Student Atlas of

World Geography

Third Edition

John L. Allen

University of Wyoming

The **McGraw·Hill** Companies

Book Team
Vice President & Publisher   *Jeffrey L. Hahn*
List Manager   *Theodore Knight*
Director of Production   *Brenda S. Filley*
Developmental Editor   *Ava Suntoke*
Designer   *Charles Vitelli*
Graphics   *Eldis Lima*
Typesetting Supervisor   *Juliana Arbo*
Proofreader   *Julie Marsh*
Cover Design   *Tom Goddard*
Cartography   *Carto-Graphics, Eau Clair, WI*

We would like to thank Digital Wisdom Incorporated for allowing us to use their Mountain High Maps cartography software. This software was used to create maps 70, 71, 73, 74, 75, 76, 77, 78, 80, 81, 82, 83, 85, 86, 87, 88, 89, 91, 92, 93, 94, 96, 97, 98, 99, 100, 101, 102, 104, 105, 106

McGraw-Hill/Dushkin
A Division of The **McGraw-Hill** *Companies*

Cover: Copyright © 2002 PhotoDisc, Inc.

The credit section for this book begins on page 226 and is considered an extension of the copyright page.

A Note to the Student

The study of geography has become an increasingly important part of the curriculum in secondary schools and institutions of higher education over the last decade. This trend, a most welcome one from the standpoint of geographers, has begun to address the massive problem of "geographic illiteracy" that has characterized the United States, almost alone among the world's developed nations. When a number of international comparative studies on world geography were undertaken, beginning in the 1970s, it became apparent that most American students fell far short of their counterparts in Europe, Russia, Canada, Australia, and Japan in their abilities to recognize geographic location, to identify countries or regions on maps, or to explain the significance of such key geographic phenomena such as population distribution, economic or urban location, or the availability of natural resources. Indeed, many American students could not even locate the United States on world maps, let alone countries like France, or Indonesia, or Nigeria. This atlas, and the texts it is intended to accompany, is a small part of the process of attempting to increase the geographic literacy of American students. As the true meaning of "the global community" becomes more apparent, such an increase in geographic awareness is not only important but necessary. If the United States has learned any lesson from the tragic events at the World Trade Center and the Pentagon on September 11, 2001, these lessons would surely include the considerations that we are not isolated from events that transpire in other parts of the world; our boundaries do not make us secure; and we ignore the conditions of political, economic, cultural, and physical geography outside those boundaries at our great peril.

The maps in the *Student Atlas of World Geography* are designed to introduce you to the patterns or "spatial distribution" of the wide variety of human and physical features of the earth's surface and to help you understand the relationships between these patterns. We call such relationships "spatial correlation" and whenever you compare the patterns made by two or more phenomena that exist at or near the earth's surface—the distribution of human population and the types of climate, for example—you are engaging in spatial correlation. Like the maps, the data sets in the atlas are intended to enable you to make comparisons between the distributions of different geographic features (for example, population growth and literacy rates) and to understand the character of the geographic variation in a single geographic feature. In many instances, the data in the tables of your atlas are the same data that have been used to produce the maps. At the very outset of your study of this atlas, you should be aware of some limitations of the data tables. In some instances, there may be data missing from a table. In such cases, the cause may represent the failure of a country to report information to a central international body (like the United Nations or the World Bank), or it may mean that the shifting of political boundaries and changed responsibility for reporting data have caused some countries (for example, those countries that made up the former Soviet Union or the former Yugoslavia) to delay their reports. It is always our aim to use the most up-to-date data that is possible. Subsequent editions of this atlas will have increased data on countries like Slovenia, Ukraine, or Uzbekistan when it becomes available. In the meantime, as events continue to restructure our world, it's an exciting time to be a student of world geography!

You will find your study of this atlas more productive if you study the maps and tables on the following pages in the context of the five distinct themes that have been developed as part of the increasing awareness of the importance of geographic education:

1. *Location: Where Is It?* This theme offers a starting point from which you discover the precise location of places in both absolute terms (the latitude and longitude of a place) and in relative terms (the location of a place in relation to the location of other places). When you think of location, you should automatically think of both forms. Knowing something about absolute location will help you to understand a variety of features of physical geography, since such key elements are so closely related to their position on the earth. But it is equally important to think of location in relative terms. The location of places in relation to other places is often more important as a determinant of social, economic, and cultural characteristics than the factors of physical geography.
2. *Place: What Is It Like?* This theme investigates the political, economic, cultural, environmental, and other characteristics that give a place its identity. You should seek to understand the similarities and differences of places by exploring their basic characteristics. Why are some places with similar environmental characteristics so very

different in economic, cultural, social, and political ways? Why are other places with such different environmental characteristics so seemingly alike in terms of their institutions, their economies, and their cultures?

3. *Human/Environment Interactions: How Is the Landscape Shaped?* This theme illustrates the ways in which people respond to and modify their environments. Certainly the environment is an important factor in influencing human activities and behavior. But the characteristics of the environment do not exert a controlling influence over human activities; they only provide a set of alternatives from which different cultures, in different times, make their choices. Observe the relationship between the basic elements of physical geography such as climate and terrain and the host of ways in which humans have used the land surfaces of the world.

4. *Movement: How Do People Stay in Touch?* This theme examines the transportation and communications systems that link people and places. Movement or "spatial interaction" is the chief mechanism for the spread of ideas and innovations from one place to another. It is spatial interaction that validates the old cliché, "the world is getting smaller." We find McDonald's restaurants in Tokyo and Honda automobiles in New York City because of spatial interaction. Advanced transportation and communications systems have transformed the world into which your parents were born. And the world your children will be born into will be very different from your world. None of this would happen without the force of movement or spatial interaction.

5. *Regions: Worlds Within a World.* This theme helps to organize knowledge about the land and its people. The world consists of a mosaic of "regions" or areas that are somehow different and distinctive from other areas. The region of Anglo-America (the United States and Canada) is, for example, different enough from the region of Western Europe that geographers clearly identify them as two unique and separate areas. Yet despite their differences, Anglo-Americans and Europeans share a number of similarities: common cultural backgrounds, comparable economic patterns, shared religious traditions, and even some shared physical environmental characteristics. Conversely, although the regions of Anglo-America and Eastern Asia are also easily distinguished as distinctive units of the earth's surface, they have a greater number of shared physical environmental characteristics. But those who live in Anglo-America and Eastern Asia have fewer similarities and more differences between them than is the case with Anglo-America and Western Europe: different cultural traditions, different institutions, different linguistic and religious patterns. An understanding of both the differences and similarities between regions like Anglo-America and Europe on the one hand, or Anglo-America and Eastern Asia on the other, will help you to understand the world around you. At the very least, an understanding of regional similarities and differences will help you to interpret what you read on the front page of your daily newspaper or view on the evening news report on your television set.

Not all of these themes will be immediately apparent on each of the maps and tables in this atlas. But if you study the contents of *Student Atlas of World Geography,* along with the reading of your text and think about the five themes, maps and tables and text will complement one another and improve your understanding of global geography.

John L. Allen

About the Author

John L. Allen is professor and chair of Geography at the University of Wyoming and emeritus professor of Geography at the University of Connecticut, where he taught from 1967 to 2000. He is a native of Wyoming. He received his bachelor's degree in 1963 and his M.A. in 1964 from the University of Wyoming, and in 1969 his Ph.D. from Clark University. His special areas of interest are perceptions of the environment and the impact of human societies on environmental systems. Dr. Allen is the author and editor of many books and articles as well as several other student atlases, including the best-selling *Student Atlas of World Politics.*

Acknowledgments

Nozar Alaolmolki
Hiram College

Barbara Batterson-Rossi
Palomar College

A. Steele Becker
University of Nebraska at Kearney

Koop Berry
Walsh University

Daniel A. Bunye
South Plains College

Winifred F. Caponigri
Holy Cross College

Femi Ferreira
Hutchinson Community College

Eric J. Fournier
Samford University

William J. Frazier
Columbus State College

Hari P. Garbharran
Middle Tennessee State University

Baher Gosheh
Edinboro University of Pennsylvania

Donald Hagan
Northwest Missouri State University

Robert Janiskee
University of South Carolina

David C. Johnson
University of Louisiana

Effie Jones
Crichton College

Cub Kahn
Marylhurst University

Artimus Keiffer
Franklin College

Leonard E. Lancette
Mercer University

Donald W. Lovejoy
Palm Beach Atlantic College

Mark Maschhoff
Harris-Stowe State College

Richard Matthews
University of South Carolina

Madolia Mills
University of Colorado–Colorado Springs

Robert Mulcahy
Providence College

Otto H. Muller
Alfred University

J. Henry Owusu
University of Connecticut

Steven Parkansky
Morehead State University

William Preston
California Polytechnic State University, San Luis Obispo

Neil Reid
The University of Toledo

A. L. Rydant
Keene State College

Deborah Berman Santana
Mills College

Steven Slakey
University of La Verne

Rolf Sternberg
Montclair State University

Richard Ulack
University of Kentucky

David Woo
California State University, Haywood

Donald J. Zeigler
Old Dominion University

Table of Contents

Part VIII Tables

Part IX Geographic Index   195

Introduction: How to Read an Atlas

An atlas is a book containing maps which are "models" of the real world. By the term "model" we mean exactly what you think of when you think of a model: a representation of reality that is generalized, usually considerably smaller than the original, and with certain features emphasized, depending on the purpose of the model. A model of a car does not contain all of the parts of the original but it may contain enough parts that it is recognizable as a car and can be used to study principles of automotive design or maintenance. A car model designed for racing, on the other hand, may contain fewer parts but would have the mobility of a real automobile. Car models come in a wide variety of types containing almost anything you can think of relative to automobiles that doesn't require the presence of a full-size car. Since geographers deal with the real world, virtually all of the printed or published studies of that world require models. Unlike a mechanic in an automotive shop, we can't roll our study subject into the shop, take it apart, put it back together. We must use models. In other words, we must generalize our subject, and the way we do that is by using maps. Some maps are designed to show specific geographic phenomena, such as the climates of the world or the relative rates of population growth for the world's countries. We call these maps "thematic maps" and Parts I through VI of this atlas contain maps of this type. Other maps are designed to show the geographic location of towns and cities and rivers and lakes and mountain ranges and so on. These are called "reference maps" and they make up many of the maps in Part VII. All of these maps, whether thematic or reference, are models of the real world that selectively emphasize the features that we want to show on the map.

In order to read maps effectively—in other words, in order to understand the models of the world presented in the following pages—it is important for you to know certain things about maps: how they are made using what are called *projections;* how the level of mathematical proportion of the map or what geographers call *scale* affects what you see; and how geographers use *generalization* techniques such as simplification and symbols where it would be impossible to draw a small version of the real world feature. In this brief introduction, then, we'll explain to you three of the most important elements of map interpretation: projection, scale, and generalization.

MAP PROJECTIONS

Perhaps the most basic problem in *cartography,* or the art and science of map-making, is the fact that the subject of maps—the earth's surface—is what is called by mathematicians "a non-developable surface." Since the world is a sphere (or nearly so—it's actually slightly flattened at the poles and bulges a tiny bit at the equator), it is impossible to flatten out the world or any part of its curved surface without producing some kind of distortion. This "near sphere" is represented by a geographic grid or coordinate system of lines of latitude or *parallels* that run east and west and are used to measure distance north and south on the globe, and lines of longitude or *meridians* that run north and south and are used to measure distance east and west. All the lines of longitude are half circles of equal length and they all converge at the poles. These meridians are numbered from 0 degrees (Prime or

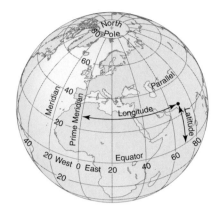

The Coordinate System

Greenwich Meridian) east and west to 180 degrees. The meridian of 0 degrees and the meridian of 180 degrees are halves of the same "great circle" or line representing a plane that bisects the globe into two equal hemispheres. All lines of longitude are halves of great circles. All the lines of latitude are complete circles that are parallel to one another and are spaced equidistant on the meridians. The circumference of these circles lessens as you move north or south from the equator. Parallels of latitude are numbered from 0 degrees at the equator north and south to 90 degrees at the North and South poles. The only line of latitude that is a great circle is the equator, which equally divides the world into a northern and southern hemisphere. In the real world, all these grid lines of latitude and longitude intersect at right angles. The problem for cartographers is to convert this spherical or curved grid into a geometrical shape that is "developable"; that is, it can be flattened (such as a cylinder or cone) or is already flat (a plane). The reason the results of the conversion process are called "projections" is that we imagine a world globe (or some part of it) that is made up of wires running north-south and east-west to represent the grid lines of latitude and longitude and other wires or even solid curved plates to represent the coastlines of continents or the continents themselves. We then imagine a light source at some location inside or outside the wire globe that can "project" or cast shadows of the wires representing grid lines onto a developable surface. Sometimes the basic geometric principles of projection may be modified by other mathematical principles to yield projections that are not truly geometric but have certain desirable features. We call these types of projections "arbitrary." The three most basic types of projections are named according to the type of developable surface: cylindrical, conic, or azimuthal (plane). Each type has certain characteristic features: they may be *equal area* projections in which the size of each area on the map is a direct proportional representation of that same area in the real world but shapes are distorted; they may be *conformal* projections in which area may be distorted but shapes are shown correctly; or they may be *compromise* projections in which both shape and area are distorted but the overall picture presented is fairly close to reality. It is important to remember that all maps distort the geographic grid and continental outlines in characteristic ways. The only representation of the world that does not distort either shape or area is a globe. You can see why we must use projections—can you imagine an atlas that you would have to carry back and forth across campus that would be made up entirely of globes?

CYLINDRICAL PROJECTIONS

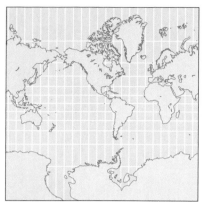

The Mercator Projection

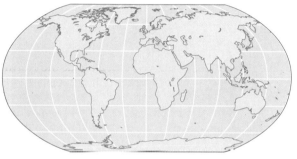

The Robinson Projection

Cylindrical projections are drawn as if the geographic grid were projected onto a cylinder. Cylindrical projections have the advantage of having all lines of latitude as true parallels or straight lines. This makes these projections quite useful for showing geographic relationships in which latitude or distance north-south is important (many physical features, such as climate, are influenced by latitude). Unfortunately, most cylindrical-type projections distort area significantly. One of the most famous is the Mercator projection shown above. This projection makes areas disproportionately large as you move toward the pole, making Greenland, which is actually about one-seventh the size of South America, appear to be as large as the southern continent. But the Mercator projection has the quality of conformality: landmasses on the map are true in shape and thus all coastlines on the map intersect lines of latitude and longitude at the proper angles. This makes the Mercator projection, named after its inventor, a sixteenth-century Dutch cartographer, ideal for its original purpose as a tool for navigation—but not a good projection for attempting to show some geographical feature in which areal relationship is important. Unfortunately, the Mercator projection has often been used for wall maps for schoolrooms and the consequence is that generations of American school children have been "tricked" into thinking that Greenland is actually larger than South America. Much better cylindrical-

type projections are those like the Robinson projection used in this atlas that is neither equal area nor conformal but a compromise that portrays the real world much as it actually looks, enough so that we can use it for areal comparisons.

CONIC PROJECTIONS

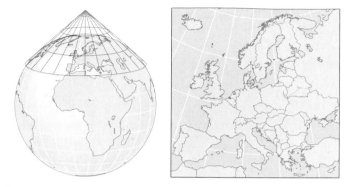

Conic Projection of Europe

Conic projections are those that are imagined as being projected onto a cone that is tangent to the globe along a standard parallel, or a series of cones tangent along several parallels or even intersecting the globe. Conic projections usually show latitude as curved lines and longitude as straight lines. They are good projections for areas with north-south extent, like the map of Europe to the right, and may be either conformal, equal area, or compromise, depending on how they are constructed. Many of the regional maps in the last map section of this atlas are conic projections.

AZIMUTHAL PROJECTIONS

Azimuthal projections are those that are imagined as being projected onto a plane or flat surface. They are named for one of their essential properties. An "azimuth" is a line of compass bearing and azimuthal projections have the property of yielding true compass directions from the center of the map. This makes azimuthal maps useful for navigation purposes, particularly air navigation. But, because they distort area and shape so greatly, they are seldom used for maps designed to show geographic relationships. When they are used as illustrative rather than navigation maps, it is often in the "polar case" projection shown below where the plane has been made tangent to the globe at the North Pole.

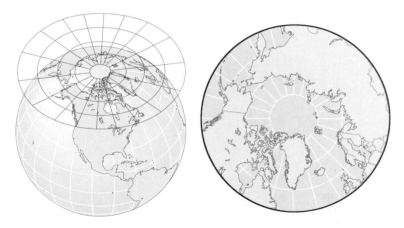

Azimuthal Projection of the North Polar Region

MAP SCALE

Since maps are models of the real world, it follows that they are not the same size as the real world or any portion of it. Every map, then, is subject to generalization, which is another way of saying that maps are drawn to certain scales. The term "scale" refers to the mathematical quality of *proportional representation,* and is expressed as a ratio between an area of the real world or the distance between places on the real world and the same area or distance on the map. We show map scale on maps in three different ways. Sometimes we simply use the proportion and write what is called a "natural scale" or representative fraction": for example, we might show on a map the mathematical proportion of 1:62,500. A map at this scale is one that is one sixty-two thousand five-hundredth the size of the same area in the real world. Other times we convert the proportion to a written description that approximates the relationship between distance on the map and distance in the real world. Since there are nearly 62,500 inches in a mile, we would refer to a map having a natural scale of 1:62,500 as having an "inch-mile" scale of "1 inch represents 1 mile." If we draw a line one inch long on this map, that line represents a distance of approximately one mile in the real world. Finally, we usually use a graphic or linear scale: a bar or line, often graduated into miles or kilometers, that shows graphically the proportional representation. A graphic scale for our 1:62,500

map might be about five inches long, divided into five equal units clearly labeled as "1 mile," "2 miles," and so on. Our examples below show all three kinds of scales.

The most important thing to keep in mind about scale, and the reason why knowing map scale is important to being able to read a map correctly, is the relationship between proportional representation and generalization. A map that fills a page but shows the whole world is much more highly generalized than a map that fills a page but shows a single city. On the world map, the city may appear as a dot. On the city map, streets and other features may be clearly seen. We call the first map, the world map, a "small scale" map because the proportional representation is a small number. A page-size map showing the whole world may be drawn at a scale of 1:150,000,000. That is a very small number indeed—hence the term "small scale" map even though the area shown is large. Conversely, the second map, a city map, may be drawn at a scale of 1:250,000. That is still a very small number but it is a great deal larger than 1:150,000,000! And so we'd refer to the city map as a "large scale" map, even though it shows only a small area. On our world map, geographical features are generalized greatly and many features can't even be shown at all. On the city map, much less generalization occurs—we can show specific features that we couldn't on the world map—but generalization still takes place. The general rule is that the smaller the map scale, the greater the degree of generalization;

Map 1 Small Scale Map of the United States

Map 2 Map of the Northeast

Map 3 Map of Southeastern New England

Map 4 Large Scale Map of Boston, MA

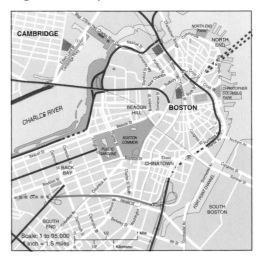

the larger the map scale, the less the degree of generalization. The only map that would not generalize would be a map at a scale of 1:1. and that map wouldn't be very handy to use. Examine the relationship between scale and generalization in the four maps on the previous page.

GENERALIZATION ON MAPS

A review of the four maps on the previous page should give you some indication of how cartographers generalize on maps. One thing that you should have noticed is that the first map, that of the United States, is much simpler than the other three and that the level of *simplification* decreases with each map. When a cartographer simplifies map data, information that is not important for the purposes of the map is just left off. For example, on the first map the objective may have been to show cities over 1 million in population. To do that clearly and effectively, it is not necessary to show and label rivers and lakes. The map has been simplified by leaving those items out. The final map, on the other hand, is more complex and shows and labels geographic features that are important to the character of the city of Boston; therefore, the Charles River is clearly indicated on the map.

Another type of generalization is *classification*. Map 1 on the previous page shows cities over 1 million in population. Map 2 shows cities of several different sizes and a different symbol is used for each size classification or category. Many of the thematic maps used in this atlas rely on classification to show data. A thematic map showing population growth rates (see Map 21 on page 32) will use different colors to show growth rates in different classification levels or what are sometimes called "class intervals." Thus, there will be one color applied to all countries with population growth rates between 1.0 percent and 1.4 percent, another color applied to all countries with population growth rates between 1.5 percent and 2.1 percent, and so on. Classification is necessary because it is impossible to find enough symbols or colors to represent precise values. Classification may also be used for qualitative data, such as the national or regional origin of migrating populations. Cartographers show both quantitative and qualitative classification levels or class intervals in important sections of maps called "legends." These legends, as in the samples shown below, make it possible for the reader of the map to interpret the patterns shown.

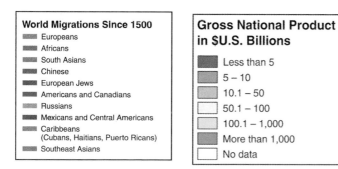

Map Legends from Maps 20 and 33

A third technique of generalization is *symbolization* and we've already noted several different kinds of symbols: those used to represent cities on the preceding maps, or the colors used to indicate population growth levels on Map 21. One general category of map symbols is quantitative in nature

and this category can further be divided into a number of different types. For example, the symbols showing city size on Maps 1 and 2 on the preceding page can be categorized as *ordinal* in that they show relative differences in quantities (the size of cities). A cartographer might also use lines of different widths to express the quantities of movement of people or goods between two or more points as on Map 20 (see page 30).

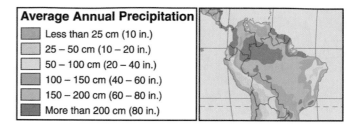

Interval Symbols

The color symbols used to show rates of population growth can be categorized as *interval* in that they express certain levels of a mathematical quantity (the percentage of population growth). Interval symbols are often used to show physical geographic characteristics such as inches of precipitation, degrees of temperature, or elevation above sea level. The sample above, for example, shows precipitation (from Map 3a, page 6).

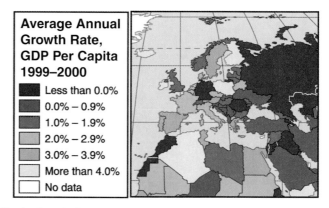

Ratio Symbols

Still another type of mathematical symbolization is the *ratio* in which sets of mathematical quantities are compared: the number of persons per square mile (population density) or the growth in gross national product per capita (per person). The map above shows GDP change per capita (from Map 34, page 46).

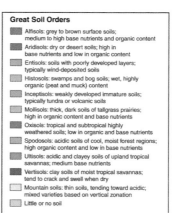

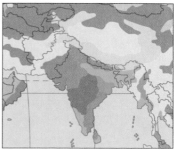

Nominal Symbols

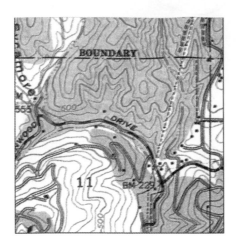

Interpolation

Finally, there are a vast number of cartographic symbols that are not mathematical but show differences in the kind of information being portrayed. These symbols are called *nominal* and they range from the simplest differences such as land and water to more complex differences such as those between different types of vegetation. Shapes or patterns or colors or iconographic drawings may all be used as nominal symbols on maps. The sample map at the bottom of page xiii uses color to show the distribution of soil types.

The final technique of generalization is what cartographers refer to as *interpolation*. Here, the maker of a map may actually show more information on the map than is actually supplied by the original data. In understanding the process of interpolation is it necessary for you to visualize the quantitative data shown on maps as being three dimensional: *x* values provide geographic location along a north-south axis of the map; *y* values provide geographic location along the east-west axis of the map; and *z* values are those values of whatever data (for example, temperature) are being shown on the map at specific points. We all can imagine a real three-dimensional surface in which the *x* and y values are directions and the *z* values are the heights of mountains and the depths of valleys. On a topographic map showing a real three-dimensional surface, contour lines are used to connect points of equal elevation above sea level. These contour lines are not measured directly; they are estimated by interpolation on the basis of the elevation points that are provided.

It is harder to imagine the statistical surface of a temperature map in which the *x* and *y* values are directions and the *z* values represent degrees of temperature at precise points. But that is just what cartographers do. And to obtain the values between two or more specific points where *z* values exist, they interpolate based on a class interval they have decided is appropriate and use *isolines* (which are statistical equivalents of a contour line) to show increases or decreases in value. The diagram below shows an example of an interpolation process. Occasionally interpolation is referred to as *induction*. By whatever name, it is one of the most difficult parts of the cartographic process.

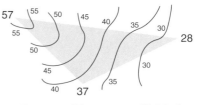

Degrees of Temperature (Celsius)
Interval = 5 degrees

And you thought all you had to do to read an atlas was look at the maps! You've now learned that it is a bit more involved than that. As you read and study this atlas, keep in mind the principles of projection and scale and generalization (including simplification, classification, symbolization, and interpolation) and you'll do just fine. Good luck and enjoy your study of the world of maps as well as maps of the world!

Part I

Global Physical Patterns

Map 1 World Countries

The international system includes the political units called "states" or countries as the most important component. The boundaries of countries are the primary source of political division in the world and for most people nationalism is the strongest source of political identity. State boundaries are an important indicator of cultural, linguistic, economic, and other geographic divisions as well, and the states themselves normally serve as the base level for which most global statistics are available. The subfield of geography known as "political geography" has as its primary concern the geographic or spatial character of this international system and its components.

Scale: 1 to 125,000,000

Note: All world maps are Robinson projection.

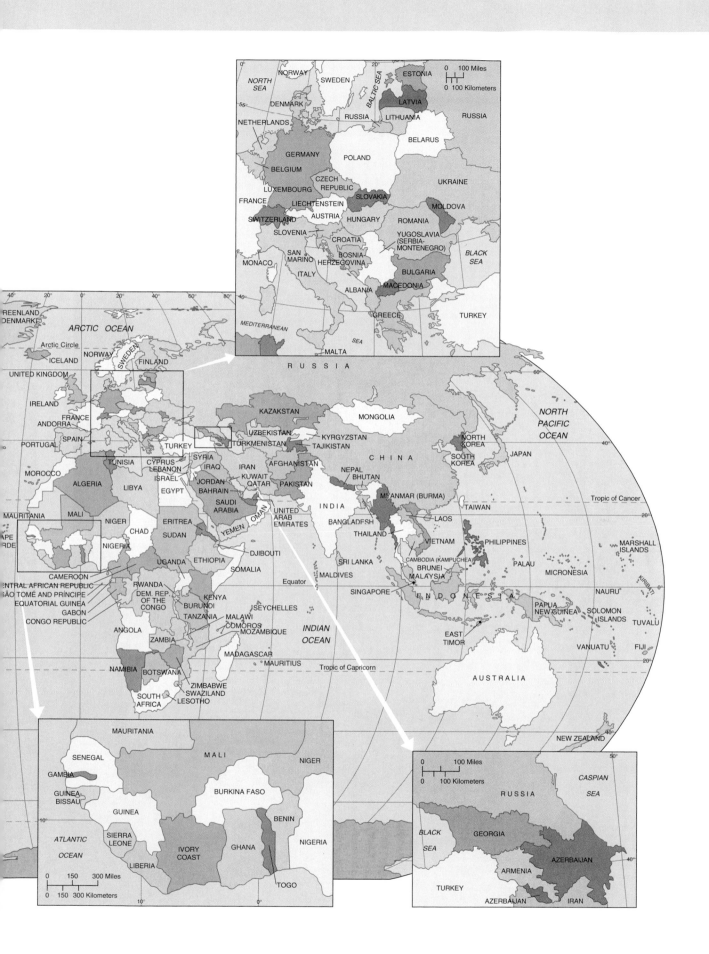

Map 2 World Physical Map

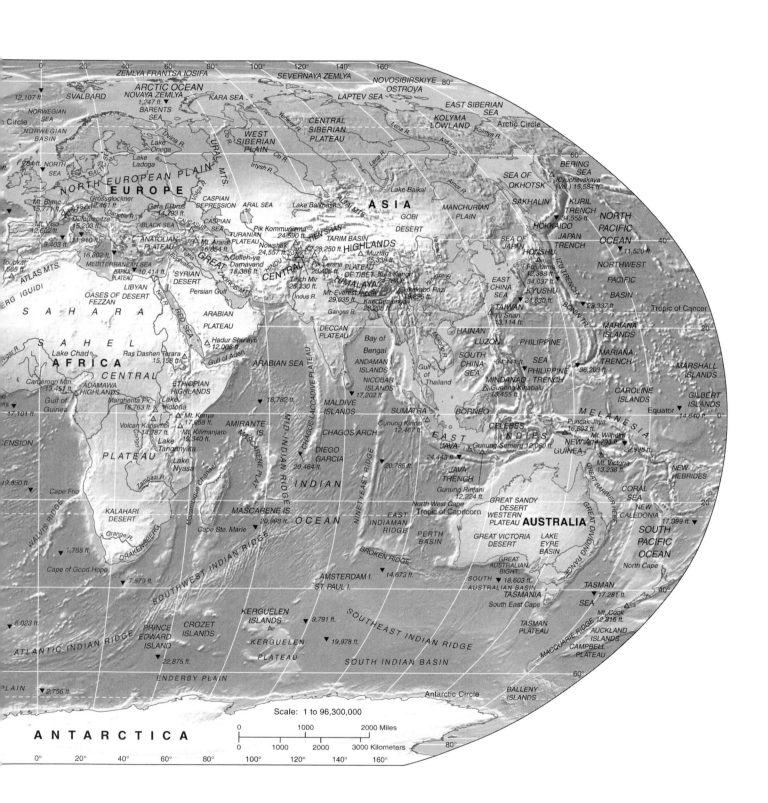

Map 3a Average Annual Precipitation

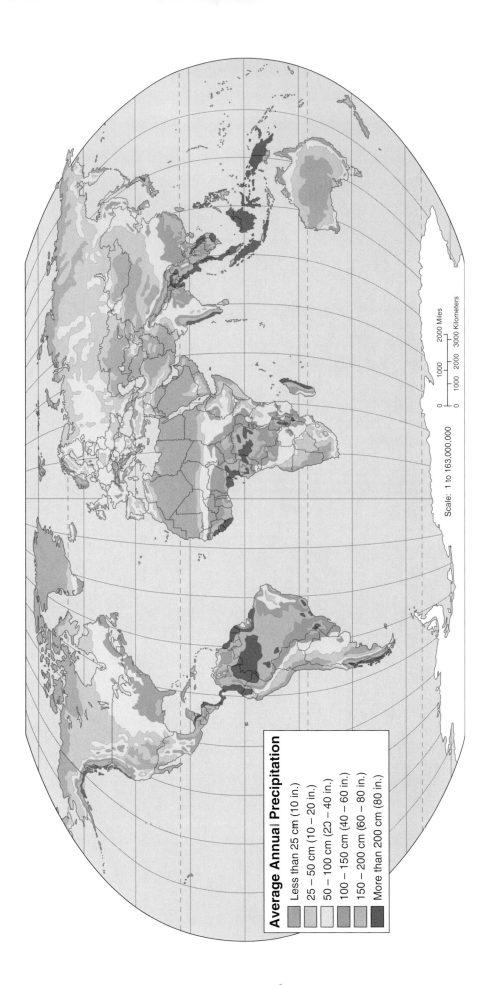

Average Annual Precipitation

- Less than 25 cm (10 in.)
- 25 – 50 cm (10 – 20 in.)
- 50 – 100 cm (20 – 40 in.)
- 100 – 150 cm (40 – 60 in.)
- 150 – 200 cm (60 – 80 in.)
- More than 200 cm (80 in.)

Scale: 1 to 163,000,000

```
0          1000          2000 Miles
0     1000   2000   3000 Kilometers
```

The two most important physical geographic variables are precipitation and temperature, the essential elements of weather and climate. Precipitation is a conditioner of both soil type and vegetation. More than any other single environmental element, it influences where people do or do not live. Water is the most precious resource available to humans, and water availability is largely a function of precipitation. Water availability is also a function of several precipitation variables that do not appear on this map: the seasonal distribution of precipitation (is precipitation or drought concentrated in a particular season?), the ratio between precipitation and temperature (how much of the water that comes to the earth in the form of precipitation is lost through mechanisms such as evaporation and transpiration that are a function of temperature?), and the annual variability of precipitation (how much do annual precipitation totals for a place or region tend to vary from the "normal" or average precipitation?). In order to obtain a complete understanding of precipitation, these variables should be examined along with the more general data presented on this map. The study of precipitation and other climatic elements is the concern of the branch of physical geography called "climatology."

-6-

Map 3b Seasonal Average Precipitation–November Through April

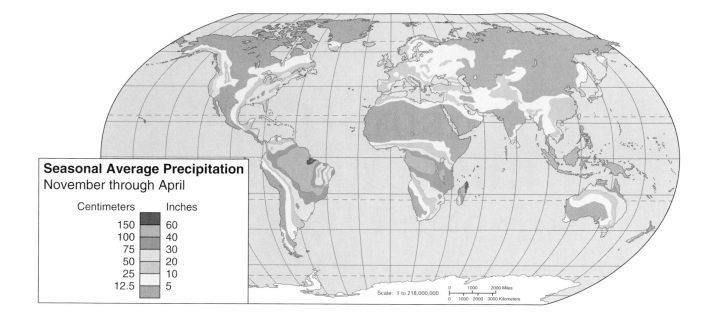

Seasonal Average Precipitation
November through April

Centimeters	Inches
150	60
100	40
75	30
50	20
25	10
12.5	5

Scale: 1 to 218,000,000

0 1000 2000 Miles
0 1000 2000 3000 Kilometers

Map 3c Seasonal Average Precipitation–May Through October

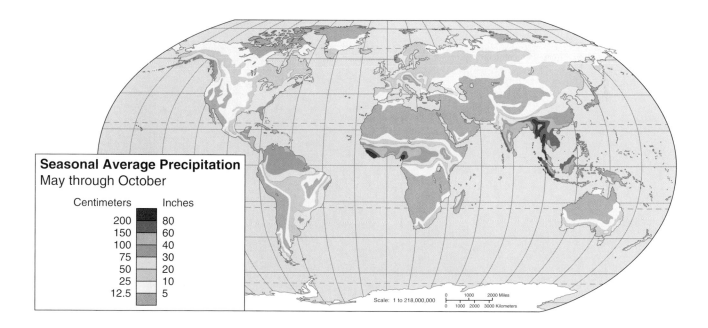

Seasonal Average Precipitation
May through October

Centimeters	Inches
200	80
150	60
100	40
75	30
50	20
25	10
12.5	5

Scale: 1 to 218,000,000

0 1000 2000 Miles
0 1000 2000 3000 Kilometers

Seasonal average precipitation is nearly as important as annual precipitation totals in determining the habitability of an area. Critical factors are such things as whether precipitation coincides with the growing season and thus facilitates agriculture or during the winter when it is less effective in aiding plant growth, and whether precipitation occurs during summer with its higher water loss through evaporation and transpiration or during the winter when more of it can go into storage. Several of the world's great climate zones have pronounced seasonal precipitation rhythms. The tropical and subtropical savanna grasslands have a long winter dry season and abundant precipitation in the summer. The Mediterranean climate is the only major climate with a marked dry season during the summer, making agriculture possible only through irrigation or other adjustments to cope with drought during the period of plant growth. And the great monsoon climates of south and southeast Asia have their winter dry season and summer rain that have conditioned the development of Asian agriculture and the rhythms of Asian life.

Map 3d Variation in Annual Precipitation

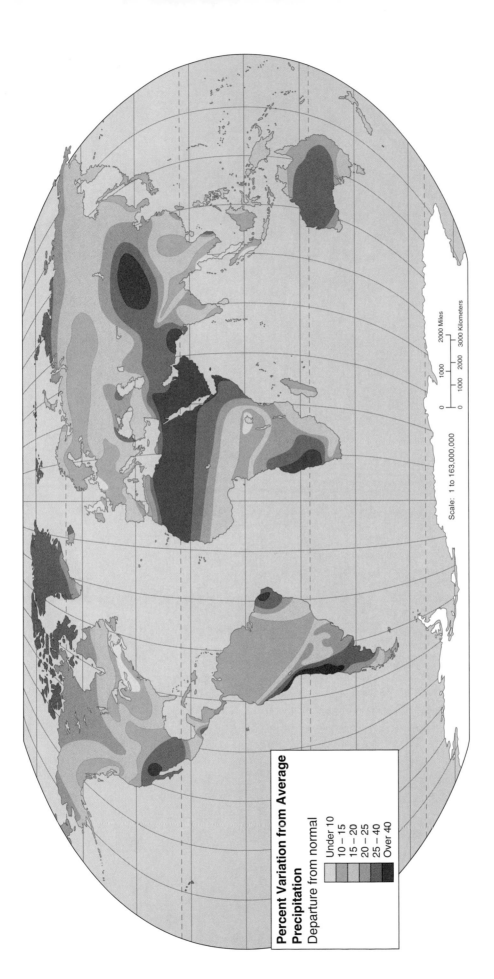

Scale: 1 to 163,000,000

Percent Variation from Average Precipitation

Departure from normal

- Under 10
- 10 – 15
- 15 – 20
- 20 – 25
- 25 – 40
- Over 40

While annual precipitation totals and seasonal distribution of precipitation are important variables, the variability of precipitation from one year to the next may be even more critical. You will note from the map that there is a general spatial correlation between the world's drylands and the amount of annual variation in precipitation. Generally, the drier the climate, the more likely it is that there will be considerable differences in rainfall and/or snowfall from one year to the next. We might determine that the average precipitation of the mid-Sahara is 2 inches per year. What this really means is that a particular location in the Sahara during one year might receive .5", during the next year 3.5", and during a third year 2". If you add these together and divide by the number of years, the "average" precipitation is 2" per year. The significance of this is that much of the world's crucial agricultural output of cereals (grains) comes from dryland climates (the Great Plains of the United States, the Pampas of Argentina, the steppes of Ukraine and Russia, for example), and variations in annual rainfall totals can have significant impacts on levels of grain production and, therefore, important consequences for both economic and political processes.

-8-

Map 4a Temperature Regions and Ocean Currents

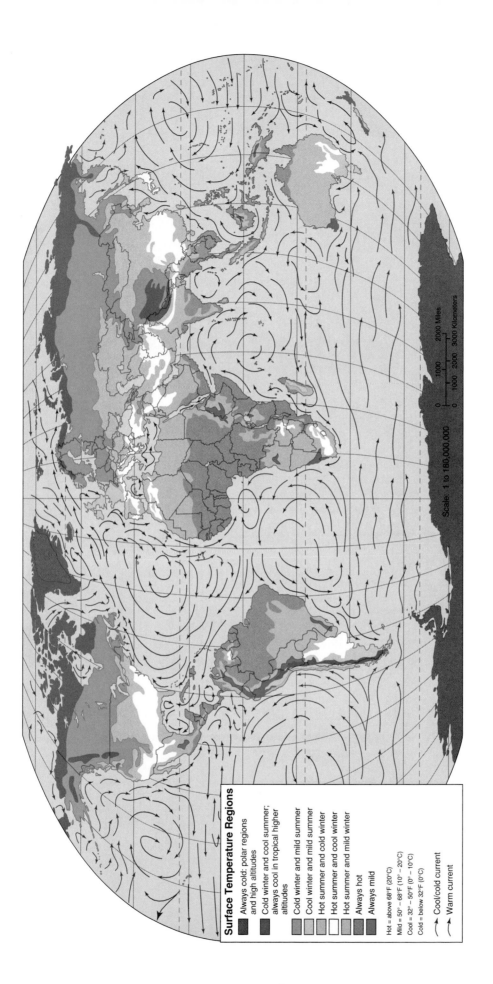

Surface Temperature Regions

- Always cold: polar regions and high altitudes
- Cold winter and cool summer; always cool in tropical higher altitudes
- Cold winter and mild summer
- Cool winter and mild summer
- Hot summer and cold winter
- Hot summer and cool winter
- Hot summer and mild winter
- Always hot
- Always mild

Hot = above 68°F (20°C)
Mild = 50° – 68°F (10° – 20°C)
Cool = 32° – 50°F (0° – 10°C)
Cold = below 32°F (0°C)

— Cool/cold current
— Warm current

Scale: 1 to 180,000,000

0 ... 1000 ... 2000 Miles

0 ... 1000 ... 2000 ... 3000 Kilometers

-9-

Along with precipitation, temperature is one of the two most important environmental variables, defining the climate conditions so essential for the distribution of such human activities as agriculture and the distribution of the human population. The seasonal rhythm of temperature, including such measures as the average annual temperature range (difference between the average temperature of the warmest month and that of the coldest month), is an additional variable not shown on the map but, like the sea-

sonality of precipitation, should be a part of any comprehensive study of climate. The ocean currents illustrated exert a significant influence over the climate of adjacent regions and are the most important mechanism for redistributing surplus heat from the equatorial region into middle and high latitudes. Physical geographers known as "climatologists" study the phenomenon of temperature and related climatic characteristics.

Map 4b Average July Temperature

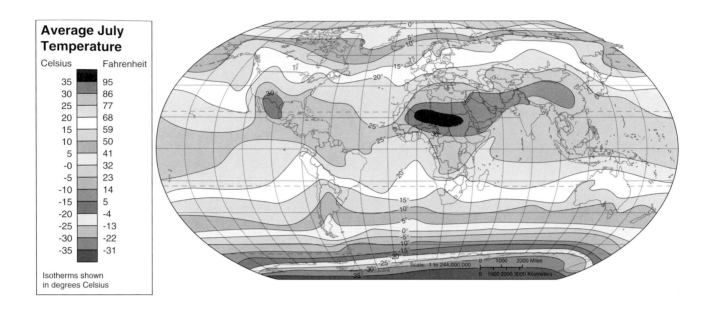

Average July Temperature

Celsius	Fahrenheit
35	95
30	86
25	77
20	68
15	59
10	50
5	41
-0	32
-5	23
-10	14
-15	5
-20	-4
-25	-13
-30	-22
-35	-31

Isotherms shown in degrees Celsius

Map 4c Average January Temperature

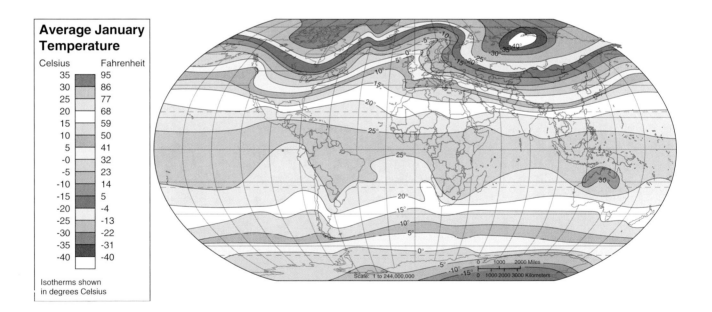

Average January Temperature

Celsius	Fahrenheit
35	95
30	86
25	77
20	68
15	59
10	50
5	41
-0	32
-5	23
-10	14
-15	5
-20	-4
-25	-13
-30	-22
-35	-31
-40	-40

Isotherms shown in degrees Celsius

Where moisture availability tends to mark the seasons in the tropics and subtropics, in the mid-latitudes, seasons are defined by temperature. Temperature is determined by latitudinal position, by altitude or elevation above sea level, and by the location of a place relative to the world's land masses and oceans. The most important of these controls is latitude, and temperatures generally become lower with increasing latitude. However, proximity to water tends to moderate temperature extremes, and "maritime" climates influenced by the oceans will be warmer in the winter and cooler in the summer than continental climates in the same general latitude. Maritime climates will also show smaller temperature ranges, the differ-

ence between January and July temperatures, while climates of continental interiors, far from the moderating influences of the oceans, will tend to have great temperature ranges. In the Northern Hemisphere, where there are both large land masses and oceans, the range is great. But in the Southern Hemisphere, dominated by water and, hence, by the more moderate maritime air masses, the temperature range is comparatively small. Significant temperature departures from the "normal" produced by latitude may also be the result of elevation. With exceptions, lower temperatures produced by topography are difficult to see on maps of this scale.

Map 5a Atmospheric Pressure and Predominant Surface Winds–January

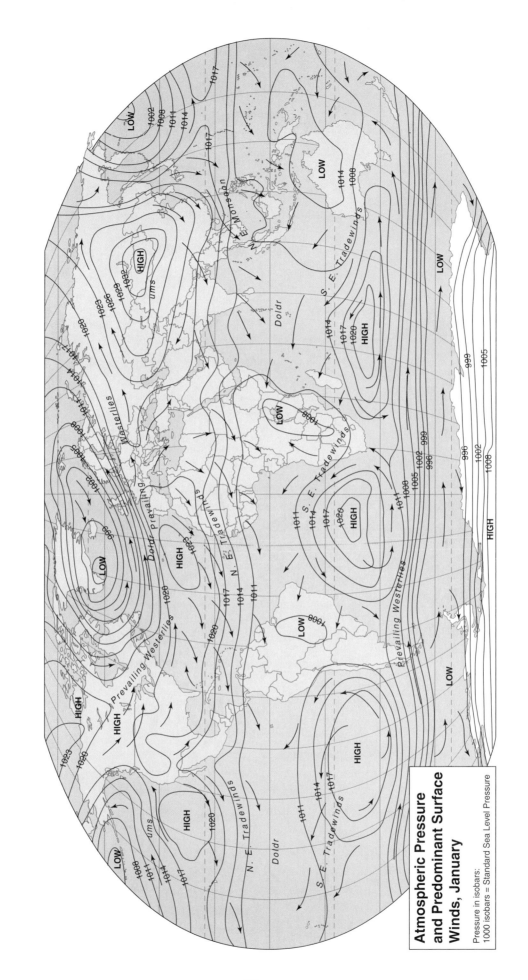

Atmospheric Pressure and Predominant Surface Winds, January

Pressure in isobars:
1000 isobars = Standard Sea Level Pressure

Map 5b Atmospheric Pressure and Predominant Surface Winds–July

Atmospheric Pressure and Predominant Surface Winds, July

Pressure in isobars:
1000 isobars = Standard Sea Level Pressure

Atmospheric pressure, or the density of air, is a function largely of air temperature: the colder the air, the denser and heavier it is, hence the higher its pressure; the warmer the air, the lighter and less stab le it is, hence the lower its pressure. Global pressure systems are the alternating low and high pressure systems that, from the equator north and south, include: the equatorial low (sometimes called the intertropical convergence) centered on the equator for much of the year; the subtropical highs with their centers near the 30th degrees of north and south latitude; the subpolar lows or polar front cen-

tered near the 60th parallel of north and south latitude; and the polar highs near the north and south poles. Air flows from high pressure to low pressure regions, and this air flow constitutes the earth's major surface winds such as the tropical tradewinds and the prevailing westerlies. This flow of air is one of the chief mechanisms by which surplus heat energy from the equatorial region is redistributed to higher latitudes. It is also the primary conditioner of the world's major precipitation belts, with rainfall and snowfall associated primarily with lower atmospheric pressure conditions.

Map 6 World Climate Regions

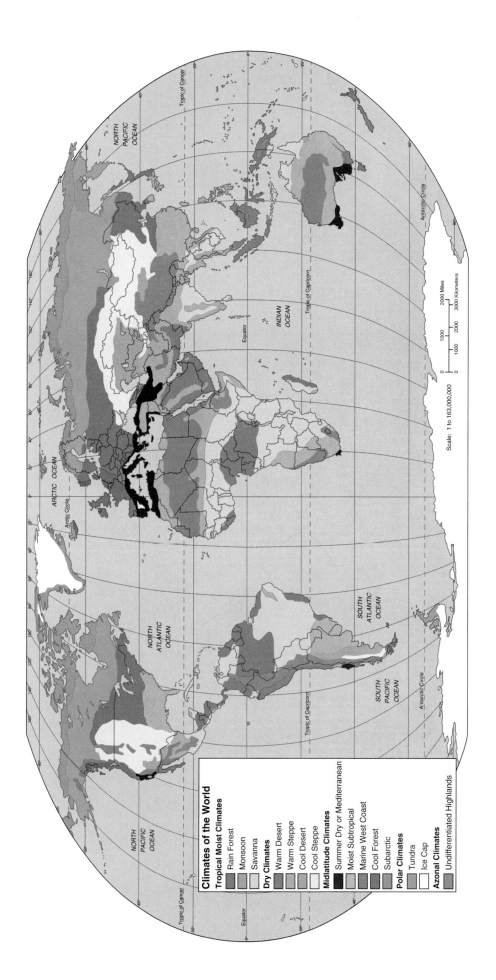

Climates of the World

Tropical Moist Climates
- Rain Forest
- Monsoon
- Savanna

Dry Climates
- Warm Desert
- Warm Steppe
- Cool Desert
- Cool Steppe

Midlatitude Climates
- Summer Dry or Mediterranean
- Moist Subtropical
- Marine West Coast
- Cool Forest
- Subarctic

Polar Climates
- Tundra
- Ice Cap

Azonal Climates
- Undifferentiated Highlands

Scale: 1 to 163,000,000

Of the world's many patterns of physical geography, climate or the long-term average of weather conditions such as temperature and precipitation is the most important. It is climate that conditions the distribution of natural vegetation and the types of soils that will exist in an area. Climate also influences the availability of our most crucial resource: water. From an economic standpoint, the world's most important activity is agriculture; no other element of physical geography is more important for agriculture than climate. Ultimately, it is agricultural production that determines where the bulk of human beings live, and therefore, climate is a basic determinant of the distribution of human populations as well. The study of climates or "climatology" is one of the most important branches of physical geography.

Map 7 Vegetation Types

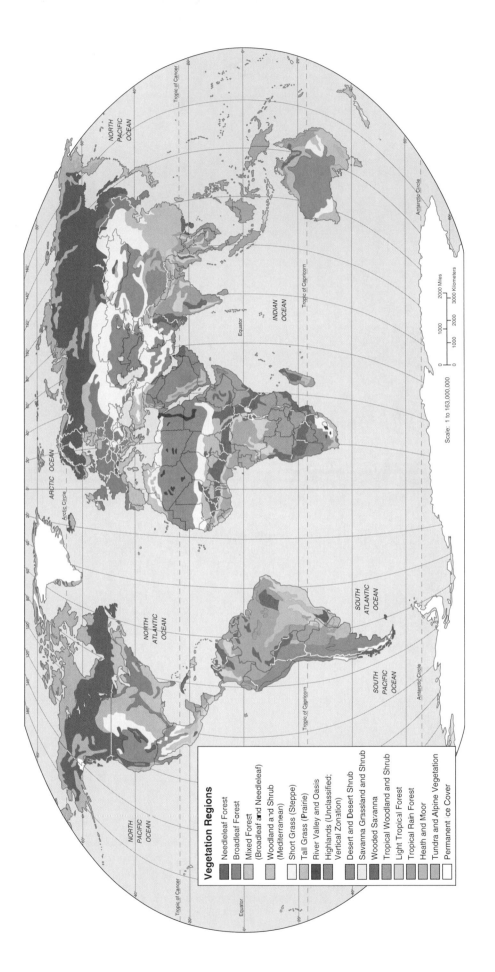

Vegetation Regions

- Needleleaf Forest
- Broadleaf Forest
- Mixed Forest (Broadleaf and Needleleaf)
- Woodland a and Shrub (Mediterranean)
- Short Grass (Steppe)
- Tall Grass (Prairie)
- River Valley and Oasis
- Highlands (Unclassified; Vertical Zonation)
- Desert and Desert Shrub
- Savanna Grassland and Shrub
- Wooded Savanna
- Tropical Woodland and Shrub
- Light Tropical Forest
- Tropical Rain Forest
- Heath and Moor
- Tundra and Alpine Vegetation
- Permanent Ice Cover

Scale: 1 to 163,000,000

Vegetation is the most visible consequence of the distribution of temperature and precipitation. The global pattern of vegetative types or "habitat classes" and the global pattern of climate are closely related and make up one of the great global spatial correlations. But not all vegetation types are the consequence of temperature and precipitation or other climatic variables. Many types of vegetation in many areas of the world are the consequence of human activities, particularly the grazing of domesticated livestock, burning, and forest clearance. This map shows the pattern of natural or "potential" vegetation, or vegetation as it might be expected to exist without significant human influences, rather than the actual vegetation that results from a combination of environmental and human factors. Physical geographers who are interested in the distribution and geographic patterns of vegetation are "biogeographers."

Map 8 World Soil Orders

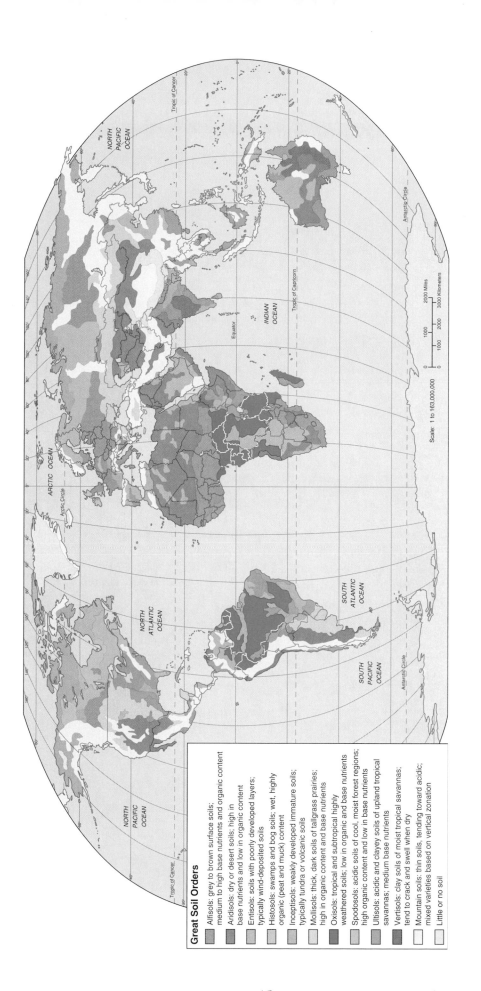

Great Soil Orders

- Alfisols: grey to brown surface soils; medium to high base nutrients and organic content
- Aridisols: dry or desert soils; high in base nutrients and low in organic content
- Entisols: soils with poorly developed layers; typically wind-deposited soils
- Histosols: swamps and bog soils; wet, highly organic (peat and muck) content
- Inceptisols: weakly developed immature soils; typically tundra or volcanic soils
- Mollisols: thick, dark soils of tallgrass prairies; high in organic content and base nutrients
- Oxisols: tropical and subtropical highly weathered soils; low in organic and base nutrients
- Spodosols: acidic soils of cool, moist forest regions; high organic content and low in base nutrients
- Ultisols: acidic and clayey soils of upland tropical savannas; medium base nutrients
- Vertisols: clay soils of moist tropical savannas; tend to crack and swell when dry
- Mountain soils: thin soils, tending toward acidic; mixed varieties based on vertical zonation
- Little or no soil

Scale: 1 to 163,000,000

0   1000   2000   3000 Kilometers
0   1000   2000 Miles

The characteristics of soil are one of the three primary physical geographic factors, along with climate and vegetation, that determine the habitability of regions for humans. In particular, soils influence the kinds of agricultural uses to which land is put. Since soils support the plants that are the primary producers of all food in the terrestrial food chain, their characteristics are crucial to the health and stability of ecosystems. Two types of soil are shown on this map: zonal soils, the characteristics of which are based on climatic patterns; and azonal soils, such as alluvial (water-deposited) or aeolian (wind-deposited) soils, the characteristics of which are derived from forces other than climate. However, many of the azonal soils, particularly those dependent upon drainage conditions, appear over areas too small to be readily shown on a map of this scale. Thus, almost none of the world's swamp or bog soils appear on this map. People who study the geographic characteristics of soils are most often "soil scientists," a discipline closely related to that branch of physical geography called "geomorphology."

Map 9 World Ecological Regions

World Ecological Regions

Arctic and Subarctic Zone

Ice Cap

Tundra Province: moss-grass and moss-lichen tundra

Tundra Altitudinal Zone: polar desert (no vegetation)

Subarctic Province: evergreen forest, needleleaf taiga; mixed coniferous and small-leafed forest

Subarctic Altitudinal Zone: open woodland; wooded tundra

Humid Temperate Zone

Moderate Continental Province: mixed coniferous and broadleaf forest

Moderate Continental Altitudinal Zone: coastal and alpine forest; open woodland

Warm Continental Province: broadleaf deciduous forest

Warm Continental Altitudinal Zone: upland broadleaf and alpine needleleaf forest

Marine Province: lowland, west-coastal humid forest

Marine Altitudinal Zone: humid coastal and alpine coniferous forest

Humid Subtropical Province: broadleaf evergreen and broadleaf deciduous forest

Humid Subtropical Altitudinal Zone: upland, subtropical broadleaf forest

Prairie Province: tallgrass and mixed prairie

Prairie Altitudinal Zone: upland mixed prairie and woodland

Mediterranean Province: sclerophyll woodland, shrub, and steppe grass

Mediterranean Altitudinal Zone: upland shrub and steppe

Humid Tropical Zone

Savanna Province: seasonally dry forest; open woodland; tallgrass savanna

Savanna Altitudinal Zone: open woodland steppe

Rain Forest Province: constantly humid, broadleaf evergreen forest

Rain Forest Altitudinal Zone: broadleaf evergreen and subtropical deciduous forest

Arid and Semiarid Zone

Tropical/Subtropical Steppe Province: dry steppe (short grass), desert shrub, semidesert savanna

Tropical/Subtropical Steppe Altitudinal Zone: upland steppe (short grass) and desert shrub

Tropical/Subtropical Desert Province: hot, lowland desert in subtropical and coastal locations; xerophytic vegetation

Tropical/Subtropical Desert Altitudinal Zone: desert shrub

Temperate Steppe Province: medium to shortgrass prairie

Temperate Steppe Altitudinal Zone: alpine meadow and coniferous woodland

Temperate Desert Province: midlatitude rainshadow desert; desert shrub

Temperate Desert Altitudinal Zone: extreme continental desert steppe; desert shrub, xerophytic vegetation, shortgrass steppe

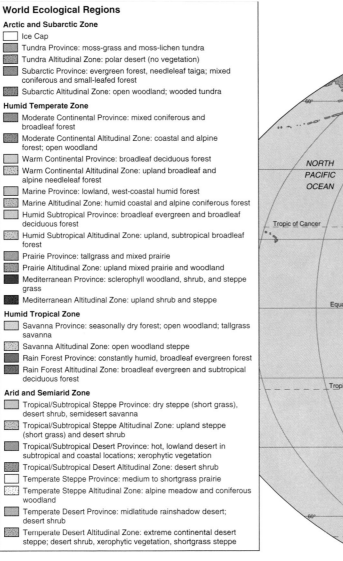

NORTH PACIFIC OCEAN

NORTH ATLANTIC OCEAN

Tropic of Cancer

Equator

Tropic of Capricorn

SOUTH PACIFIC OCEAN

SOUTH ATLANTIC OCEAN

Antarctic Circle

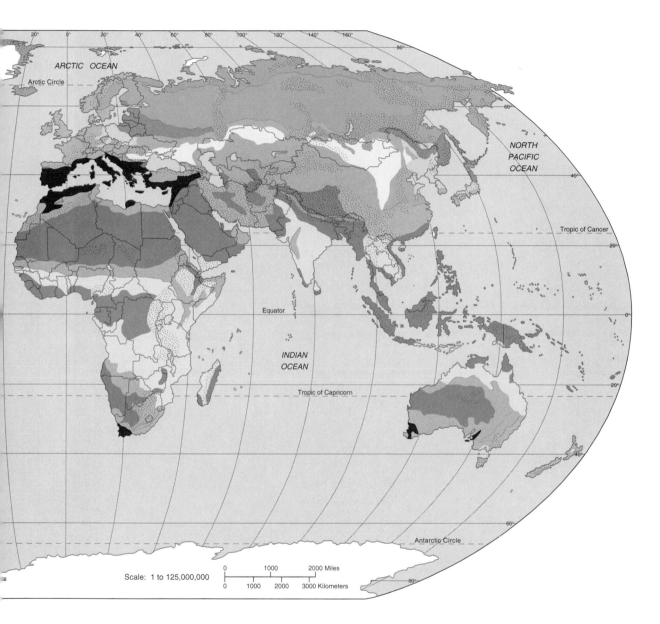

Ecological regions are distinctive areas within which unique sets of organisms and environments are found. We call the study of the relationships between organisms and their environmental surroundings "ecology." Within each of the ecological regions portrayed on the map, a particular combination of vegetation, wildlife, soil, water, climate, and terrain defines that region's habitability, or ability to support life, including human life. Like climate and landforms, ecological relationships are crucial to the existence of agriculture, the most basic of our economic activities, and important for many other kinds of economic activity as well. Biogeographers are especially concerned with the concept of ecological regions since such regions so clearly depend upon the geographic distribution of plants and animals in their environmental settings.

Map 10 Plate Tectonics

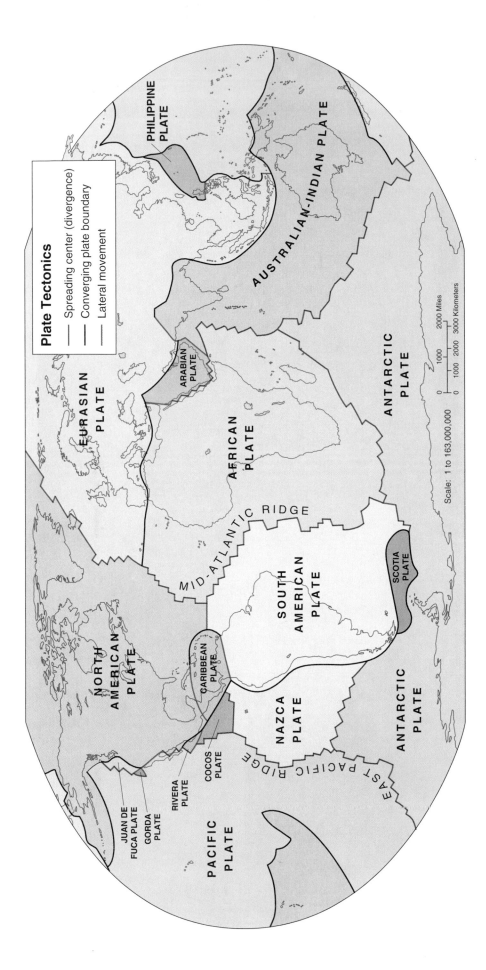

Plate Tectonics

— Spreading center (divergence)
— Converging plate boundary
— Lateral movement

PHILIPPINE PLATE

AUSTRALIAN-INDIAN PLATE

EURASIAN PLATE

ARABIAN PLATE

AFRICAN PLATE

ANTARCTIC PLATE

MID-ATLANTIC RIDGE

NORTH AMERICAN PLATE

SOUTH AMERICAN PLATE

SCOTIA PLATE

CARIBBEAN PLATE

COCOS PLATE

RIVERA PLATE

JUAN DE FUCA PLATE

GORDA PLATE

NAZCA PLATE

ANTARCTIC PLATE

PACIFIC PLATE

EAST PACIFIC RIDGE

Scale: 1 to 163,000,000

0   1000   2000 Miles
0   1000   2000   3000 Kilometers

An understanding of the forces that shape the primary features of the earth's surface—the continents and ocean basins—requires a view of the earth's crust as fragments or "lithospheric plates" that shift position relative to one another. There are three dominant types of plate movement: *convergence*, in which plates move together, compressing former ocean floor or continental rocks together to produce mountain ranges, or producing mountain ranges through volcanic activity if one plate slides beneath another; *divergence*, in which the plates move away from one another, producing rifts in the earth's crust through which molten material wells up to produce new sea floors and mid-oceanic ridges; and *lateral shift*, in which plates move horizontally relative to one another, causing significant earthquake activity. All the major forms of these types of shifts are extremely slow and take place over long periods of geologic time. The movement of crustal plates, or what is known as "plate tectonics," is responsible for the present shape and location of the continents but is also the driving force behind some much shorter-term earth phenomena like earthquakes and volcanoes. A comparison of the map of plates with maps of hazards and terrain will reveal some interesting relationships.

-18-

Map 11 World Topography

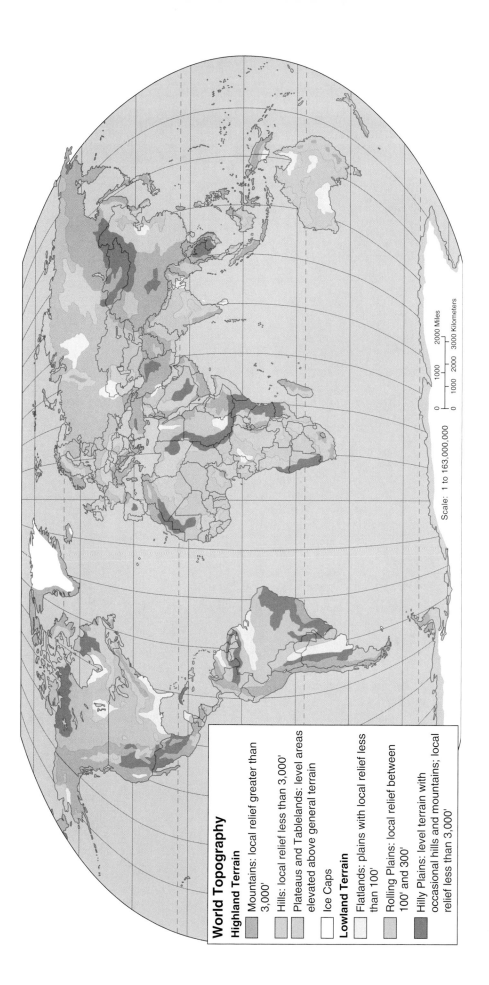

World Topography

Highland Terrain

- Mountains: local relief greater than 3,000'
- Hills: local relief less than 3,000'
- Plateaus and Tablelands: level areas elevated above general terrain
- Ice Caps

Lowland Terrain

- Flatlands: plains with local relief less than 100'
- Rolling Plains: local relief between 100' and 300'
- Hilly Plains: level terrain with occasional hills and mountains; local relief less than 3,000'

Scale: 1 to 163,000,000

0   1000   2000 Miles

0   1000   2000   3000 Kilometers

Topography or terrain, also called "landforms," is second only to climate as a conditioner of human activity, particularly agriculture but also the location of cities and industry. A comparison of this map of mountains, valleys, plains, plateaus, and other features of the earth's surface with a map of land use (Map 15) shows that most of the world's productive agricultural zones are located in lowland and relatively level regions. Where large regions of agricultural productivity are found, we also tend to find urban concentrations and, with cities, we find industry. There is also a good spatial correlation between the map of topography and the map showing the distribution and density of the human population (Map 14). Normally the world's major landforms are the result of extremely gradual primary geologic activity such as the long-term movement of crustal plates. This activity occurs over hundreds of millions of years. Also important is the more rapid (but still slow by human standards) geomorphological or erosional activity of water, wind, glacial ice, and waves, tides, and currents. Some landforms may be produced by abrupt or "cataclysmic" events such as a major volcanic eruption or a meteor strike, but such events are relatively rare and their effects are usually too minor to show up on a map of this scale. The study of the processes that shape topography is known as "geomorphology" and is an important branch of physical geography.

-19-

Map 12a World Resources: Mineral Fuels

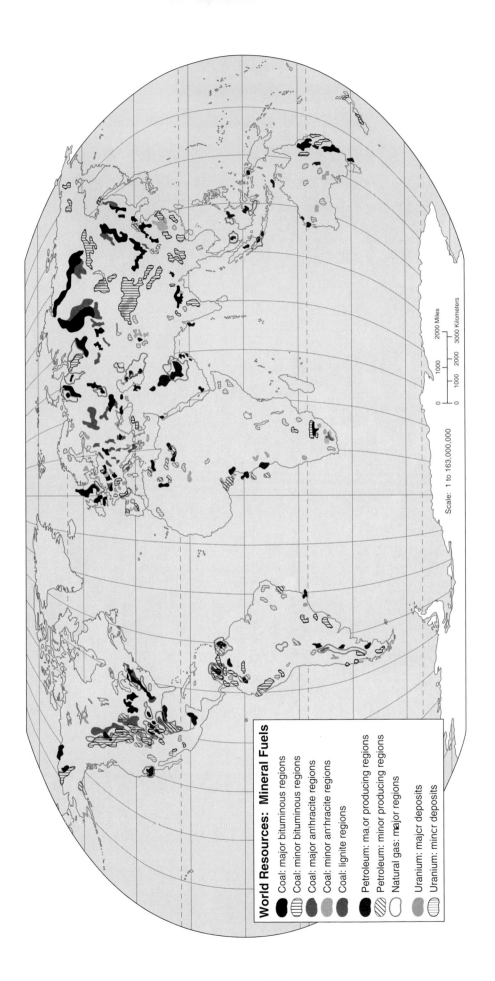

World Resources: Mineral Fuels

- Coal: major bituminous regions
- Coal: minor bituminous regions
- Coal: major anthracite regions
- Coal: minor anthracite regions
- Coal: lignite regions
- Petroleum: major producing regions
- Petroleum: minor producing regions
- Natural gas: major regions
- Uranium: major deposits
- Uranium: minor deposits

Scale: 1 to 163,000,000

```
0        1000        2000 Miles
0   1000   2000   3000 Kilometers
```

The extraction and transportation of mineral fuels rank with agriculture and forestry as "primary" human activities that impact on the environment at a global scale. Nearly all of the most highly publicized environmental disasters of recent decades—the Prince William Sound oil spill or the Chernobyl nuclear accident, for example—have involved mineral fuels that were being stored, transported, or used. And the continuing extraction of mineral fuels like oil, natural gas, coal, and uranium produces high levels of atmospheric, soil, and water pollution. The location of mineral fuels tells us a great deal about where environmental degradation is likely to be occurring or to occur in the future. One need only look at the levels of atmospheric pollution and vegetative disruption in central and eastern Europe to recognize the damaging consequences of heavy reliance on coal as a domestic and industrial fuel. The location of mineral fuels also tells us something about existing or potential levels of economic development with those countries possessing abundant reserves of mineral fuels having more of a chance to maintain or attain higher levels of prosperity.

Map 12b World Resources: Critical Metals

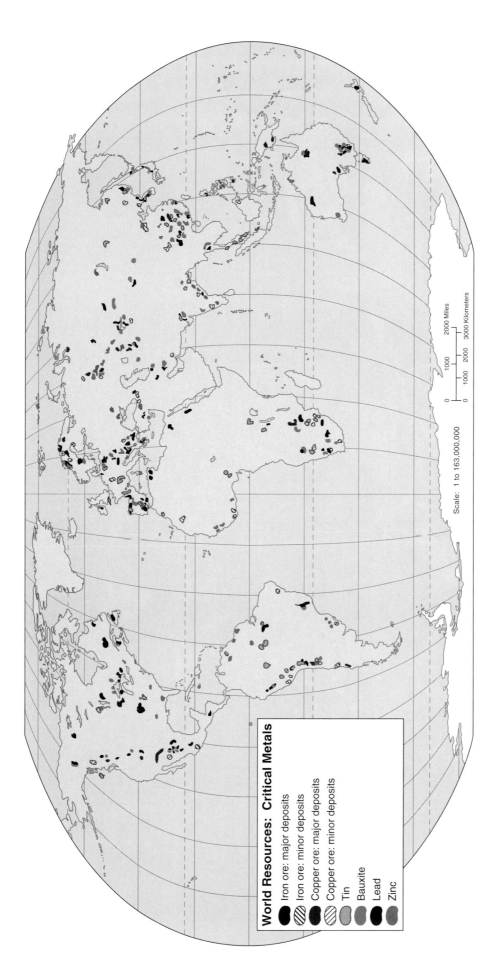

Scale: 1 to 163,000,000

World Resources: Critical Metals

- Iron ore: major deposits
- Iron ore: minor deposits
- Copper ore: major deposits
- Copper ore: minor deposits
- Tin
- Bauxite
- Lead
- Zinc

The location of deposits of critical metals such as iron, copper, tin, and others is an important determinant of the location of mining activities, like mineral fuel extraction, a "primary" economic activity. Also like mineral fuel extraction, mining for critical metallic ores makes significant environmental impact, particularly on vegetation, soils, and water resources. Some of the world's most dramatic examples of human modification of environments are located in areas of metallic ore extraction: the open pit copper mining areas of Arizona and Utah, for example. Environmental impact aside, those countries with significant critical metal deposits tend to stand a better chance of reaching higher levels of economic development, as long as they can extract and market the ores

themselves rather than having the extraction process controlled by outside concerns. The average Bolivian, for example, does not benefit greatly from the fact that his/her country is an important producer of tin and other metals. Bolivia is a "colonial dependency" country and the wealth generated by metallic ore production there tends to flow out of the country to Europe and North America. On the other hand, another South American country, Brazil, is paying for much of its own current economic development by utilizing its reserves of iron and other metals and more of the wealth from the extraction of those resources stays within the country.

Map 13 World Natural Hazards

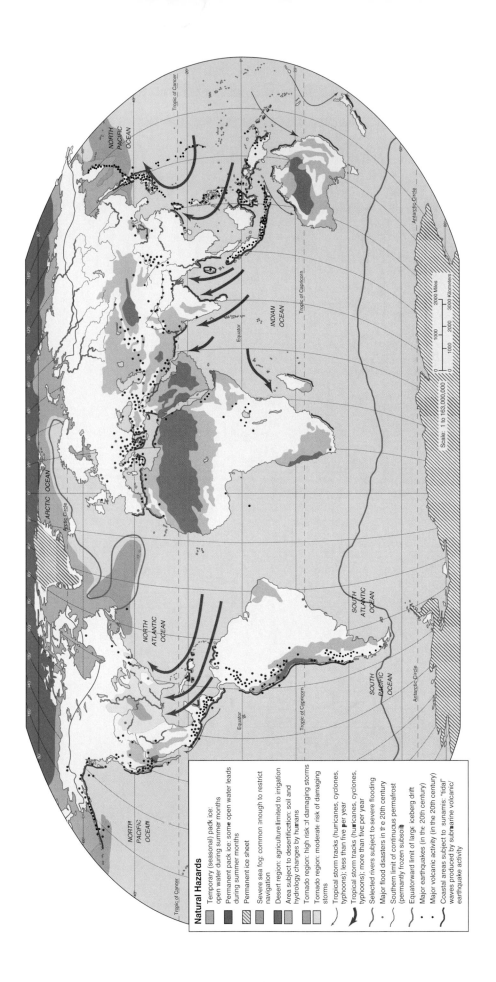

Natural Hazards

- Temporary (seasonal) pack ice: open water during summer months
- Permanent pack ice: some open water leads during summer months
- Permanent ice sheet
- Severe sea fog: common enough to restrict navigation
- Desert region: agriculture limited to irrigation
- Area subject to desertification: soil and hydrology changes by humans
- Tornado region: high risk of damaging storms
- Tornado region: moderate risk of damaging storms
- Tropical storm tracks (hurricanes, cyclones, typhoons): less than five per year
- Tropical storm tracks (hurricanes, cyclones, typhoons); more than five per year
- Selected rivers subject to severe flooding
- Major flood disasters in the 20th century
- Southern limit of continuous permafrost (permanly frozen subsoil)
- Equatorward limit of large iceberg drift
- Major earthquakes (in the 20th century)
- Major volcanic activity (in the 20th century)
- Coastal areas subject to sunamis: "tidal" waves produced by submarine volcanic/earthquake activity

Scale: 1 to 163,000,000

Unlike other elements of physical geography, most natural hazards are unpredictable. However, there are certain regions where the probability of the occurrence of a particular natural hazard is high. This map shows regions affected by major natural hazards at rates that are higher than the global norm. The presence of persistent natural hazards may influence the types of modifications that people make in the environment and certainly influence the styles of housing and other elements of cultural geography. Natural hazards may also undermine the utility of an area for economic purposes and some scholars suggest that regions of environmental instability may be regions of political instability as well. The study of natural hazards has become an important activity for "resource geographers" whose areas of interest overlap both human and physical fields of geography.

Part II

The Human Environment: Global Patterns

Map 14 World Population Density

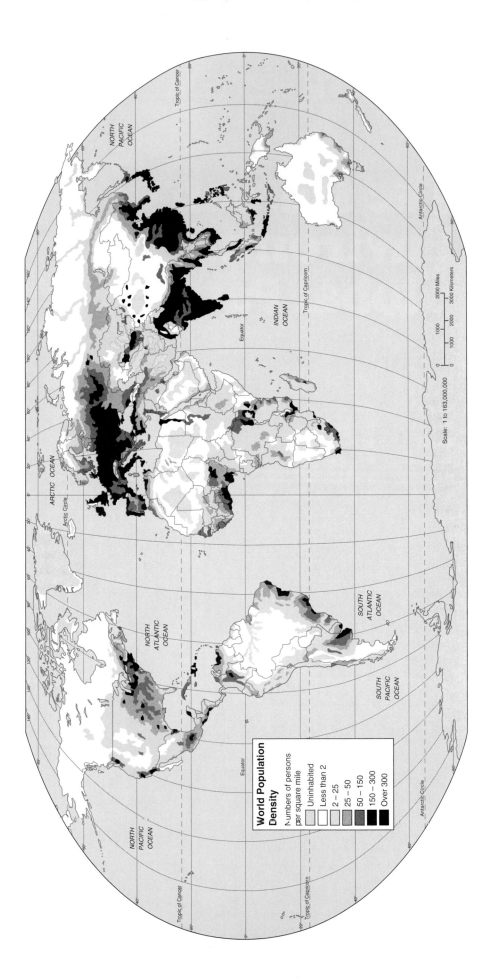

World Population Density

Numbers of persons per square mile

- Uninhabited
- Less than 2
- 2 – 25
- 25 – 50
- 50 – 150
- 150 – 300
- Over 300

Scale: 1 to 163,000,000

No feature of human activity is more reflective of geographic relationships than where people live. In the areas of densest populations, a mixture of natural and human factors has combined to allow maximum food production, maximum urbanization, and maximum centralization of economic activities. Three great concentrations of human population appear on the map—East Asia, South Asia, and Europe—with a fourth, lesser concentration in eastern North America. While population growth is relatively slow in three of these population clusters, in the fourth—South Asia—growth is still rapid and South Asia is expected to become even more densely populated in the early years of the twenty-first century, while density of the other regions is expected to remain about

as it now appears. In Europe and North America, the relatively stable population growth rates are the result of economic development that has caused population growth to level off within the last century. In East Asia, the growth rates have also begun to decline. In the case of Japan, Taiwan, the Koreas, and other more highly developed nations of the Pacific Rim, the reduced growth is the result of economic development. In China, at least until recently, lowered population growth rates have resulted from strict family planning. The areas of future high density of population, in addition to those already existing, are likely to be in Middle and South America and in Central Africa, where population growth rates are well above the world average.

Map 15 World Economic Activity

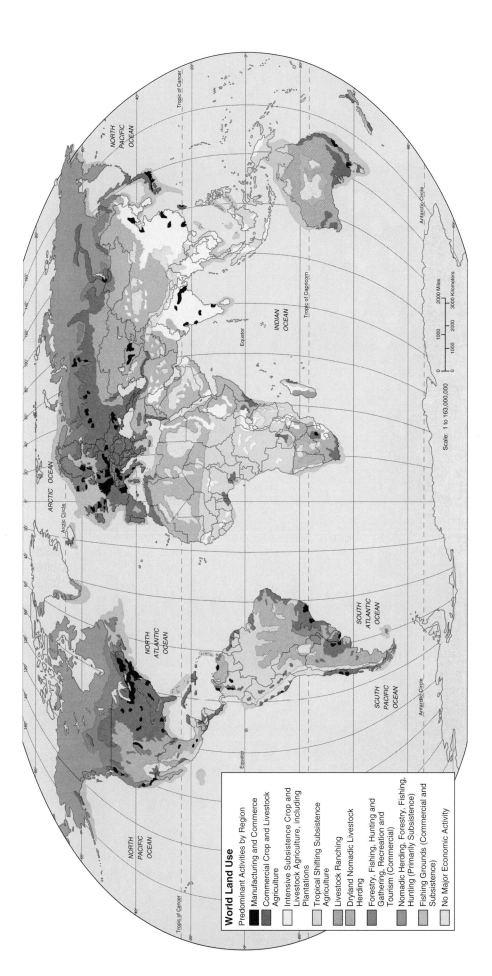

World Land Use

Predominant Activities by Region

- Manufacturing and Commerce
- Commercial Crop and Livestock Agriculture
- Intensive Subsistence Crop and Livestock Agriculture, including Plantations
- Tropical Shifting Subsistence Agriculture
- Livestock Ranching
- Dryland Nomadic Livestock Herding
- Forestry, Fishing, Hunting and Gathering, Recreation and Tourism (Commercial)
- Nomadic Herding, Forestry, Fishing, Hunting (Primarily Subsistence)
- Fishing Grounds (Commercial and Subsistence)
- No Major Economic Activity

Scale: 1 to 163,000,000

Land uses can be categorized as lying somewhere on a scale between extensive uses in which human activities are dispersed over relatively large areas and intensive uses in which human activities are concentrated in relatively small areas. Many of the most important land use patterns of the world (such as urbanization, industry, mining, or transportation) are intensive and therefore relatively small in area and not easily seen on maps of this scale. Hence, even in the areas identified as "Manufacturing and Commerce" on the map there are many land uses that are not strictly industrial or commercial in nature and, in fact, more extensive land uses (farming, residential, open space)

may actually cover more ground than the intensive industrial or commercial activities. On the other hand, the more extensive land uses, like agriculture and forestry, tend to dominate the areas in which they are found. Thus, primary economic activities such as agriculture and forestry tend to dominate the world map of land use because of their extensive character. Much of this map is, therefore, a map that shows the global variations in agricultural patterns. Note, among other things, the differences between land use patterns in the more developed countries of the temperate zones and the less developed countries of the tropics.

Map 16 World Urbanization

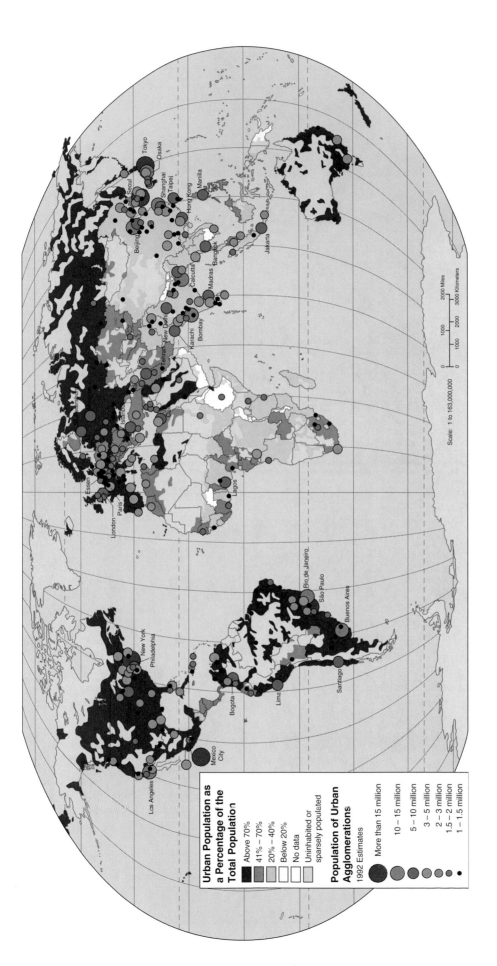

Urban Population as a Percentage of the Total Population

- Above 70%
- 41% – 70%
- 20% – 40%
- Below 20%
- No data
- Uninhabited or sparsely populated

Population of Urban Agglomerations

1992 Estimates

- More than 15 million
- 10 – 15 million
- 5 – 10 million
- 3 – 5 million
- 2 – 3 million
- 1.5 – 2 million
- 1 – 1.5 million

Scale: 1 to 163,000,000

The degree to which a region's population is concentrated in urban areas is a major indicator of a number of things: the potential for environmental impact, the level of economic development, and the problems associated with human concentrations. Urban dwellers are rapidly becoming the norm among the world's people and rates of urbanization are increasing worldwide, with the greatest increases in urbanization taking place in developing or developing regions. Whether in developed or developing countries, those who live in cities exert an influence on the environment, politics, economics, and social systems that go far beyond the confines of the city itself. Acting as the focal points for the flow of goods and ideas, cities draw resources and people not just from their immediate hinterland but from the entire world. This process creates far-reaching impacts as resources are extracted, converted through industrial processes, and transported over great distances to metropolitan regions, and as ideas spread or *diffuse* along with the movements of people to cities and the flow of communication from them. The significance of urbanization can be most clearly seen, perhaps, in North America where, in spite of vast areas of relatively unpopulated land, well over 90 percent of the population lives in urban areas.

Map 17 World Transportation Patterns

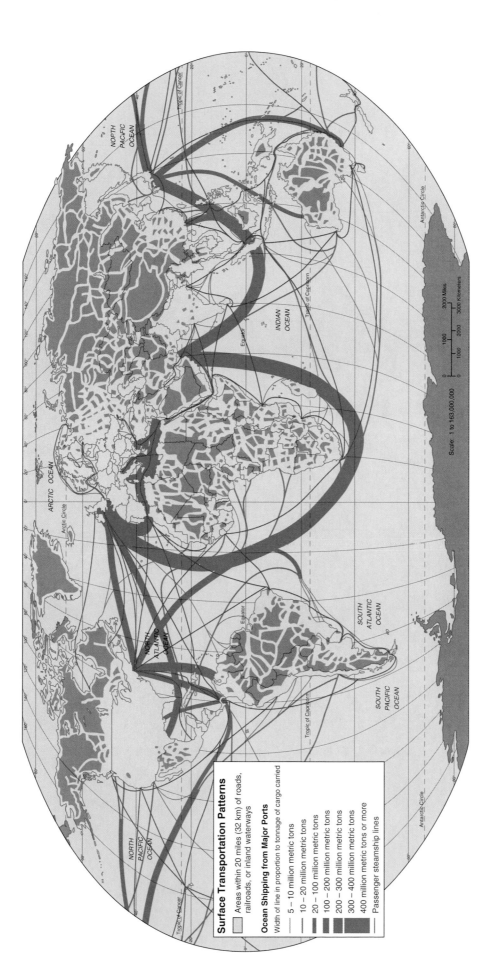

Surface Transportation Patterns

Areas within 20 miles (32 km) of roads, railroads, or inland waterways

Ocean Shipping from Major Ports

Width of line in proportion to tonnage of cargo carried

5 – 10 million metric tons

10 – 20 million metric tons

20 – 100 million metric tons

100 – 200 million metric tons

200 – 300 million metric tons

300 – 400 million metric tons

400 million metric tons or more

Passenger steamship lines

Scale: 1 to 163,000,000

As a form of land use, transportation is second only to agriculture in its coverage of the earth's surface and is one of the clearest examples in the human world of a *network*, a linked system of lines allowing flows from one place to another. The global transportation network and its related communication web is responsible for most of the *spatial interaction* or movement of goods, people, and ideas between places. As the chief mechanism of spatial interaction, transportation is linked firmly with the concept of a shrinking world and the development of a global community and economy. Because transportation systems require significant modification of the earth's surface, transportation is also responsible for massive alterations in the quantity and quality of water, for major soil degradations and erosion, and (indirectly) for the air pollution that emanates from vehicles utilizing the transportation system. In addition, as improved transportation technology draws together places on the earth that were formerly remote, it allows people to impact environments a great distance away from where they live.

Map 18 World Religions

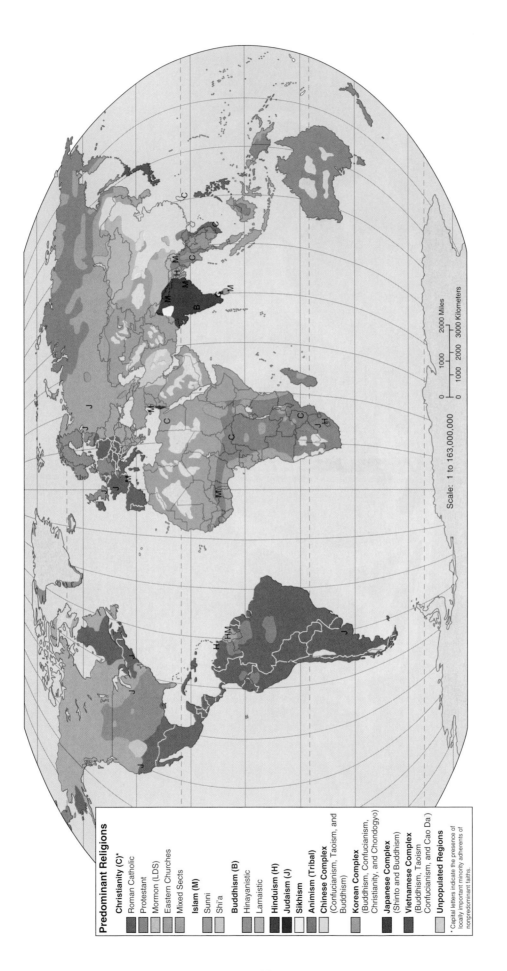

Predominant Religions

Christianity (C)*
- Roman Catholic
- Protestant
- Mormon (LDS)
- Eastern Churches
- Mixed Sects

Islam (M)
- Sunni
- Shi'a

Buddhism (B)
- Hinayanistic
- Lamaistic

Hinduism (H)

Judaism (J)

Sikhism

Animism (Tribal)

Chinese Complex
(Confucianism, Taoism, and Buddhism)

Korean Complex
(Buddhism, Confucianism, Christianity, and Chondogyo)

Japanese Complex
(Shinto and Buddhism)

Vietnamese Complex
(Buddhism, Taoism Confucianism, and Cao Da)

Unpopulated Regions

* Capital letters indicate the presence of locally important minority adherents of nonpredominant faiths.

Scale: 1 to 163,000,000

0   1000   2000 Miles
0   1000 2000 3000 Kilometers

-28-

Religious adherence is one of the fundamental defining characteristics of human culture, the style of life adopted by a people and passed from one generation to the next. Because of the importance of religion for culture, a depiction of the spatial distribution of religions is as close as we can come to a map of cultural patterns. More than just a set of behavioral patterns having to do with worship and ceremony, religion is a vital conditioner of the ways that people deal with one another, with their institutions, and with the environments they occupy. In many areas of the world, the ways in which people make a living, the patterns of occupation that they create on the land, and the impacts that they make on ecosystems are the direct consequences of their adherence to a religious faith. An examination of the map in the context of international and intranational conflict will also show that tension between countries and the internal stability of states is also a function of the spatial distribution of religion.

Map 19 World Languages

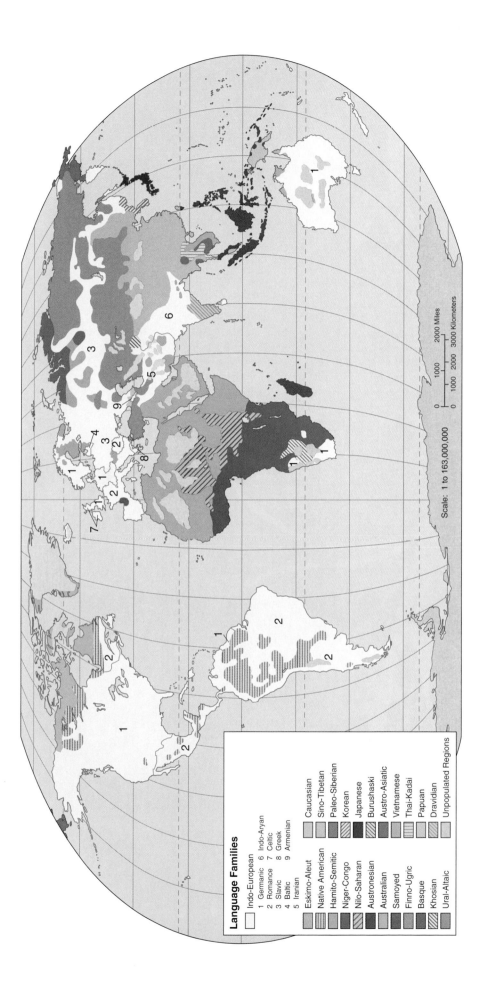

Language Families

Indo-European

1 Germanic 6 Indo-Aryan
2 Romance   7 Celtic
3 Slavic   8 Greek
4 Baltic   9 Armenian
5 Iranian

- Eskimo-Aleut
- Native American
- Hamito-Semitic
- Niger-Congo
- Nilo-Saharan
- Austronesian
- Australian
- Samoyed
- Finno-Ugric
- Basque
- Khosian
- Ural-Altaic
- Caucasian
- Sino-Tibetan
- Paleo-Siberian
- Korean
- Japanese
- Burushaski
- Austro-Asiatic
- Vietnamese
- Thai-Kadai
- Papuan
- Dravidian
- Unpopulated Regions

Scale: 1 to 163,000,000

0   1000   2000 Miles
0 1000 2000 3000 Kilometers

Language, like religion, is an important identifying characteristic of culture. Indeed, it is perhaps the most durable of all those identifying characteristics or *cultural traits*: language, religion, institutions, material technologies, and ways of making a living. After centuries of exposure to other languages or even conquest by speakers of other languages, the speakers of a specific tongue will often retain their own linguistic identity. As a geographic element, language helps us to locate areas of potential conflict, particularly in regions where two or more languages overlap. Many, if not most, of the world's conflict zones are also areas of linguistic diversity and knowing the distribution of languages helps us to understand some of the reasons behind important current events: for example, linguistic identity differences played an important part in the disintegration of the Soviet Union in the 1990s; and in areas emerging from recent colonial rule, such as Africa, the participants in conflicts over territory and power are often defined in terms of linguistic groups. Language distributions also help us to comprehend the nature of the human past by providing clues that enable us to chart the course of human migrations, as shown in the distribution of Indo-European, Austronesian, or Hamito-Semitic languages. Finally, because languages have a great deal to do with the way people perceive and understand the world around them, linguistic patterns help to explain the global variations in the ways that people interact with their environments.

Map 20 World External Migrations in Modern Times

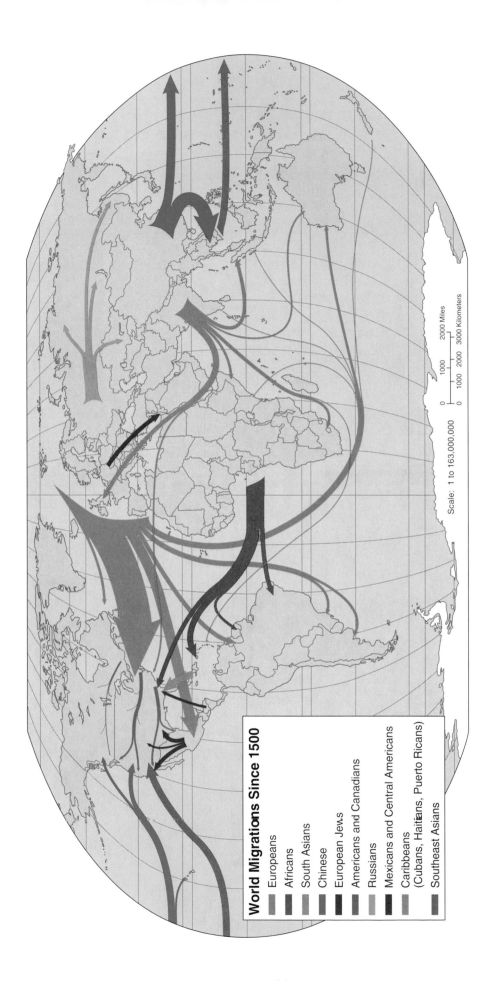

World Migrations Since 1500

- Europeans
- Africans
- South Asians
- Chinese
- European Jews
- Americans and Canadians
- Russians
- Mexicans and Central Americans
- Caribbeans (Cubans, Haitians, Puerto Ricans)
- Southeast Asians

Scale: 1 to 163,000,000

| 0 | 1000 | 2000 Miles |

| 0 | 1000 | 2000 | 3000 Kilometers |

Migration has had a significant effect on world geography, contributing to cultural change and development, to the diffusion of ideas and innovations, and to the complex mixture of people and cultures found in the world today. *Internal migration* occurs within the boundaries of a country; *external migration* is movement from one country or region to another. Over the last 50 years, the most important migrations in the world have been internal, largely the rural-to-urban migration that has been responsible for the recent rise of global urbanization. Prior to the mid-twentieth century, three types of external migrations were most important: *voluntary*, most often in search of better con-

ditions and opportunities; *involuntary* or *forced*, involving people who have been driven from their homelands by war, political unrest, or environmental disasters, or who have been transported as slaves or prisoners; and *imposed*, not entirely forced but which conditions make highly advisable. Human migrations in recorded history have been responsible for major changes in the patterns of languages, religions, ethnic composition, and economies. Particularly during the last 500 years, migrations of both the voluntary and involuntary or forced type have literally reshaped the human face of the earth.

Part III

Global Demographic Patterns

Map 21 Population Growth Rates

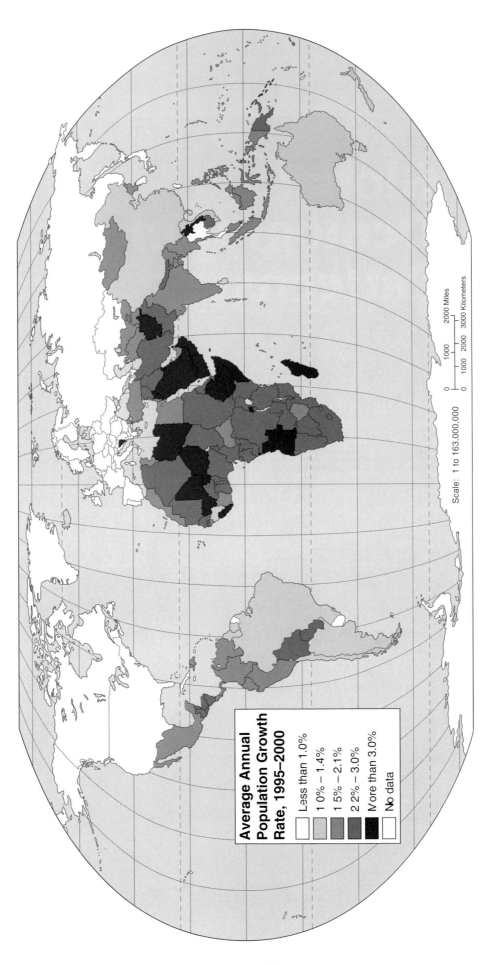

Average Annual Population Growth Rate, 1995–2000

- Less than 1.0%
- 1.0% – 1.4%
- 1.5% – 2.1%
- 2.2% – 3.0%
- More than 3.0%
- No data

Scale: 1 to 163,000,000

0   1000   2000 Miles
0   1000   2000   3000 Kilometers

Of all the statistical measurements of human population, that of the rate of population growth is the most important. The growth rate of a population is a combination of natural change (births and deaths), in-migration, and out-migration; it is obtained by adding the number of births to the number of immigrants during a year and subtracting from that total the sum of deaths and emigrants for the same year. For a specific country, this figure will determine many things about the country's future ability to feed, house, educate, and provide medical services to its citizens. Some of the countries with the largest populations (such as India) also have high growth rates. Since these countries tend to be in developing regions, the combination of high population and high growth rates poses special problems for continuing economic development and carries heightened risks of environmental degradation. Many people believe that the rapidly expanding world population is a potential crisis that may cause environmental and human disaster by the middle of the twenty-first century.

Map 22 Total Fertility Rates

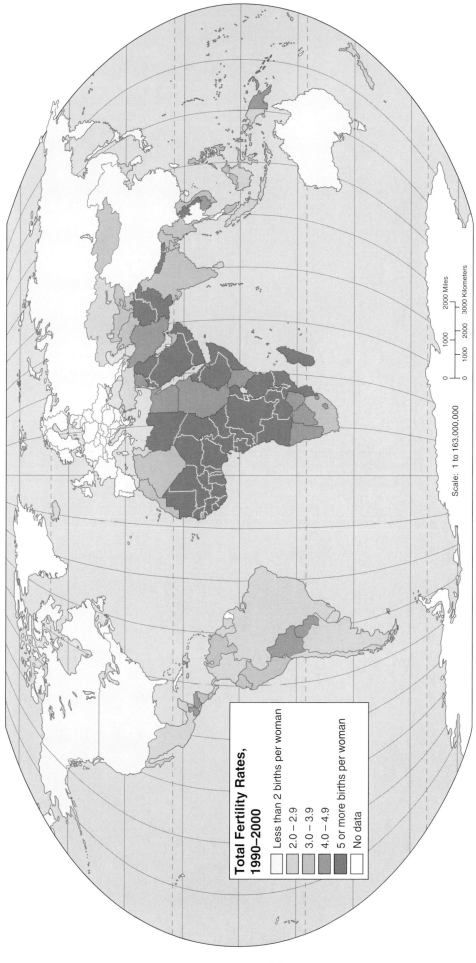

Scale: 1 to 163,000,000

Total Fertility Rates, 1990–2000

- Less than 2 births per woman
- 2.0 – 2.9
- 3.0 – 3.9
- 4.0 – 4.9
- 5 or more births per woman
- No data

0   1000   2000 Miles
0   1000   2000   3000 Kilometers

The fertility rate measures the number of children that a woman is expected to bear during her lifetime, based on the age-specific fertility figures of women between 15 and 40 (the normal childbearing years). While fertility rates tell us a great deal about present population growth, with high fertility rates indicating high population growth rates, they are also indicative of potential or projected growth. A country whose women can be expected to bear many children is a country with enormous potential for population growth in the future. Given present fertility rates, for example, the number of offspring from the average German woman over the next three generations (the total number of children, grandchildren, and great-grandchildren) will be 7. During the same three generations, the average American woman will have a total of 17 children, grandchildren, and great-grandchildren. But during this time, assuming that present fertility rates are maintained, the average woman in sub-Saharan Africa will have [&em]258[&stop] children, grandchildren, and great-grandchildren. You might be interested in working out some potential population growth rates over two or three generations, using the data as presented on the map.

-33-

Map 23 Infant Mortality Rates

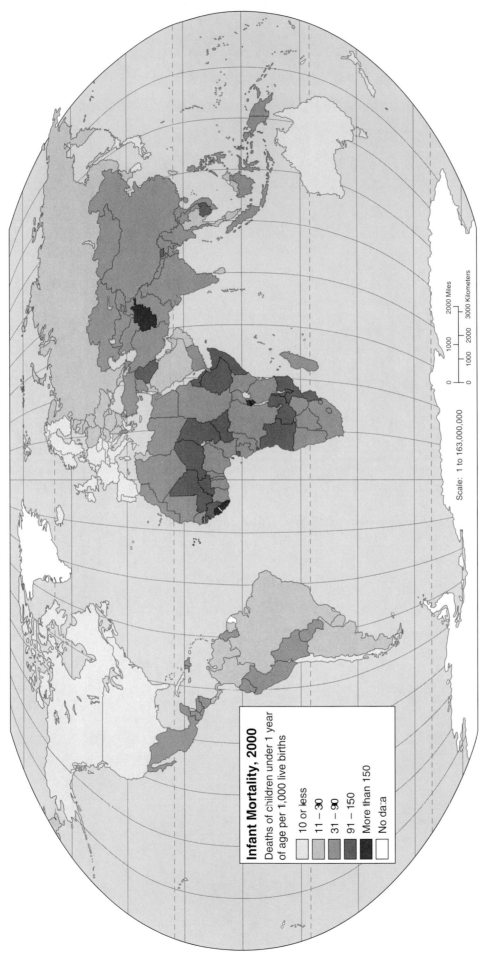

Infant Mortality, 2000

Deaths of children under 1 year
of age per 1,000 live births

	10 or less
	11 – 30
	31 – 90
	91 – 150
	More than 150
	No data

Scale: 1 to 163,000,000

0   1000   2000 Miles
0   1000   2000   3000 Kilometers

Infant mortality rates are calculated by dividing the number of children born in a given year who die before their first birthday by the total number of children born that year—and then multiplying by 1,000; this shows how many infants have died for every 1,000 births. Infant mortality rates are prime indicators of economic development. In highly developed economies, with advanced medical technologies, sufficient diets, and adequate public sanitation, infant mortality rates tend to be quite low. By contrast, in less developed countries, with the disadvantages of poor diet, limited access to medical technology, and the other problems of poverty, infant mortality rates tend to be high.

Although worldwide infant mortality has decreased significantly during the last 2 decades, many regions of the world still experience infant mortality above the 10 percent level (100 deaths per 1,000 live births). Such infant mortality rates represent not only human tragedy at its most basic level, but also are powerful inhibiting factors for the future of human development. Comparing infant mortality rates in the midlatitudes and the tropics shows that children in most African countries are more than 10 times as likely to die within a year of birth as children in European countries.

Map 24 Child Mortality Rate

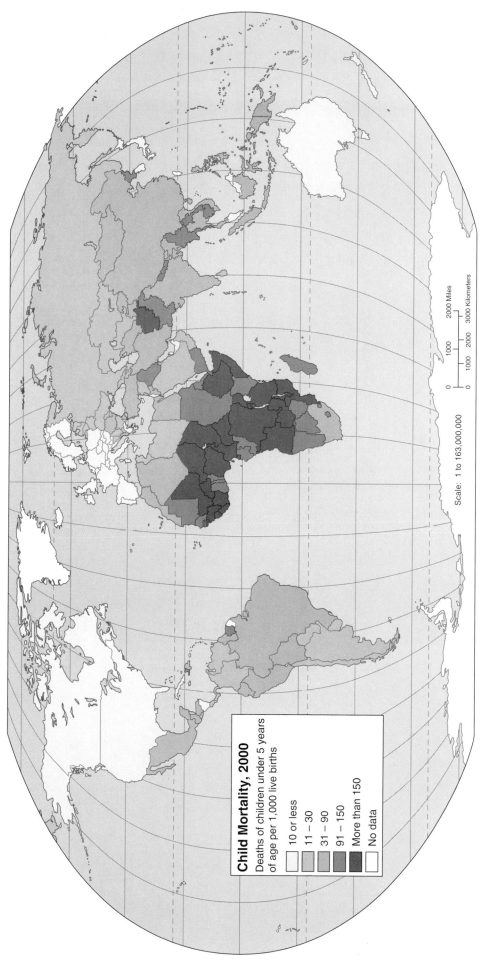

Child Mortality, 2000

Deaths of children under 5 years
of age per 1,000 live births

- 10 or less
- 11 – 30
- 31 – 90
- 91 – 150
- More than 150
- No data

Scale: 1 to 163,000,000

0	1000	2000 Miles
0	1000 2000	3000 Kilometers

Child mortality rates are calculated by determining the probability that a child born in a specified year will die before reaching the age 5, using current age-specific mortality rates for a population. The major sources of mortality rates are vital registration systems and estimates made from surveys and/or census reports. Along with infant mortality and average life-expectancy rates, child mortality rates, according to the World Bank, "are probably the best general indicators of a community's current health status and are often cited as overall measures of a population's welfare or quality of life." Where infant mortality often reflects health care conditions, child mortality is usually a reflection of the inadequacy of nutrition, leading to early deaths from nutritionally related diseases. In some less developed countries in Africa and Asia, child mortality is also an indicator of the widespread presence of infectious diseases such as malaria, tuberculosis, and HIV/AIDS.

Map 25 Average Life Expectancy at Birth

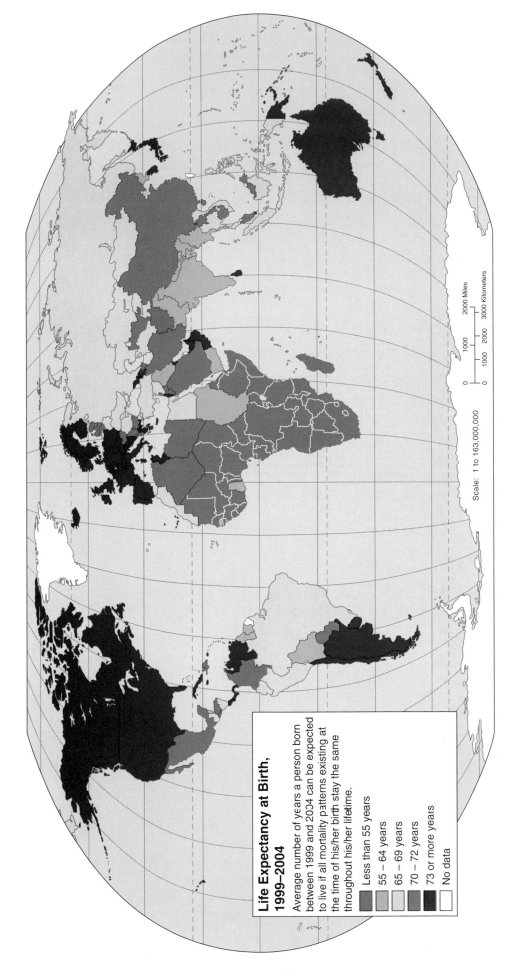

Life Expectancy at Birth, 1999–2004

Average number of years a person born between 1999 and 2004 can be expected to live if all mortality patterns existing at the time of his/her birth stay the same throughout his/her lifetime.

- Less than 55 years
- 55 – 64 years
- 65 – 69 years
- 70 – 72 years
- 73 or more years
- No data

Scale: 1 to 163,000,000

0   1000   2000 Miles

0   1000   2000   3000 Kilometers

Average life expectancy at birth is a measure of the average longevity of the population of a country. Like all average measures, it is distorted by extremes. For example, a country with a high mortality rate among children will have a low average life expectancy. Thus, an average life expectancy of 45 years does not mean that everyone can be expected to die at the age of 45. More normally, what the figure means is that a substantial number of children die between birth and 5 years of age, thus reducing the average life expectancy for the entire population. In spite of the dangers inherent in misinterpreting the data, average life expectancy (along with infant mortality and several other measures) is a valid way of judging the relative health of a population. It reflects the nature of the health care system, public sanitation and disease control, nutrition, and a number of other key human need indicators. As such, it is a measure of well-being that is significant in indicating economic development.

Map 26 Population by Age Group

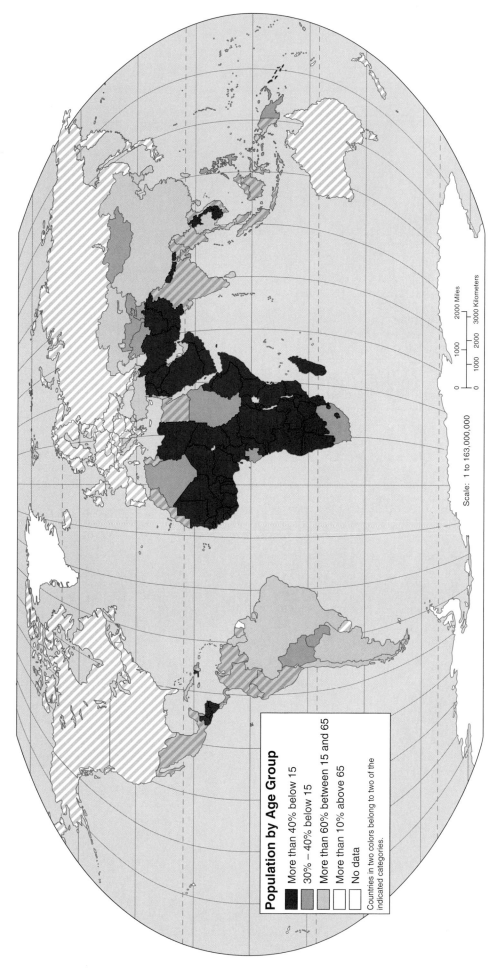

Population by Age Group

- More than 40% below 15
- 30% – 40% below 15
- More than 60% between 15 and 65
- More than 10% above 65
- No data

Countries in two colors belong to two of the indicated categories.

Scale: 1 to 163,000,000

0   1000   2000 Miles

0   1000   2000   3000 Kilometers

Of all the measurements that illustrate the dynamics of a population, age distribution may be the most significant, particularly when viewed in combination with average growth rates. The particular relevance of age distribution is that it tells us what to expect from a population in terms of growth over the next generation. If, for example, approximately 40–50 percent of a population is below the age of 15, that suggests that in the next generation about one-quarter of the total population will be women of childbearing age. When age distribution is combined with fertility rates (the average number of children born per woman in a population), an especially valid mea-

surement of future growth potential may be derived. A simple example: Nigeria, with a 1995 population of 127 million, has 47 percent of its population below the age of 15 and a fertility rate of 6.4; the United States, with a 1995 population of 263 million, has 22 percent of its population below the age of 15 and a fertility rate of 2.1. During the period in which those women presently under the age of 15 are in their childbearing years, Nigeria can be expected to add a total of approximately 197 million persons to its total population. Over the same period, the United States can be expected to add only 61 million.

Map 27 World Daily Per Capita Food Supply

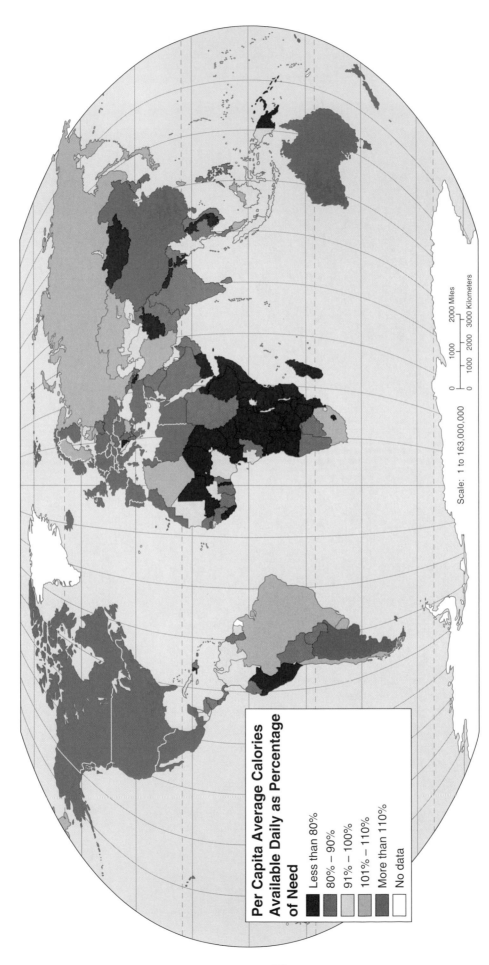

Per Capita Average Calories Available Daily as Percentage of Need

- Less than 80%
- 80% – 90%
- 91% – 100%
- 101% – 110%
- More than 110%
- No data

Scale: 1 to 163,000,000

0   1000   2000 Miles
0   1000   2000   3000 Kilometers

The data shown on this map indicate the presence or absence of critical food shortages. While they do not necessarily indicate the presence of starvation or famine, they certainly do indicate potential problem areas for the next decade. The measurements are in calories from all food sources: domestic production, international trade, drawdown on stocks or food reserves, and direct fore gn contributions or aid. The quantity of calories available is that amount, estimated by the UN's Food and Agriculture Organization (FAO), that actually reaches consumers. The calories actually consumed may be lower than the figures shown, depending on how much is lost in a variety of ways: in home storage (to pests such as rats and mice), in preparation and cooking, through consumption by pets and domestic animals, and as discarded foods, for example. The former practice in such maps was to evaluate available calories as a percentage of "need" or minimum daily requirements to maintain health. Such a statistical measure was virtually impossible to standardize for the variety of human types in the world and for such variables as age and sex distribution. A newer form of measure—available calories as a percentage of the world average available—eliminates many of the problems of the former set of numbers while still maintaining a good relative picture of global hunger.

Map 28 Illiteracy Rate

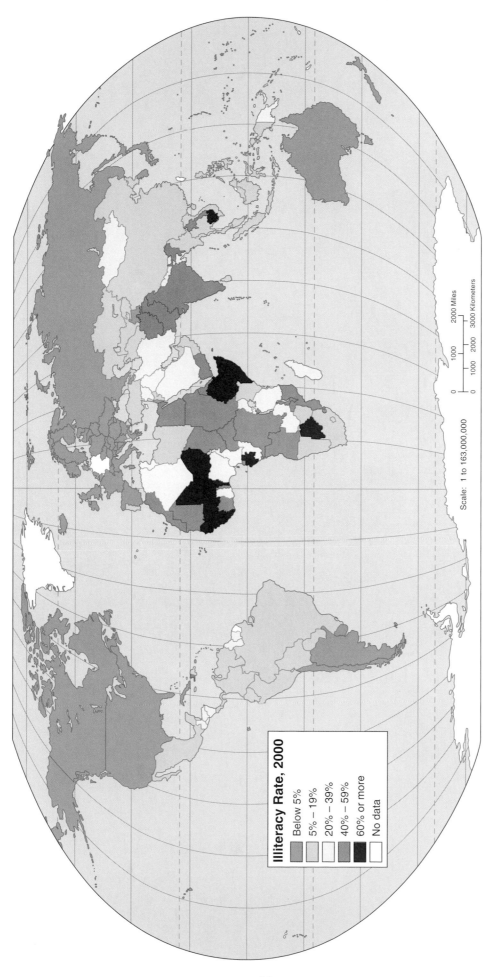

Illiteracy Rate, 2000

- Below 5%
- 5% – 19%
- 20% – 39%
- 40% – 59%
- 60% or more
- No data

Scale: 1 to 163,000,000

0   1000   2000 Miles

0   1000   2000   3000 Kilometers

The gains in living standards that developing countries have experienced during the last two decades are manifested in two major areas: life expectancy and literacy. The increase in global literacy is largely the consequence of an increase in primary school enrollment, particularly throughout Middle and South America, Africa, and Asia. Worldwide, education is perceived as a way to advance economic status. Unfortunately, although gains have been made, there are still countries where illiteracy rates—particularly among the females of the population—are well above global norms. The long-term potential of these countries is severely compromised as a result. [&b]A word of caution:[&stop] Most countries view their literacy or illiteracy rates as hallmarks of their status in the world community and there is, therefore, a tendency to overstate or overestimate literacy (or, conversely, to underestimate illiteracy). In the United States for example, the stated illiteracy rate is less than 2%. Yet many experts indicate that somewhere between 10% and 15% of the U.S. population may, in fact, be functionally illiterate—that is, unable to read street signs, advertisements, or newspapers.

Map 29 Female/Male Inequality in Education and Employment

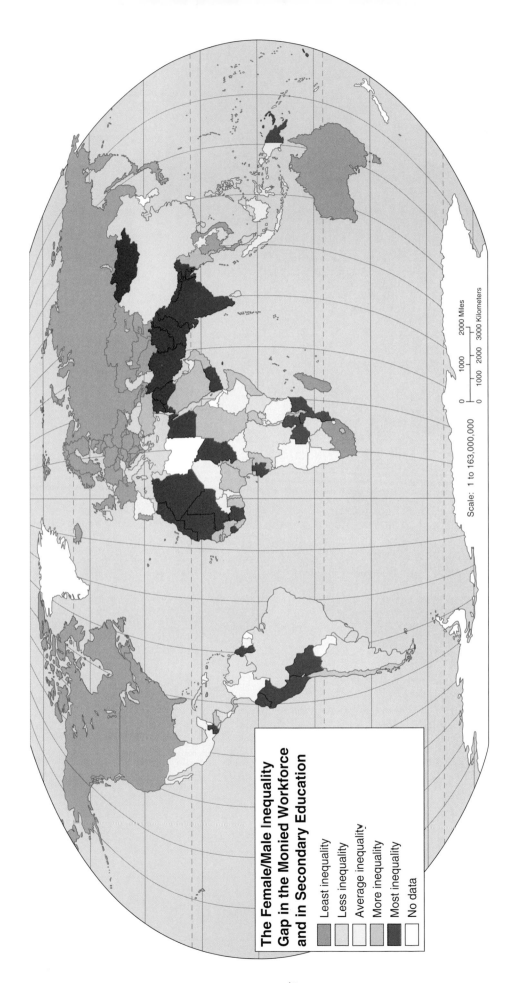

The Female/Male Inequality Gap in the Monied Workforce and in Secondary Education

- Least inequality
- Less inequality
- Average inequality
- More inequality
- Most inequality
- No data

Scale: 1 to 163,000,000

0   1000   2000 Miles

0   1000   2000   3000 Kilometers

While women in developed countries, particularly in North America and Europe, have made significant advances in socioeconomic status in recent years, in most of the world females suffer from significant inequality when compared with their male counterparts. Although women have received the right to vote in most of the world's countries, in over 90 percent of these countries that right has only been granted in the last 50 years. In most regions, literacy rates for women still fall far short of those for men; In Africa and Asia, for example, only about half as many women are as literate as men. Women marry considerably younger than men and attend school for shorter periods of time. Inequalities in education and employment are perhaps the most telling indicators of the unequal status of women in most of the world. Lack of secondary education in comparison with men prevents women from entering the workforce with equally high-paying jobs. Even where women are employed in positions similar to those held by men, they still tend to receive less compensation. The gap between rich and poor involves not only a clear geographic differentiation, but a clear gender differentiation as well.

-40-

Map 30 Global Scourges: Major Infectious Diseases

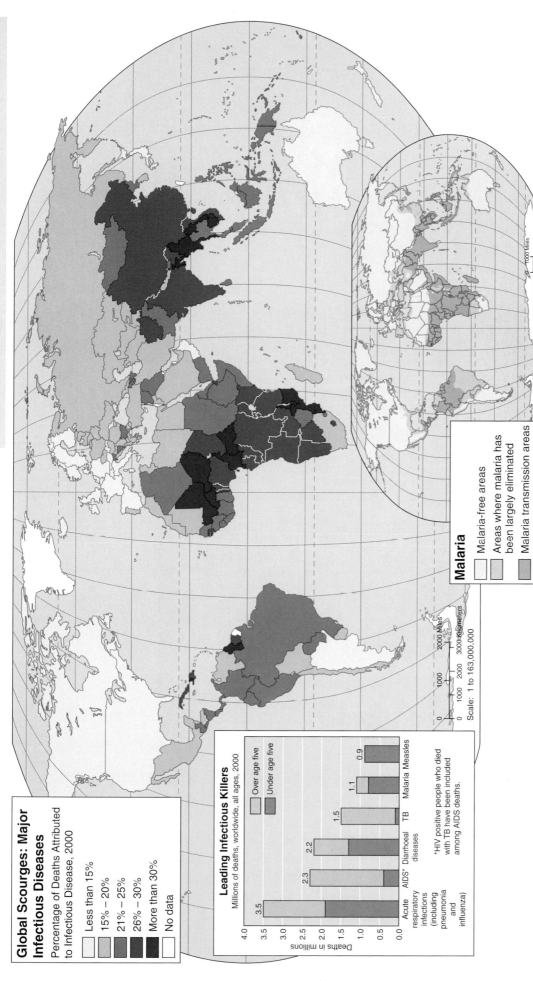

Global Scourges: Major Infectious Diseases

Percentage of Deaths Attributed to Infectious Disease, 2000

- Less than 15%
- 15% – 20%
- 21% – 25%
- 26% – 30%
- More than 30%
- No data

Leading Infectious Killers

Millions of deaths, worldwide, all ages, 2000

- Over age five
- Under age five

	Deaths in millions
Acute respiratory infections (including pneumonia and influenza)	3.5
AIDS*	2.3
Diarrhoeal diseases	2.2
TB	1.5
Malaria	1.1
Measles	0.9

*HIV positive people who died with TB have been included among AIDS deaths.

Malaria

- Malaria-free areas
- Areas where malaria has been largely eliminated
- Malaria transmission areas

Scale: 1 to 360,000,000

0 1000 Kilometers
0 1000 Miles

Scale: 1 to 163,000,000

0 1000 2000 3000 Kilometers
0 1000 2000 Miles

Infectious diseases are the world's leading cause of premature death and at least half of the world's population is, at any time, at risk of contracting an infectious disease. Although we often think of infectious diseases as being restricted to the tropical world (malaria, dengue fever), many if not most of them have attained global proportions. A major case in point is HIV/AIDS, which quite probably originated in Africa but has, over the last two decades, spread throughout the entire world. Major diseases of the nineteenth century, such as cholera and tuberculosis, are making a major comeback in many parts of the world, in spite of being preventable or treatable. Part of the problem with infectious diseases is that they tend to be associated with poverty (poor nutrition, poor sanitation, substandard housing, and so on) and, therefore, are seen as a problem of undeveloped countries, with the consequent lack of funding for prevention and treatment. Infectious diseases are also tending to increase because lifesaving drugs, such as antibiotics and others used in the fight against diseases, are losing their effectiveness as bacteria develop genetic resistance to them. The problem of global warming is also associated with a spread of infectious diseases as many disease vectors (certain species of mosquito, for example) are spreading into higher latitudes with increasingly warm temperatures and are spreading disease into areas where populations have no resistance to them. Infectious diseases have become something greater than simply a health issue of poor countries. They are now major social problems with potentially enormous consequences for the entire world.

Map 31 The Quality of Life: The Index of Human Development

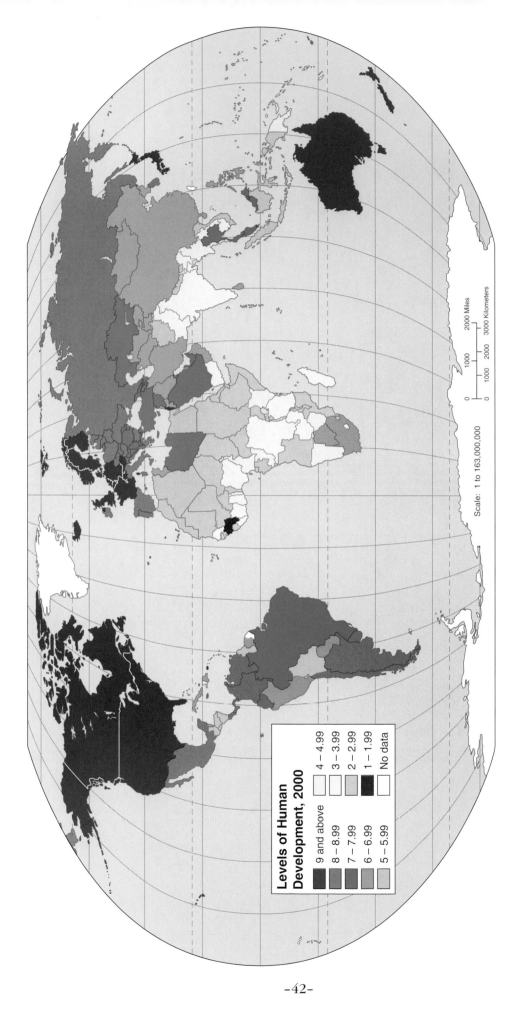

Levels of Human Development, 2000

- 9 and above
- 8 – 8.99
- 7 – 7.99
- 6 – 6.99
- 5 – 5.99
- 4 – 4.99
- 3 – 3.99
- 2 – 2.99
- 1 – 1.99
- No data

Scale: 1 to 163,000,000

0   1000   2000   3000 Kilometers

0   1000   2000 Miles

The development index upon which this map is based takes into account a wide variety of demographic, health, and educational data, including population growth, per capita gross domestic income, longevity, literacy, and years of schooling. The map reveals significant improvement in the quality of life in Middle and South America, although it is questionable whether the gains made in those regions can be maintained in the face of the dramatic population increases expected over the next 30 years. More clearly than anything else, the map illustrates the near-desperate situation in Africa and South Asia. In those regions, the unparalleled growth in population threatens to overwhelm all efforts to improve the quality of life. In Africa, for example, the population is increasing by 20 million persons per year. With nearly 45 percent of the continent's population aged 15 years or younger, this growth rate will accelerate as the women reach childbearing age. Africa, along with South Asia, faces the very difficult challenge of providing basic access to health care, education, and jobs for a rapidly increasing population. The map also illustrates the striking difference in quality of life between those who inhabit the world's equatorial and tropical regions and those fortunate enough to live in the temperate zones, where the quality of life is significantly higher.

Part IV

Global Economic Patterns

Map 32 Rich and Poor Countries: Gross National Income

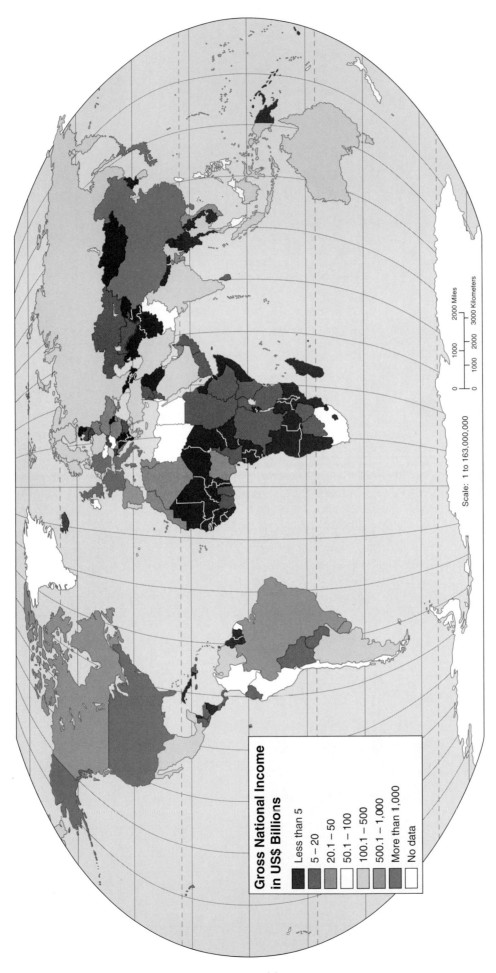

Scale: 1 to 163,000,000

Gross National Income
in US$ Billions

- Less than 5
- 5 – 20
- 20.1 – 50
- 50.1 – 100
- 100.1 – 500
- 500.1 – 1,000
- More than 1,000
- No data

Gross National Income (GNI) is the broadest measure of national income and measures the total claims of a country's residents to all income from domestic and foreign products during a year. Although GNI is often misleading and commonly incomplete, it is often used by economists, geographers, political scientists, policy makers, development experts, and others not only as a measure of relative well-being but also as an instrument of assessing the effectiveness of economic and political policies. What is wrong with GNI? First of all, it does not take into account a number of real economic factors, such as, environmental deterioration, the accumulation or degradation of human and social capital, or the value of household work. Yet in spite of these deficiencies, GNI is still a reasonable way to assess the relative wealth of nations: the vast differences in wealth that separate the poorest countries from the richest. One of the more striking features of the map is the evidence that such a small number of countries possess so many of the world's riches (keeping in mind that GNI provides no measure of the distribution of wealth within a country).

Map 33 Gross National Income Per Capita

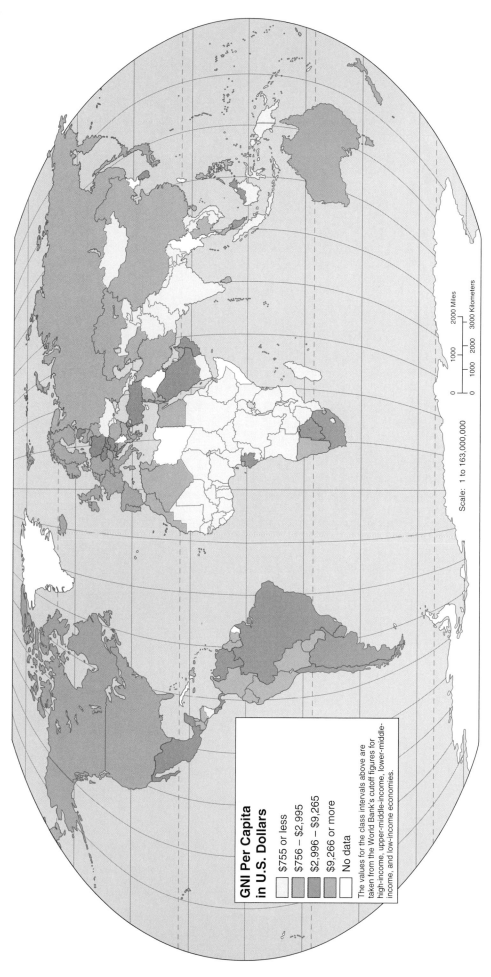

GNI Per Capita
in U.S. Dollars

- $755 or less
- $756 – $2,995
- $2,996 – $9,265
- $9,266 or more
- No data

The values for the class intervals above are taken from the World Bank's cutoff figures for high-income, upper-middle-income, lower-middle-income, and low-income economies.

Scale: 1 to 163,000,000

0   1000   2000 Miles

0   1000   2000   3000 Kilometers

Gross National Income in either absolute or per capita form should be used cautiously as a yardstick of economic strength, because it does not measure the distribution of wealth among a population. There are countries (most notably, the oil-rich countries of the Middle East) where per capita GNI is high but where the bulk of the wealth is concentrated in the hands of a few individuals, leaving the remainder in poverty. Even within countries in which wealth is more evenly distributed (such as those in North America or western Europe), there is a tendency for dollars or pounds sterling or francs or marks to concentrate in the bank accounts of a relatively small percentage of the population. Yet the maldistribution of wealth tends to be greatest in the less developed countries, where the per capita GNI is far lower than in North America and western Europe, and poverty is widespread. In fact, a map of GNI per capita offers a reasonably good picture of comparative economic well-being. It should be noted that a low per capita GNI does not automatically condemn a country to low levels of basic human needs and services. There are a few countries, such as Costa Rica and Sri Lanka, that have relatively low per capita GNI figures but rank comparatively high in other measures of human well-being, such as average life expectancy, access to medical care, and literacy.

Map 34 Economic Growth: GDP Change Per Capita

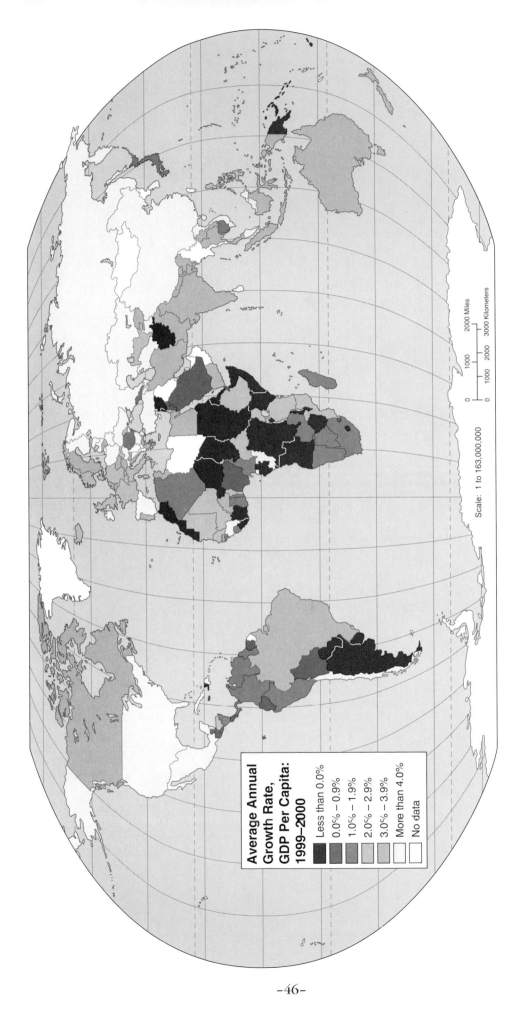

Average Annual Growth Rate, GDP Per Capita: 1999–2000

- Less than 0.0%
- 0.0% – 0.9%
- 1.0% – 1.9%
- 2.0% – 2.9%
- 3.0% – 3.9%
- More than 4.0%
- No data

Scale: 1 to 163,000,000

Gross Domestic Product or GDP is Gross National Income (GNI) less receipts of primary income from foreign sources. While the calculations of GDP growth per capita are complex, the growth rate is considered by the World Bank and international economists to be a particularly good measure of economic growth. One of the worldwide tendencies measured by GDP growth per capita is for continued economic development in Africa, and in South, Southeast, and East Asia where GDP grew at rates higher than the growth rates of "richer" countries in Europe and North and South America. This should not necessarily be viewed as a case of the poor catching up with the rich; in fact, it shows the huge impact that even relatively small production increases will have in countries with small GNIs and GDPs. Nevertheless, in spite of the continuing low economic growth through most of sub-Saharan Africa, the GDP growth rate of some of the world's poorer countries is an encouraging trend.

Map **35** Total Labor Force

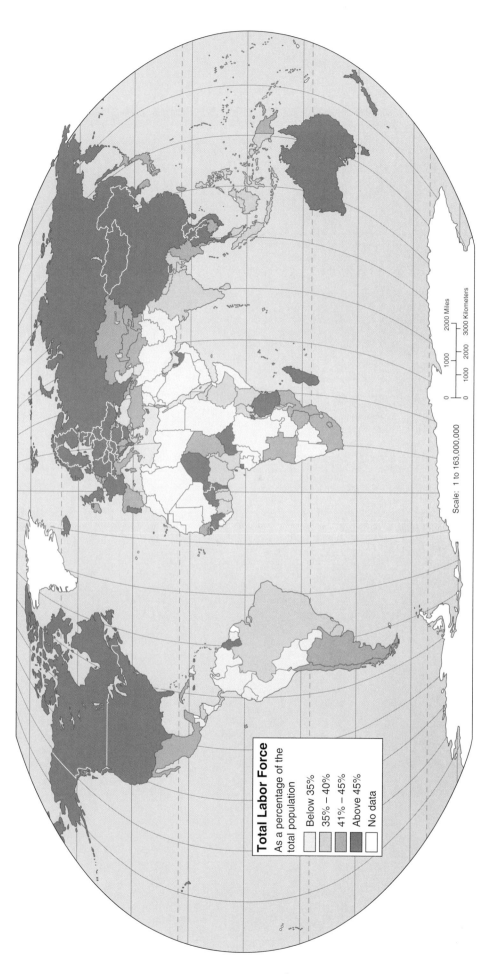

Scale: 1 to 163,000,000

Total Labor Force
As a percentage of the total population
Below 35%
35% – 40%
41% – 45%
Above 45%
No data

The term *labor force* refers to the economically active portion of a population, that is, all people who work or are without work but are available for and are seeking work to produce economic goods and services. The total labor force thus includes both the employed and the unemployed (as long as they are actively seeking employment). Labor force is considered a better indicator of economic potential than employment/unemployment figures, since unemployment figures will include experienced workers with considerable potential who are temporarily out of work. Unemployment figures will also incorporate persons seeking employment for the first time (many recent college graduates, for example). Generally, countries with higher percentages of total pop-

ulation within the labor force will be countries with higher levels of economic development. This is partly a function of levels of education and training and partly a function of the age distribution of populations. In developing countries, substantial percentages of the total population are too young to be part of the labor force. Also in developing countries a significant percentage of the population consist of women engaged in household activities or subsistence cultivation. These people seldom appear on lists of either employed or unemployed seeking employment and are the world's forgotten workers.

Map 36 Employment by Sector

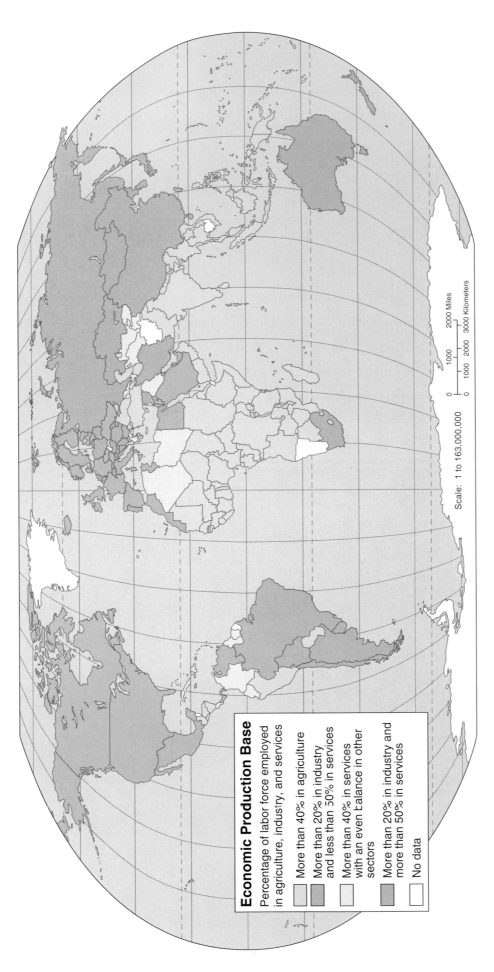

Economic Production Base

Percentage of labor force employed in agriculture, industry, and services

- More than 40% in agriculture
- More than 20% in industry and less than 50% in services
- More than 40% in services with an even balance in other sectors
- More than 20% in industry and more than 50% in services
- No data

Scale: 1 to 163,000,000

0   1000   2000 Miles

0 1000 2000 3000 Kilometers

The employment structure of a country's population is one of the best indicators of the country's position on the scale of economic development. At one end of the scale are those countries with more than 40 percent of their labor force employed in agriculture. These are almost invariably the least developed, with high population growth rates, poor human services, significant environmental problems, and so on. In the middle of the scale are two types of countries: those with more than 20 percent of their labor force employed in industry and those with a fairly even balance among agricultural, industrial, and service employment but with at least 40 percent of their labor force employed in service activities. Generally, these countries have undergone the industrial revolution fairly recently and are still developing an industrial base while building up their service activities. This category also includes countries with a disproportionate share of their economies in service activities primarily related to resource extraction. On the other end of the scale are countries with more than 20 percent of their labor force employed in industry and more than 50 percent in service activities. These countries are, for the most part, those with a highly automated industrial base and a highly mechanized agricultural system (the "postindustrial," developed countries). They also include, particularly in Middle and South America and Africa, industrializing countries that are also heavily engaged in resource extraction as a service activity.

Map 37 Agricultural Production Per Capita

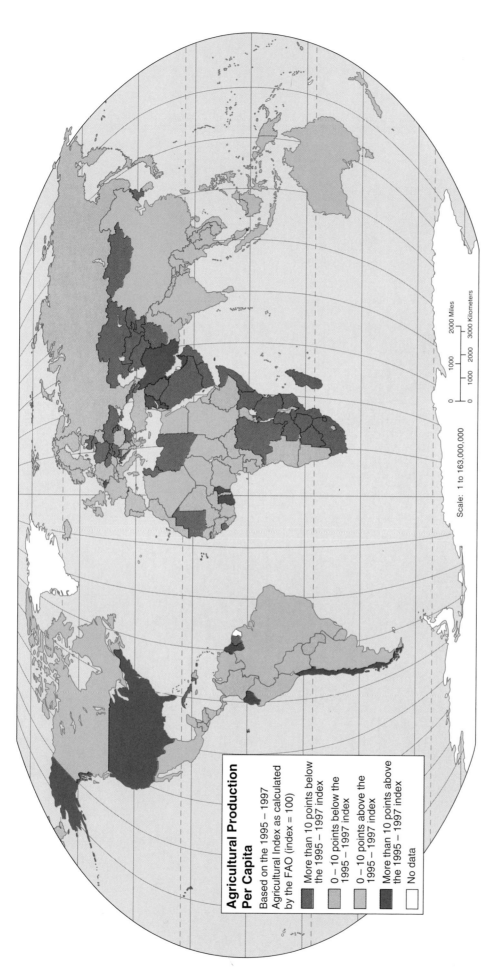

Agricultural Production Per Capita

Based on the 1995 – 1997
Agricultural Index as calculated
by the FAO (index = 100)

- More than 10 points below the 1995 – 1997 index
- 0 – 10 points below the 1995 – 1997 index
- 0 – 10 points above the 1995 – 1997 index
- More than 10 points above the 1995 – 1997 index
- No data

Scale: 1 to 163,000,000

2000 Miles

1000   2000   3000 Kilometers
0   1000   2000

Agricultural production includes the value of all crop and livestock products originating within a country for the base period of 1995–1997. The index value portrays the disposable output (after deductions for livestock feed and seed for planting) of a country's agriculture in comparison with the base period 1989–1991. Thus, the production values show not only the relative ability of countries to produce food but also show whether or not that ability has increased or decreased over a 10-year period. In general, global food production has kept up with or very slightly exceeded population growth. However, there are significant regional variations in the trend of food production keeping up with or surpassing population growth. For example, agricultural production in Africa and in Middle America has fallen, while production in South America, Asia, and Europe has risen. In the case of Africa, the drop in production reflects a population

growing more rapidly than agricultural productivity. Where rapid increases in food production per capita exist (as in certain countries in South America, Asia, and Europe), most often the reason is the development of new agricultural technologies that have allowed food production to grow faster than population. In much of Asia, for example, the so-called Green Revolution of new, highly productive strains of wheat and rice made positive index values possible. Also in Asia, the cessation of major warfare allowed some countries (Cambodia, Laos, and Vietnam) to show substantial increases over the 1982–1984 index. In some cases, a drop in production per capita reflects government decisions to limit production in order to maintain higher prices for agricultural products. The United States and Japan fall into this category.

Map 38 Exports of Primary Products

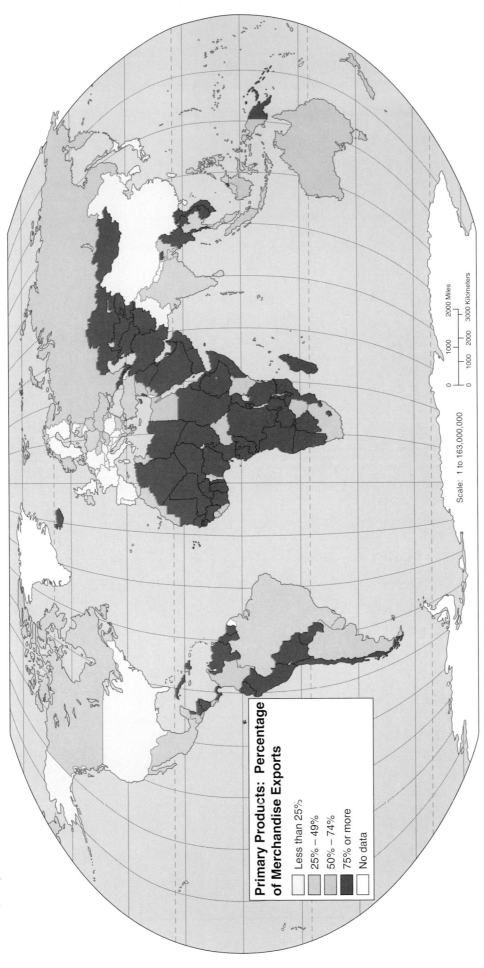

Primary Products: Percentage of Merchandise Exports

- Less than 25%
- 25% – 49%
- 50% – 74%
- 75% or more
- No data

Scale: 1 to 163,000,000

0   1000   2000   2000 Miles

0   1000   2000   3000 Kilometers

Primary products are those that require additional processing before they enter the consumer market: metallic ores that must be converted into metals and then into metal products such as automobiles or refrigerators; forest products such as timber that must be converted to lumber before they become suitable for construction purposes; and agricultural products that require further processing before being ready for human consumption. It is an axiom in international economics that the more a country relies on primary products for its export commodities, the more vulnerable its economy is to market fluctuations. Those countries with only primary products to export are hampered in their economic growth. A country dependent on only one or two products for export revenues is unprotected from economic shifts, particularly a changing market demand for its products. Imagine what would happen to the thriving economic status of the oil-exporting states of the Persian Gulf, for example, if an alternate source of cheap energy were found. A glance at this map shows that those countries with the lowest levels of economic development tend to be concentrated on primary products and, therefore, have economies that are especially vulnerable to economic instability.

Map 39 Dependence on Trade

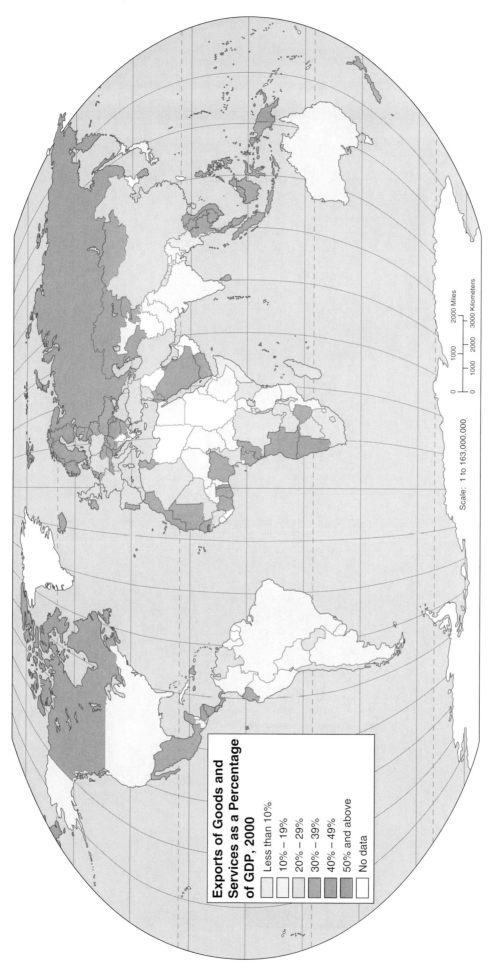

Exports of Goods and Services as a Percentage of GDP, 2000

- Less than 10%
- 10% – 19%
- 20% – 29%
- 30% – 39%
- 40% – 49%
- 50% and above
- No data

Scale: 1 to 163,000,000

0   1000   2000 Miles
0   1000   2000   3000 Kilometers

As the global economy becomes more and more a reality, the economic strength of virtually all countries is increasingly dependent upon trade. For many developing nations, with relatively abundant resources and limited industrial capacity, exports provide the primary base upon which their economies rest. Even countries like the United States, Japan, and Germany, with huge and diverse economies, depend on exports to generate a significant percentage of their employment and wealth. Without imports, many products that consumers want would be unavailable or more expensive; without exports, many jobs would be eliminated. But exports alone do not provide the full story on trade

dependence; part of what a map such as this masks is "what kind of exports?" For the more developed parts of the world, exports tend to be industrial products, perhaps along with some few raw materials (the United States and Russia are exceptions here in that they export a significant quantity of raw materials). But for the lesser developed countries, the exports are largely in the raw materials category. While this, as noted, provides jobs, it also means that countries may not have the necessary quantity of raw materials with which to develop their own industries and further their economic development.

Map 40 Global Flow of Investment Capital

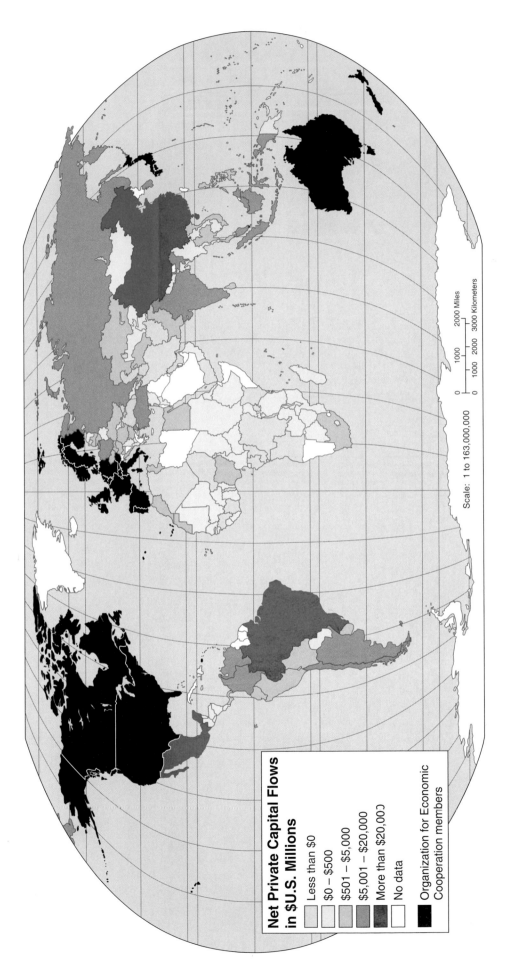

Scale: 1 to 163,000,000

0   1000   2000 Miles

0   1000   2000   3000 Kilometers

Net Private Capital Flows in $U.S. Millions

Less than $0

$0 – $500

$501 – $5,000

$5,001 – $20,000

More than $20,000

No data

Organization for Economic Cooperation members

International capital flows include private debt and nondebt flows from one country to another, shown on the map as flows into a country. Nearly all of the capital comes from those countries that are members of the Organization for Economic Cooperation and Development (OECD), shown in black on the map. Capital flows include commercial bank lending, bonds, other private credits, foreign direct investment, and portfolio investment. Most of these flows are indicators of the increasing influence developed countries exert over the developing economies. Foreign direct investment or FDI, for example, is a measure of the net inflow of investment monies used to acquire long-term management interest in businesses located somewhere other than in the economy of the investor. Usually this means the acquisition of at least 10 percent of the stock of a company by a foreign investor and is, then, a measure of what might be termed "economic colonialism": control of a region's economy by foreign investors that could, in the world of the future, be as significant as colonial political control was in the past. International capital flows have increased greatly in the last decade as the result of the increasing liberalization of developing countries, the strong economic growth exhibited by many developing countries, and the falling costs and increased efficiency of communication and transportation services.

Map 41 Central Government Expenditures Per Capita

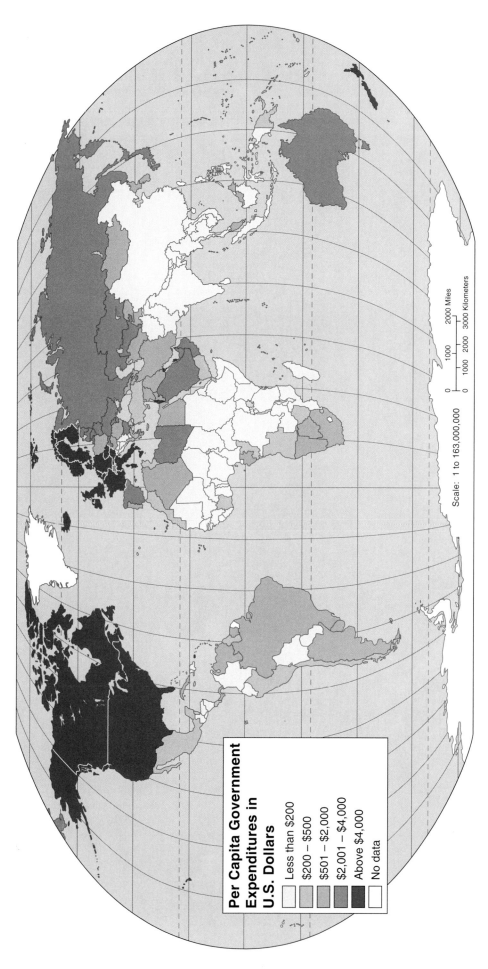

Per Capita Government Expenditures in U.S. Dollars

- Less than $200
- $200 – $500
- $501 – $2,000
- $2,001 – $4,000
- Above $4,000
- No data

Scale: 1 to 163,000,000

0   1000   2000 Miles

0   1000   2000   3000 Kilometers

The amount of money that the central government of a country spends upon a variety of essential governmental functions is a measure of relative economic development, particularly when it is viewed on a per-person basis. These functions include such governmental responsibilities as agriculture, communications, culture, defense, education, fishing and hunting, health, housing, recreation, religion, social security, transportation, and welfare. Generally, the higher the level of economic development, the greater the per capita expenditures on these services. However, the data do mask some internal variations. For example, countries that spend 20 percent or more of their central gov-

ernment expenditures on defense will often show up in the more developed category when, in fact, all that the figures really show is that a disproportionate amount of the money available to the government is devoted to purchasing armaments and maintaining a large standing military force. Thus, the fact that Libya spends $2,937 per capita—more than 10 times the average for Africa—does not suggest that the average Libyan is 10 times better off than the average Tanzanian. Nevertheless, this map—particularly when compared with Map 44, Energy Consumption Per Capita—does provide a reasonable approximation of economic development levels.

Map 42 Purchasing Power Parity

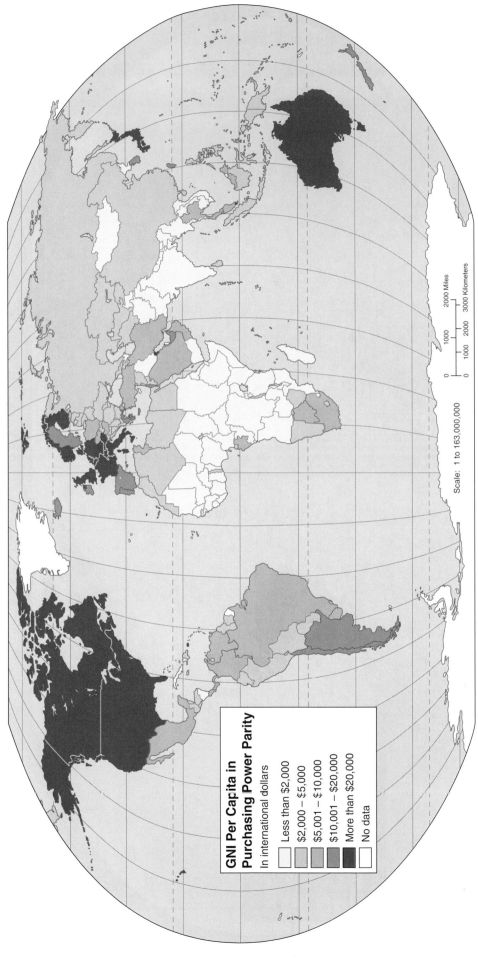

GNI Per Capita in Purchasing Power Parity

In international dollars

- Less than $2,000
- $2,000 – $5,000
- $5,001 – $10,000
- $10,001 – $20,000
- More than $20,000
- No data

Scale: 1 to 163,000,000

0 1000 2000 3000 Kilometers
0   1000   2000 Miles

Among all the economic measures that separate the "haves" from the "have-nots," per capita Purchasing Power Parity (PPP) may be the most meaningful. Per capita GNP and GDP (Gross Domestic Product) figures, and even per capita income, have the limitation of seldom reflecting the true purchasing power of a country's currency at home. In order to get around this limitation, international economists seeking to compare national currencies developed the PPP measure, which shows the level of goods and services that holders of a country's money can acquire locally. By converting all currencies to the "international dollar," the World Bank and other organizations using PPP can now show more truly comparative values, since the new currency value shows the number of units of a country's currency required to buy the same quantity of goods and services in the local market as one U.S. dollar would buy in an average country. The use of PPP currency values can alter the perceptions about a country's true comparative position in the world economy. PPP provides a valid measurement of the ability of a country's population to provide for itself the things that people in the developed world take for granted: adequate food, shelter, clothing, education, and access to medical care. A glance at the map shows a clear-cut demarcation between temperate and tropical zones, with most of the countries with a PPP above $5,000 in the midlatitude zones and most of those with lower PPPs in the tropical and equatorial regions.

Map 43 Energy Production

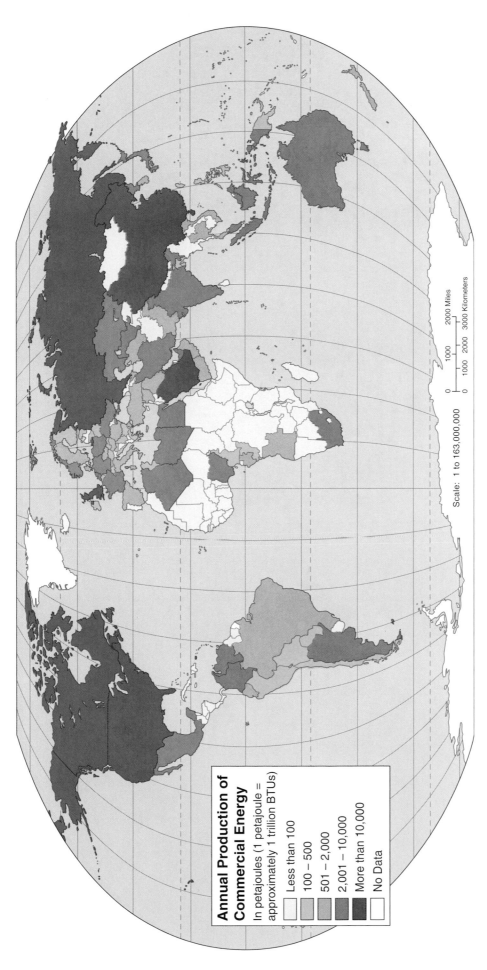

Annual Production of Commercial Energy

In petajoules (1 petajoule = approximately 1 trillion BTUs)

- Less than 100
- 100 – 500
- 501 – 2,000
- 2,001 – 10,000
- More than 10,000
- No Data

Scale: 1 to 163,000,000

0   1000   2000 Miles

0   1000   2000   3000 Kilometers

The production of commercial energy in all its forms—solid fuels (primarily coal), liquid fuels (primarily petroleum), natural gas, geothermal, wind, solar, hydroelectric, and nuclear—is a good measure of a country's ability to produce sufficient quantities of energy to meet domestic demands or to provide a healthy export commodity—or, in some instances, both. Commercial energy production is also a measure of the level of economic development, although a fairly subjective one. With exceptions, wealthier countries produce more energy from all sources than do poorer countries. Countries such as Japan and many European states rank among the world's wealthiest, but are energy-poor and produce relatively little of their own energy. They have the ability, however, to pay for it. On the other hand, countries such as those of the Persian Gulf or the oil-producing states of Middle and South America may rank relatively low on the scale of economic development but rank high as producers of energy. The map does not show the enormous amounts of energy from noncommercial sources (traditional fuels like firewood and animal dung) used by the world's poor, particularly in Middle and South America, Africa, South Asia, and East Asia. In these regions, firewood and animal dung may account for more actual energy production than coal or oil. Indeed, for many in the developing world, the real energy crisis is a shortage of wood for cooking and heating.

Map 44 Energy Consumption Per Capita

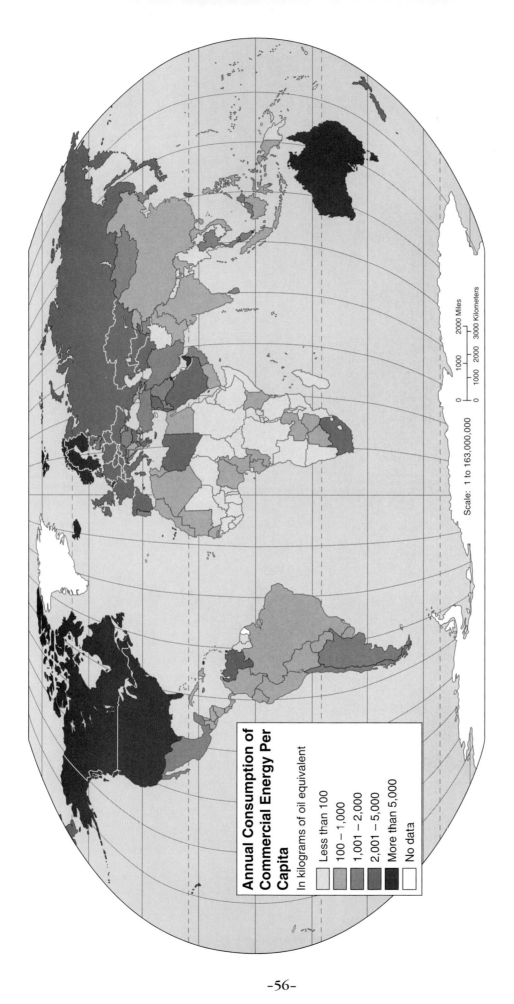

Annual Consumption of Commercial Energy Per Capita

In kilograms of oil equivalent

- Less than 100
- 100 – 1,000
- 1,001 – 2,000
- 2,001 – 5,000
- More than 5,000
- No data

Scale: 1 to 163,000,000

2000 Miles

1000

2000   3000 Kilometers

Of all the quantitative measures of economic well-being, energy consumption per capita may be the most expressive. All of the countries defined by the World Bank as having high incomes consume at least 100 gigajoules of commercial energy (the equivalent of about 3.5 metric tons of coal) per person per year, with some, such as the United States and Canada, having consumption rates in the 300 gigajoule range (the equivalent of more than 10 metric tons of coal per person per year). With the exception of the oil-rich Persian Gulf states, where consumption figures include the costly "burning off" of excess energy in the form of natural gas flares at wellheads, most of the highest-consuming countries are in the Northern Hemisphere, concentrated in North America and Western Europe. At the other end of the scale are low-income countries, whose consumption rates are often less than 1 percent of those of the United States and other high consumers. These figures do not, of course, include the consumption of noncommercial energy—the traditional fuels of firewood, animal dung, and other organic matter—widely used in the less developed parts of the world.

-56-

Part V

Global Patterns of Environmental Disturbance

Map 45 Global Air Pollution: Sources and Wind Currents

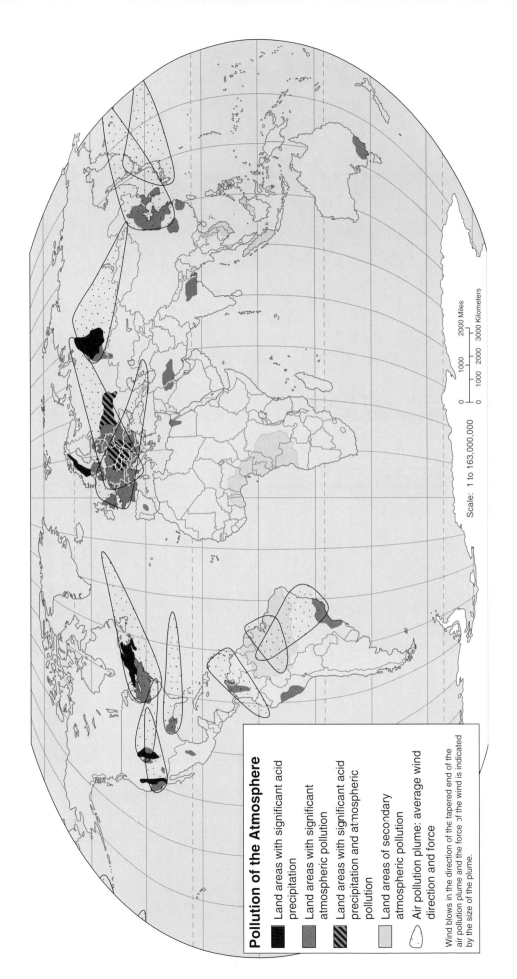

Pollution of the Atmosphere

- ⬛ Land areas with significant acid precipitation
- ⬛ Land areas with significant atmospheric pollution
- ▨ Land areas with significant acid precipitation and atmospheric pollution
- ⬜ Land areas of secondary atmospheric pollution
- ⬭ Air pollution plume: average wind direction and force

Wind blows in the direction of the tapered end of the air pollution plume and the force of the wind is indicated by the size of the plume.

Scale: 1 to 163,000,000

| 0 | 1000 | 2000 Miles |
| 0 | 1000 | 2000 | 3000 Kilometers |

Almost all processes of physical geography begin and end with the flows of energy and matter among land, sea, and air. Because of the primacy of the atmosphere in this exchange system, air pollution is potentially one of the most dangerous human modifications in environmental systems. Pollutants such as various oxides of nitrogen or sulfur cause the development of acid precipitation, which damages soil, vegetation, and wildlife and fish. Air pollution in the form of smog is often dangerous for human health. And

most atmospheric scientists believe that the efficiency of the atmosphere in retaining heat—the so-called "greenhouse effect"—is being enhanced by increased carbon dioxide, methane, and other gases produced by agricultural and industrial activities. The result, they fear, will be a period of global warming that will dramatically alter climates in all parts of the world.

Map 46 The Acid Deposition Problem: Air, Water, Soil

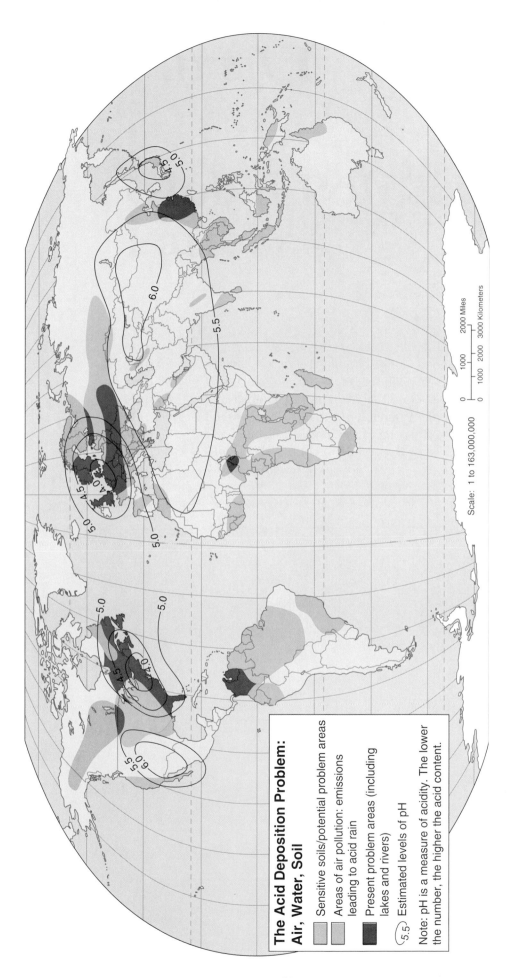

The Acid Deposition Problem: Air, Water, Soil

- Sensitive soils/potential problem areas
- Areas of air pollution: emissions leading to acid rain
- Present problem areas (including lakes and rivers)
- 5.5 — Estimated levels of pH

Note: pH is a measure of acidity. The lower the number, the higher the acid content.

Scale: 1 to 163,000,000

0   1000   2000 Miles
0   1000   2000   3000 Kilometers

The term "acid precipitation" refers to increasing levels of acidity in snowfall and rainfall caused by atmospheric pollution. Oxides of nitrogen and sulfur resulting from incomplete combustion of fossil fuels (coal, oil, and natural gas) combine with water vapor in the atmosphere to produce weak acids that then "precipitate" or fall along with water or ice crystals. Some atmospheric acids formed by this process are known as "dry-acid" precipitates and they too will fall to earth, although not necessarily along with rain or snow. In some areas of the world, the increased acidity of streams and lakes stemming from high levels of acid precipitation or dry acid fallout has damaged or destroyed aquatic life. Acid precipitation and dry acid fallout also harms soil systems and vegetation, producing a characteristic burned appearance in forests that lends the same quality to landscapes that forest fires would. The region most dramatically impacted by acid precipitation is Central Europe where decades of destructive environmental practices, including the burning of high sulfur coal for commercial, industrial, and residential purposes, has produced the destruction of hundreds of thousands of acres of woodlands—a phenomenon described by the German foresters who began their study of the area following the lifting of the Iron Curtain as "Waldsterben": Forest Death.

Map 47 Major Polluters and Common Pollutants

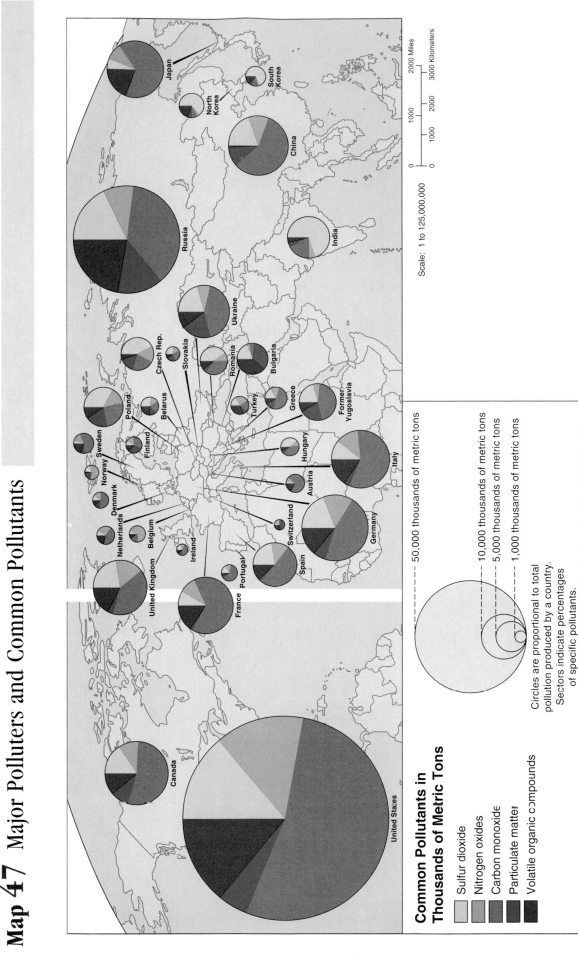

Common Pollutants in Thousands of Metric Tons

- Sulfur dioxide
- Nitrogen oxides
- Carbon monoxide
- Particulate matter
- Volatile organic compounds

50,000 thousands of metric tons

10,000 thousands of metric tons

5,000 thousands of metric tons

1,000 thousands of metric tons

Circles are proportional to total pollution produced by a country. Sectors indicate percentages of specific pollutants.

Scale: 1 to 125,000,000

2000 Miles

3000 Kilometers

2000

1000

1000

0

0

More than 90 percent of the world's total of anthropogenic (human-generated) air pollutants come from the heavily populated industrial regions of North America, Europe, South Asia (primarily in India), and East Asia (mainly in China, Japan, and the two Koreas). This map shows the origins of the five most common pollutants: sulfur dioxide, nitrogen oxide, carbon monoxide, particulate matter, and volatile organic compounds. These substances are produced both by industry and by the combustion of fossil fuels that generate electricity and power trains, planes, automobiles, buses, and trucks. In addition to combining with other components of the atmosphere and with one another to produce smog, they are the chief ingredients in acid accumulations in the atmosphere, which ultimately result in acid deposition, either as acid precipitation or dry acid fallout. Like other forms of pollutants, these air pollutants do not recognize political boundaries, and regions downwind of major polluters receive large quantities of pollutants from areas over which they often have no control.

-60-

Map **48** Global Carbon Dioxide Emissions

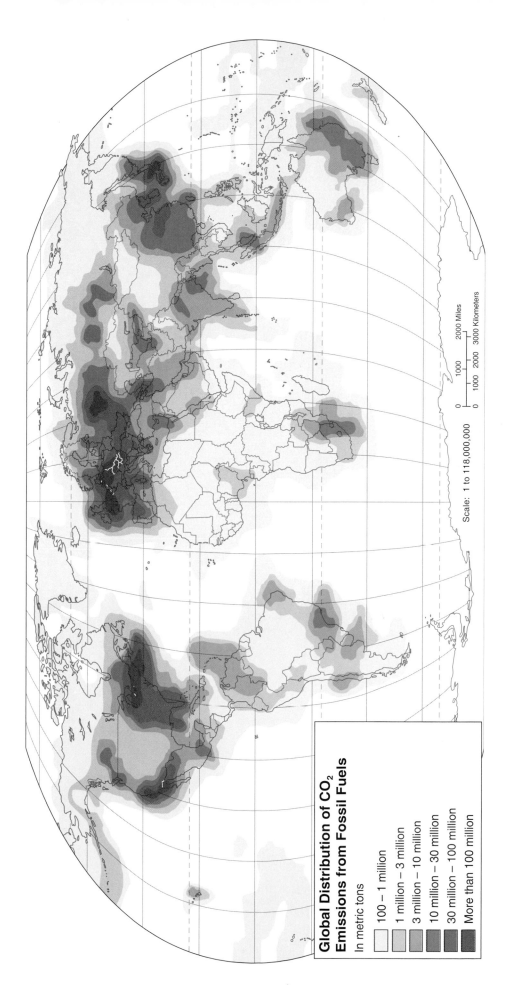

Global Distribution of CO$_2$ Emissions from Fossil Fuels

In metric tons

- 100 – 1 million
- 1 million – 3 million
- 3 million – 10 million
- 10 million – 30 million
- 30 million – 100 million
- More than 100 million

Scale: 1 to 118,000,000

0   1000   2000 Miles
0 1000 2000 3000 Kilometers

One of the most important components of the atmosphere is the gas carbon dioxide, the byproduct of animal respiration, of decomposition, and of combustion. During the past 200 years, atmospheric carbon dioxide has risen dramatically, largely as the result of the tremendous increase in fossil fuel combustion brought on by the industrialization of the world's economy and the burning and clearing of forests by the expansion of farming. While carbon dioxide by itself is relatively harmless, it is an important "greenhouse gas." The gases in the atmosphere act like the panes of glass in a greenhouse roof, allowing light in but preventing heat from escaping. The greenhouse capacity of the atmosphere is crucial for organic life and is a purely natural component of the global energy cycle. But too much carbon dioxide and other greenhouse gases such as methane could cause the earth's atmosphere to warm up too much, producing the global warming that atmospheric scientists are concerned about. Researchers estimate that if greenhouse gases such as carbon dioxide continue to increase at their present rates, the earth's mean temperature could rise between 1.5 and 4.5 degrees Celsius by the middle of the next century. Such a rise in global temperatures would produce massive alterations in the world's climate patterns.

Map 49 Potential Global Temperature Change

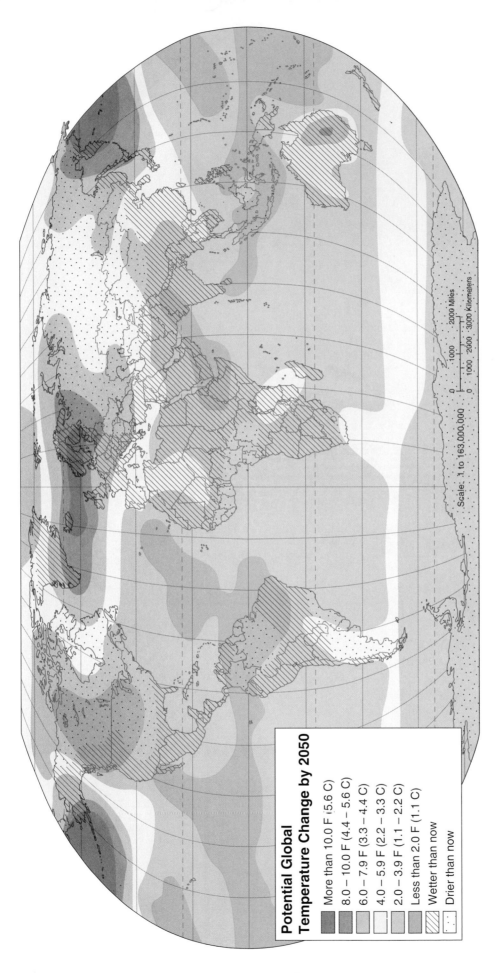

Potential Global Temperature Change by 2050

- More than 10.0 F (5.6 C)
- 8.0 – 10.0 F (4.4 – 5.6 C)
- 6.0 – 7.9 F (3.3 – 4.4 C)
- 4.0 – 5.9 F (2.2 – 3.3 C)
- 2.0 – 3.9 F (1.1 – 2.2 C)
- Less than 2.0 F (1.1 C)
- Wetter than now
- Drier than now

Scale: 1 to 163,000,000

0   1000   2000 Miles

0   1000   2000   3000 Kilometers

According to atmospheric scientists, one of the major problems of the twenty-first century will be "global warming," produced as the atmosphere's natural ability to trap and retain heat is enhanced by increased percentages of carbon dioxide, methane, chlorinated fluorocarbons or "CFCs," and other "greenhouse gases" in the earth's atmosphere. Computer models based on atmospheric percentages of carbon dioxide resulting from present use of fossil fuels show that warming is not just a possibility but a probability. Increased temperatures would cause precipitation patterns to alter significantly as well and would produce a number of other harmful effects, including a rise in the level of the world's oceans that could flood most coastal cities. International conferences on the topic of the enhanced greenhouse effect have resulted in several interna-tional agreements to reduce the emission of carbon dioxide or to maintain it at present levels. Unfortunately, the solution is not that simple since reduction of carbon dioxide emissions is, in the short run, expensive—particularly as long as the world's energy systems continue to be based on fossil fuels. Chief among the countries that could be hit by serious international mandates to reduce emissions are those highest on the development scale who use the highest levels of fossil fuels and, therefore, produce the highest emissions, and those on the lowest end of the development scale whose efforts to industrialize could be severely impeded by the more expensive energy systems that would replace fossil fuels.

-62-

Map 50 World Water Resources: Availability of Renewable Water Per Capita

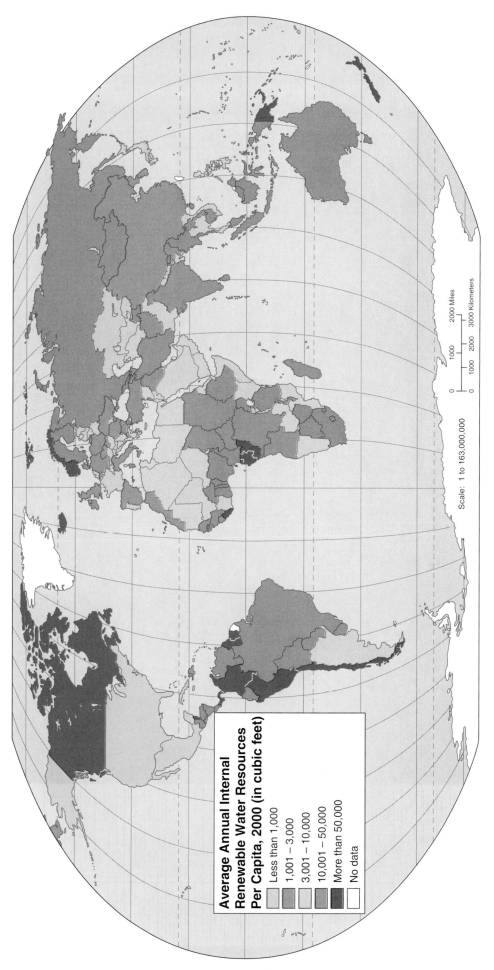

Average Annual Internal Renewable Water Resources Per Capita, 2000 (in cubic feet)

- Less than 1,000
- 1,001 – 3,000
- 3,001 – 10,000
- 10,001 – 50,000
- More than 50,000
- No data

Scale: 1 to 163,000,000

	1000	2000 Miles
0	1000 2000	3000 Kilometers

Renewable water resources are usually defined as the total water available from streams and rivers (including flows from other countries), ponds and lakes, and groundwater storage or aquifers. Not included in the total of renewable water would be water that comes from such nonrenewable sources as desalinization plants or melted icebergs. While the concept of renewable or flow resources is a traditional one in resource management, in fact, few resources, including water, are truly renewable when their use is

excessive. The water resources shown here are indications of that principle. A country like the United States possesses truly enormous quantities of water. But the United States also uses enormous quantities of water. The result is that, largely because of excessive use, the availability of renewable water is much less than in many other parts of the world where the total supply of water is significantly less.

-63-

Map 51 World Water Resources: Annual Withdrawal Per Capita

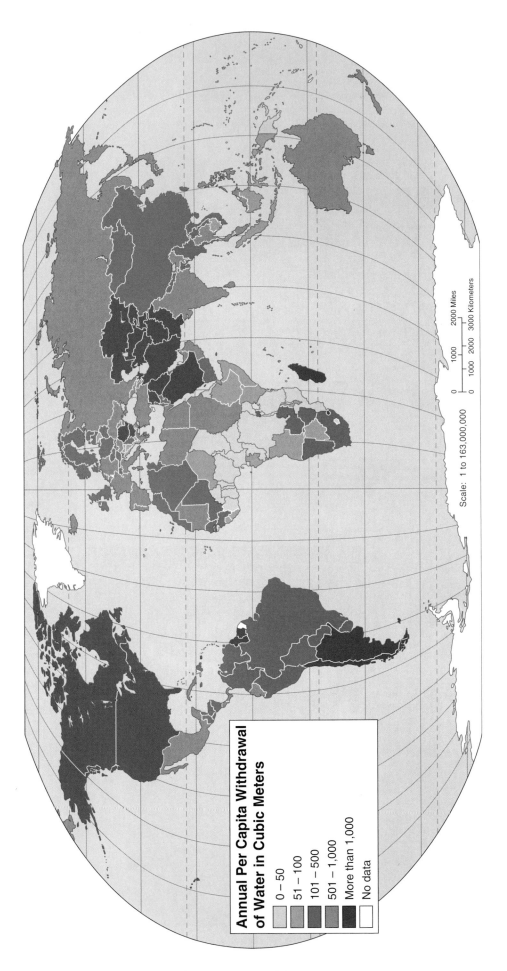

Annual Per Capita Withdrawal of Water in Cubic Meters

- 0 – 50
- 51 – 100
- 101 – 500
- 501 – 1,000
- More than 1,000
- No data

Scale: 1 to 163,000,000

0   1000   2000 Miles

0   1000   2000   3000 Kilometers

Water resources must be viewed like a bank account in which deposits and withdrawals are made. As long as the deposits are greater than the withdrawals, a positive balance remains. But when the withdrawals begin to exceed the deposits, sooner or later (depending on the relative sizes of the deposits and withdrawals) the account becomes overdrawn. For many of the world's countries, annual availability of water is insufficient to cover the demand. In these countries, reserves stored in groundwater are being

tapped, resulting in depletion of the water supply (think of this as shifting money from a savings account to a checking account). The water supply can maintain its status as a renewable resource only if deposits continue to be greater than withdrawals, and that seldom happens. In general, countries with high levels of economic development and countries that rely on irrigation agriculture are the most spendthrift when it comes to their water supplies.

Map 52 Pollutions of the Oceans

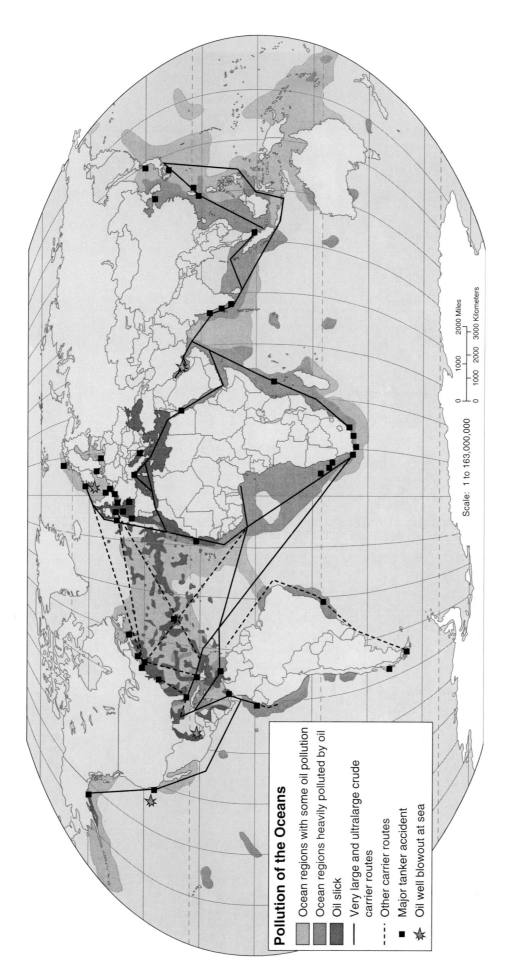

Pollution of the Oceans

- Ocean regions with some oil pollution
- Ocean regions heavily polluted by oil
- Oil slick
- —— Very large and ultralarge crude carrier routes
- ---- Other carrier routes
- ■ Major tanker accident
- ✹ Oil well blowout at sea

Scale: 1 to 163,000,000

0   1000   2000   Miles

0   1000   2000   3000 Kilometers

The pollution of the world's oceans has long been a matter of concern to physical geographers, oceanographers, and other environmental scientists. The great circulation systems of the ocean are one of the controlling factors of the earth's natural environment, and modifications to those systems have unknown consequences. This map is based on what we can measure: (1) areas of oceans where oil pollution has been proven to have inflicted significant damage to ocean ecosystems and life-forms (including phytoplankton, the oceans' primary food producers, equivalent to land-based vegetation) and (2) areas of oceans where unusually high concentrations of hydrocarbons from oil spills may have inflicted some damage to the oceans' biota. A glance at the map shows that there are few areas of the world's oceans where some form of pollution is not a part of the environmental system. What the map does not show in detail, because of the scale, are the dramatic consequences of large individual pollution events: the wreck of the *Exxon Valdez* and the polluting of Prince William Sound or the environmental devastation produced by the Gulf War in the Persian Gulf.

Map 53 Food Supply From the World's Marine and Freshwater Systems

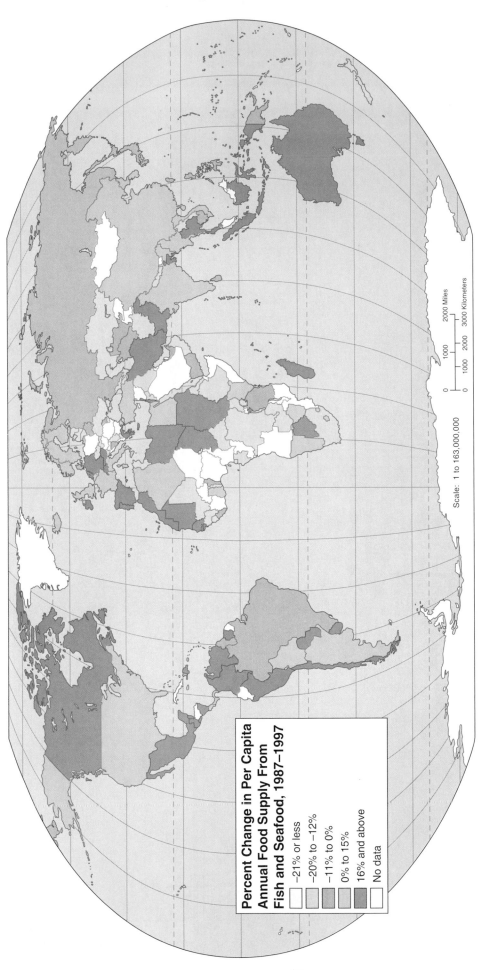

Percent Change in Per Capita Annual Food Supply From Fish and Seafood, 1987–1997

- −21% or less
- −20% to −12%
- −11% to 0%
- 0% to 15%
- 16% and above
- No data

Scale: 1 to 163,000,000

0   1000   2000 Miles
0   1000   2000   3000 Kilometers

Not that many years ago, food supply experts were confidently predicting that the "starving millions" of the world of the future could be fed from the unending bounty of the world's oceans. While the annual catch from the sea helped to keep hunger at bay for a time, by the late 1980s it had become apparent that without serious human intervention in the form of aquaculture, the supply of fish would not be sufficient to offset the population/food imbalance that was beginning to affect so many of the world's regions. The development of factory-fishing with advanced equipment to locate fish and process them before they went to market increased the supply of food from the ocean, but in that increase was sown the seeds of future problems. The factory-fishing system, efficient in terms of economics, was costly in terms of fish populations. In some well-fished areas, the stock of fish that was viewed as near infinite just a few decades ago has dwindled nearly to the point of disappearance. This map shows both increases and decreases in the amount of individual countries' food supplies from the ocean. The increases are often the result of more technologically advanced fishing operations. The decreases are usually the result of the same thing: increased technology has brought increased harvests, which has reduced the supply of fish and shellfish and that, in turn, has increased prices. Most of the countries that have experienced sharp decreases in their supply of food from the world's oceans are simply no longer able to pay for an increasingly scarce commodity.

Map **54** Changes in Cropland Per Capita, 1987–1997

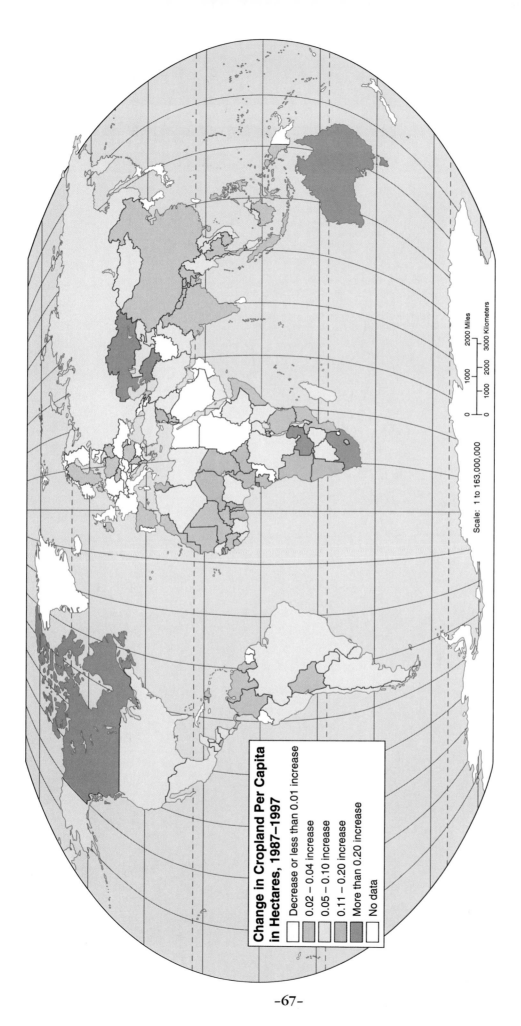

Change in Cropland Per Capita in Hectares, 1987–1997

- Decrease or less than 0.01 increase
- 0.02 – 0.04 increase
- 0.05 – 0.10 increase
- 0.11 – 0.20 increase
- More than 0.20 increase
- No data

Scale: 1 to 163,000,000

0   1000   2000 Miles

0   1000   2000   3000 Kilometers

As population has increased rapidly throughout the world, area in cultivated land has increased at the same time; in fact, the amount of farmland per person has gone up slightly. Unfortunately, the figures that show this also tell us that since most of the best (or even good) agricultural land in 1985 was already under cultivation, most of the agricultural area added since the early 1980s involves land that would have been viewed as marginal by the fathers and grandfathers of present farmers—marginal in that it was too dry, too wet, too steep to cultivate, too far from a market, and so on. The continued expansion of agricultural area is one reason that serious famine and starvation have struck only a few regions of the globe. But land, more than any other resource we deal with, is finite, and the expansion cannot continue indefinitely. Future gains in agricultural production are most probably going to come through more intensive use of existing cropland, heavier applications of fertilizers and other agricultural chemicals, and genetically engineered crops requiring heavier applications of energy and water, than from an increase in the amount of the world's cropland.

-67-

Map **55** Annual Change in Forest Cover, 1990–1995

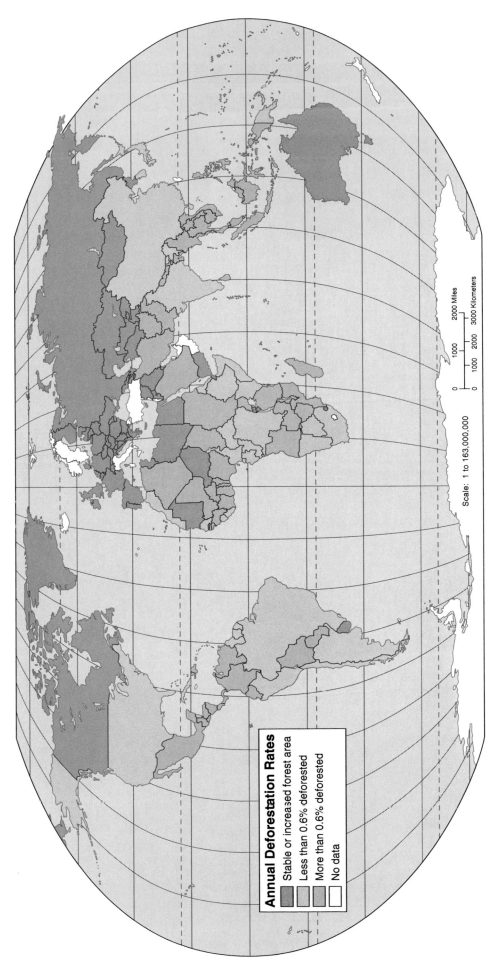

Annual Deforestation Rates
- Stable or increased forest area
- Less than 0.6% deforested
- More than 0.6% deforested
- No data

Scale: 1 to 163,000,000

0   1000   2000 Miles

0   1000   2000   3000 Kilometers

One of the most discussed environmental problems is that of deforestation. For most people, deforestation means clearing of tropical rain forests for agricultural purposes. Yet nearly as much forest land per year—much of it in North America, Europe, and Russia—is impacted by commercial lumbering as is cleared by tropical farmers and ranchers. Even in the tropics, much of the forest clearance is undertaken by large corporations producing high-value tropical hardwoods for the global market in furniture, ornaments, and other fine wood products. Still, it is the agriculturally driven clear-ing of the great rain forests of the Amazon Basin, west and central Africa, Middle America, and Southeast Asia that draws public attention. Although much concern over forest clearance focuses on the relationship between forest clearance and the reduction in the capacity of the world's vegetation system to absorb carbon dioxide (and thus delay global warming), of just as great concern are issues having to do with the loss of biodiversity (large numbers of plants and animals), the near-total destruction of soil systems, and disruptions in water supply that accompany clearing.

-68-

Map 56 The Loss of Biodiversity: Globally Threatened Animal Species

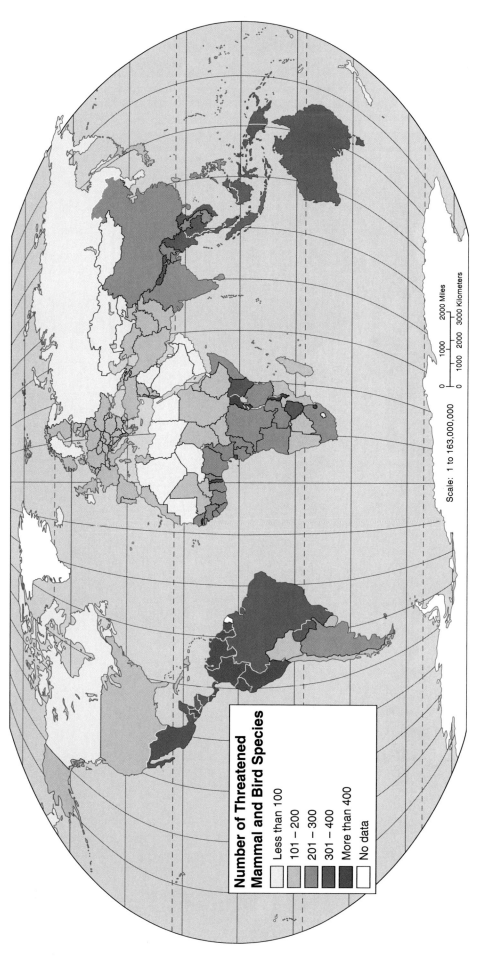

Scale: 1 to 163,000,000

Number of Threatened
Mammal and Bird Species

- Less than 100
- 101 – 200
- 201 – 300
- 301 – 400
- More than 400
- No data

Threatened species are those in grave danger of going extinct. Their populations are becoming restricted in range, and the size of the populations required for sustained breeding is nearing a critical minimum. *Endangered species* are in immediate danger of becoming extinct. Their range is already so reduced that the animals may no longer be able to move freely within an ecozone, and their populations are at the level where the species may no longer be able to sustain breeding. Most species become threatened first and then endangered as their range and numbers continue to decrease. When people think of animal extinction, they think of large herbivorous species like the rhinoc-

eros or fierce carnivores like lions, tigers, or grizzly bears. Certainly these animals make almost any list of endangered or threatened species. But there are literally hundreds of less conspicuous animals that are equally threatened. Extinction is normally nature's way of informing a species that it is inefficient. But conditions in this century are controlled more by human activities than by natural evolutionary processes. Species that are endangered or threatened fall into that category because, somehow, they are competing with us or with our domesticated livestock for space and food. And in that competition the animals are always going to lose.

Map 57 The Loss of Biodiversity: Globally Threatened Plant Species

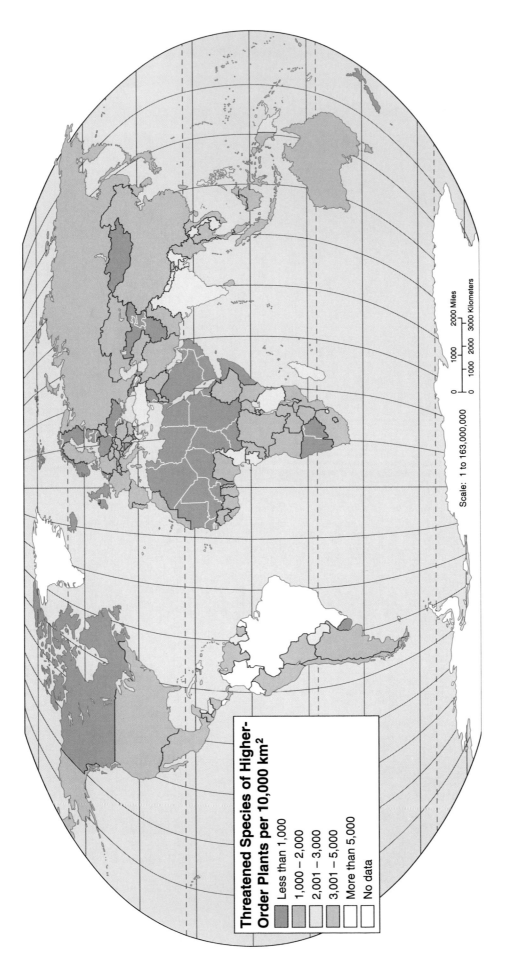

Threatened Species of Higher-Order Plants per 10,000 km²

- Less than 1,000
- 1,000 – 2,000
- 2,001 – 3,000
- 3,001 – 5,000
- More than 5,000
- No data

Scale: 1 to 163,000,000

0   1000   2000 Miles
0   1000   2000   3000 Kilometers

While most people tend to be more concerned about the animals on threatened and endangered species lists, the fact is that many more plants are in jeopardy, and the loss of plant life is, in all ecological regions, a more critical occurrence than the loss of animal populations. Plants are the primary producers in the ecosystem; that is, plants produce the food upon which all other species in the food web, including human beings, depend for sustenance. It is plants from which many of our critical medicines come, and it is plants that maintain the delicate balance between soil and water in most of the world's regions. When biogeographers and other environmental scientists speak of a loss of "biodiversity," what they are most often describing is a loss of the richness and complexity of plant life that lends stability to ecosystems. Systems with more plant life tend to be more stable than those with less. For these and other reasons, the scientific concern over extinction is greater when applied to plants than to animals. It is difficult for people to become as emotional over a teak tree as they would over an elephant. But as great a tragedy as the loss of the elephant would be, the loss of the teak would be greater.

Map 58 Global Hotspots of Biodiversity

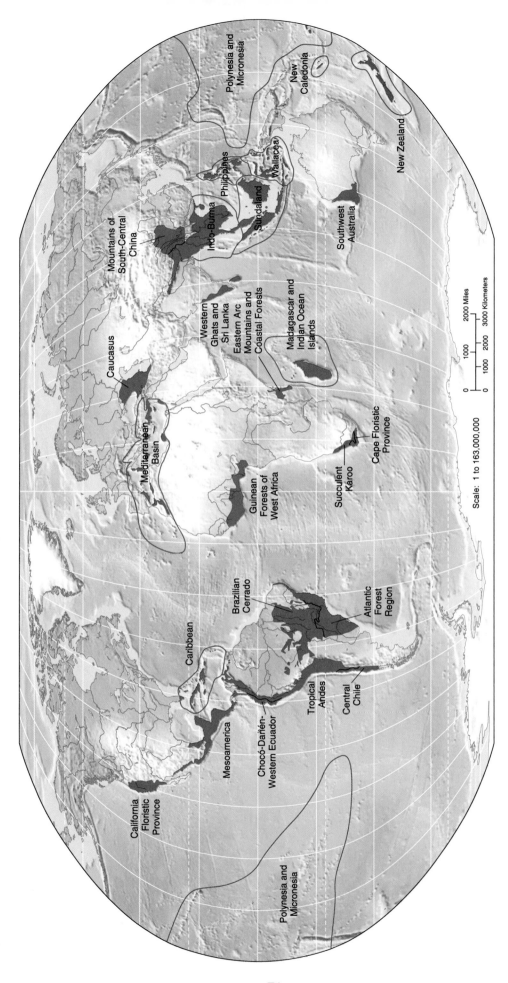

Scale: 1 to 163,000,000

0   1000   2000   3000 Kilometers
0   1000   2000 Miles

Where we have normally thought of tropical forest basins such as Amazonia as the world's most biologically diverse ecosystems, recent research has discovered the surprising fact that a number of hotspots of biological diversity exist outside the major tropical forest regions. These hotspot regions contain slightly less than 2 percent of the world's total land area but may contain up to 60 percent of the total world's terrestrial species of plants and animals. Geographically, the hotspot areas are characterized by vertical zonation (that is, they tend to be hilly to mountainous regions), long known to

be a factor in biological complexity. They are also in coastal locations or near large bodies of water, locations that stimulate climatic variability and, hence, biological complexity. Although some of the hotspots are sparsely populated, others, such as "Sundaland," are occupied by some of the world's densest populations. Protection of the rich biodiversity of these hotspots is, most biologists feel, of crucial importance to the preservation of the world's biological heritage.

-71-

Map 59 The Risks of Desertification

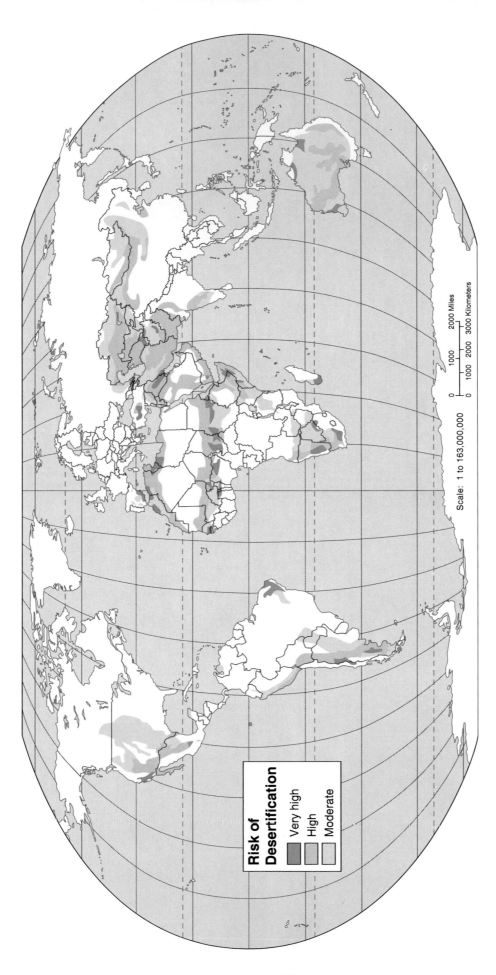

Risk of Desertification

- Very high
- High
- Moderate

Scale: 1 to 163,000,000

0   1000   2000 Miles

0   1000   2000   3000 Kilometers

The awkward-sounding term "desertification" refers to a reduction in the food-producing capacity of drylands through vegetation, soil, and water changes that culminate in either a drier climate or in soil and plant systems that are less efficient in their use of water. Most of the world's existing drylands—the shortgrass steppes, the tropical savannas, the bunchgrass regions of the desert fringe—are fairly intensively used for agriculture and are, therefore, subject to the kinds of pressures that culminate in desertification. Most desertification is a natural process that occurs near the margins of desert regions. It is caused by dehydration of the soil's surface layers during periods of drought and by high water loss through evaporation in an environment of high temperature and high winds. This natural process is greatly enhanced by human agricultural

activities that expose topsoil to wind and water erosion. Among the most important practices that cause desertification are (1) overgrazing of rangelands, resulting from too many livestock on too small an area of land; (2) improper management of soil and water resources in irrigation agriculture, leading to accelerated erosion and to salt buildup in the soil; (3) cultivation of marginal terrain with soils and slopes that are unsuitable for farming; (4) surface disturbances of vegetation (clearing of thorn scrub, mesquite, chaparral, and similar vegetation) without soil protection efforts being made or replanting being done; and (5) soil compaction by agricultural implements, domesticated livestock, and rain falling on an exposed surface.

Map 60 Global Soil Degradation

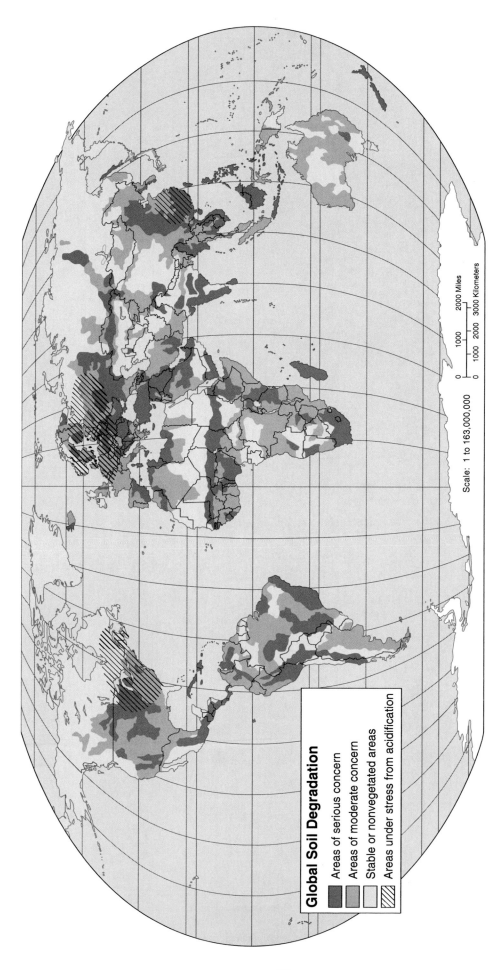

Global Soil Degradation

- Areas of serious concern
- Areas of moderate concern
- Stable or nonvegetated areas
- Areas under stress from acidification

Scale: 1 to 163,000,000

0   1000   2000 Miles

0   1000   2000   3000 Kilometers

Recent research has shown that more than 3 billion acres of the world's surface suffer from serious soil degradation, with more than 22 million acres so severely eroded or poisoned with chemicals that they can no longer support productive crop agriculture. Most of this soil damage has been caused by poor farming practices, overgrazing of domestic livestock, and deforestation. These activities strip away the protective cover of natural vegetation forests and grasslands, allowing wind and water erosion to remove the topsoil that contains necessary nutrients and soil microbes for plant growth. But millions of acres of topsoil have been degraded by chemicals as well. In some instances

these chemicals are the result of overapplication of fertilizers, herbicides, pesticides, and other agricultural chemicals. In other instances, chemical deposition from industrial and urban wastes and from acid precipitation has poisoned millions of acres of soil. As the map shows, soil erosion and pollution are not problems just in developing countries with high population densities and increasing use of marginal lands. They also afflict the more highly developed regions of mechanized, industrial agriculture. While many methods for preventing or reducing soil degradation exist, they are seldom used because of ignorance, cost, or perceived economic inefficiency.

Map **61** Degree of Human Disturbance

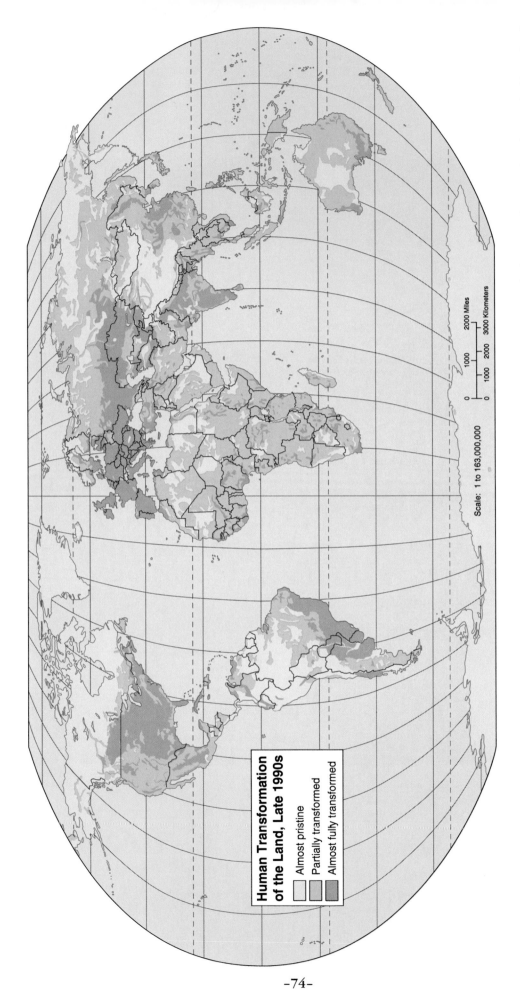

Human Transformation of the Land, Late 1990s

- Almost pristine
- Partially transformed
- Almost fully transformed

Scale: 1 to 163,000,000

0   1000   2000   3000 Kilometers

0   1000   2000 Miles

The data on human disturbance have been gathered from a wide variety of sources, some of them conflicting and not all of them reliable. Nevertheless, at a global scale this map fairly depicts the state of the world in terms of the degree to which humans have modified its surface. The almost pristine areas, covered with natural vegetation, generally have population densities under 10 persons per square mile. These areas are, for the most part, in the most inhospitable parts of the world: too high, too dry, too cold for permanent human habitation in large numbers. The partially transformed areas are normally agricultural areas, either subsistence (such as shifting cultivation) or extensive (such as livestock grazing). They often contain areas of secondary vegetation, regrown after removal of original vegetation by humans. They are also often marked by a density of livestock in excess of carrying capacity, leading to overgrazing, which further alters the condition of the vegetation. The almost fully transformed areas are those of permanent and intensive agriculture and urban settlement. The primary vegetation of these regions has been removed, with no evidence of regrowth or with current vegetation that is quite different from natural (potential) vegetation. Soils are in a state of depletion and degradation, and, in drier lands, desertification is a factor of human occupation. The disturbed areas match closely those areas of the world with the densest human populations.

-74-

Part VI

Global Political Patterns

Map **62** Political Systems

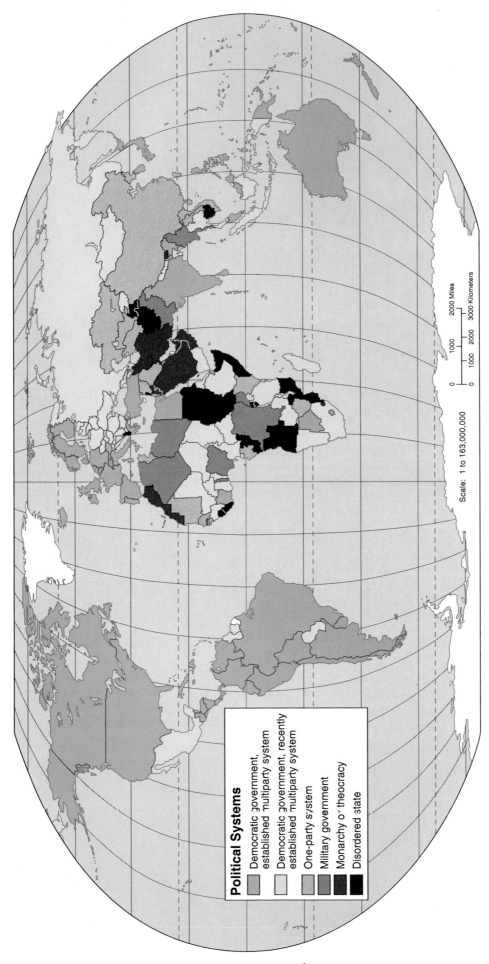

Political Systems

- Democratic government, established multiparty system
- Democratic government, recently established multiparty system
- One-party system
- Military government
- Monarchy o' theocracy
- Disordered state

Scale: 1 to 163,000,000

2000 Miles

0   1000   2000

0   1000   2000   3000 Kilometers

-76-

The world map of political systems has changed dramatically during the 1990s. The categories of political systems shown are subject to some interpretation: established multiparty democracies are those in which elections by secret ballot with adult suffrage are long-term features of the political landscape; recently established multiparty democracies are those in which this distinguishing feature has only recently emerged. The former Soviet satellites of eastern Europe and the republics that formerly constituted the USSR are in this category; so are states in emerging regions that are beginning to throw off the single-party rule that often followed the violent upheavals of the immediate postcolonial governmental transitions. The other categories are more or less obvious. One-party systems define states where single party rule provided for by constitution is a fact of political life. Monarchies are countries with heads of state who are members of a royal family. Some countries with monarchs do not fall into this category because their monarchs are titular heads of state only. Theocracies are countries in which rule is within the hands of a priestly or clerical class; for example, fundamentalist Islamic countries. Military governments are often organized around a junta that has seized control of the government from civil authority; such states are often technicially transitional—the military claims that it will return the reins of government to civil authority "when order is restored." Finally, "disordered states" are countries so beset by civil war or widespread ethnic conflict that no organized government can be said to exist within them.

Map 63 Sovereign States: Duration of Independence

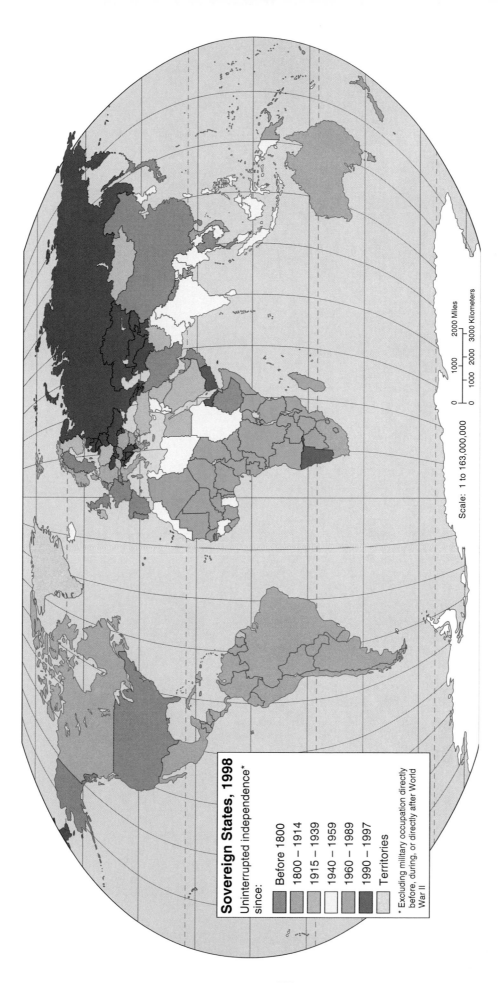

Sovereign States, 1998

Uninterrupted independence* since:

- Before 1800
- 1800 – 1914
- 1915 – 1939
- 1940 – 1959
- 1960 – 1989
- 1990 – 1997
- Territories

* Excluding military occupation directly before, during, or directly after World War II

Scale: 1 to 163,000,000

0   1000   2000 Miles
0   1000   2000   3000 Kilometers

Most countries of the modern world, including such major states as Germany and Italy, became independent after the beginning of the nineteenth century. Of the world's current countries, only 27 were independent in 1800. (Ten of the 27 were in Europe; the others were Afghanistan, China, Colombia, Ethiopia, Haiti, Iran, Japan, Mexico, Nepal, Oman, Paraguay, Russia, Taiwan, Thailand, Turkey, the United States, and Venezuela). Following 1800, there have been five great periods of national independence. During the first of these (1800–1914), most of the mainland countries of the Americas achieved independence. During the second period (1915–1939), the countries of Eastern Europe emerged as independent entities. The third period (1940–1959) includes World War II and the years that followed, when independence for African and Asian nations that had been under control of colonial powers first began to occur. During the fourth period (1960–1989), independence came to the remainder of the colonial African and Asian nations, as well as to former colonies in the Caribbean and the South Pacific. More than half of the world's countries came into being as independent political entities during this period. Finally, in the last few years (1990–1997), the breakup of the existing states of the Soviet Union, Yugoslavia, and Czechoslovakia created 22 countries where only 3 had existed before.

–77–

Map 64 Post-Cold War International Alliances

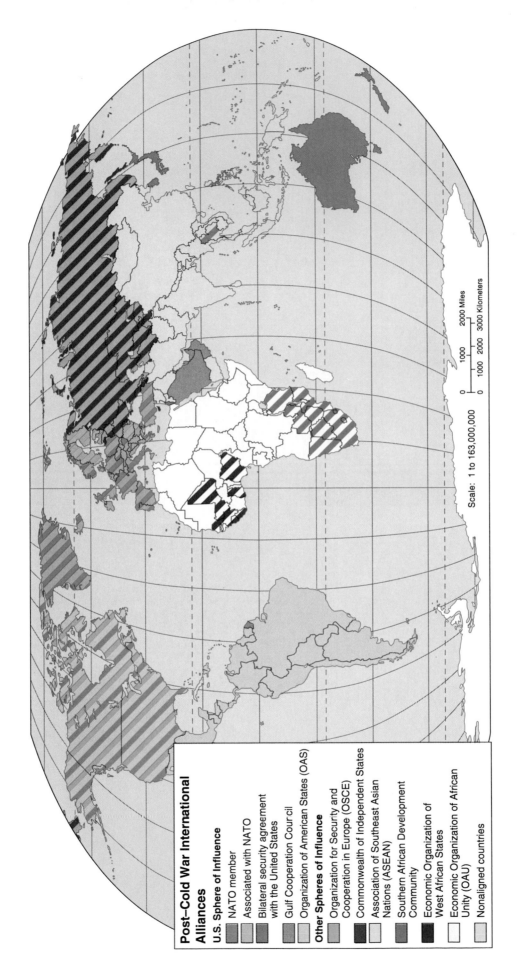

Post-Cold War International Alliances

U.S. Sphere of Influence
- NATO member
- Associated with NATO
- Bilateral security agreement with the United States
- Gulf Cooperation Council
- Organization of American States (OAS)

Other Spheres of Influence
- Organization for Security and Cooperation in Europe (OSCE)
- Commonwealth of Independent States
- Association of Southeast Asian Nations (ASEAN)
- Southern African Development Community
- Economic Organization of West African States
- Economic Organization of African Unity (OAU)
- Nonaligned countries

Scale: 1 to 163,000,000

0   1000   2000 Miles

0   1000   2000   3000 Kilometers

When the Warsaw Pact dissolved in 1992, the North Atlantic Treaty Organization (NATO) was left as the only major military alliance in the world. Some former Warsaw Pact members (Czech Republic, Hungary, and Poland) have joined NATO and others are petitioning for entry. The bipolar division of the world into two major military alliances is over, at least temporarily, leaving the United States alone as the world's dominant political and military power. But other international alliances such as the Commonwealth of Independent States (including most of the former republics of the Soviet Union) will continue to be important. It may well be that, during the first few decades of the twenty-first century, economic alliances will begin to overshadow military ones in their relevance for the world's peoples.

Map 65 International Conflicts in the Post–World War II World

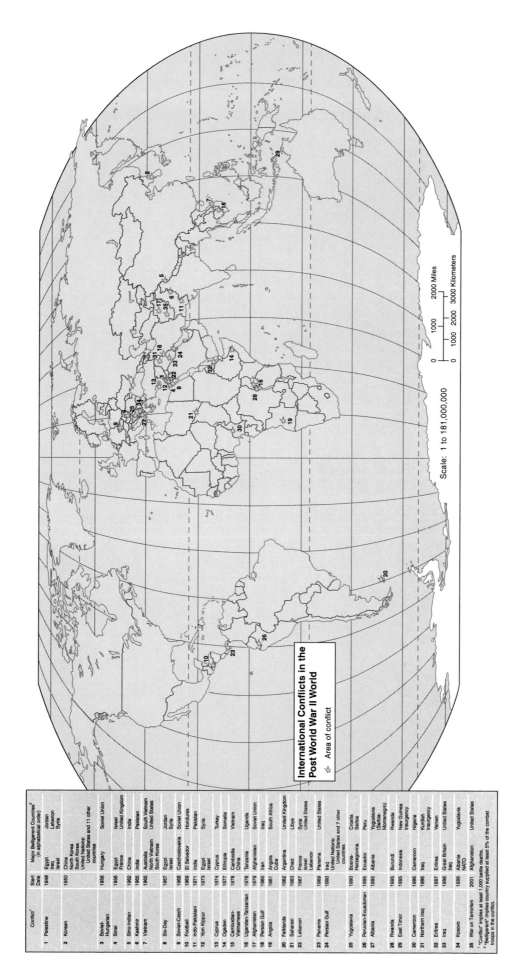

International Conflicts in the Post World War II World

☆ Area of conflict

Scale: 1 to 181,000,000

| 0   1000   2000 Miles |
| 0   1000   2000   3000 Kilometers |

	Conflict[1]	Start Date	Major Belligerent Countries[2] (in alphabetical order)
1	Palestine	1948	Egypt, Jordan, Iraq, Lebanon, Israel, Syria
2	Korean	1950	China, North Korea, South Korea, United Nations: United States and 11 other countries
3	Soviet-Hungarian	1956	Hungary, Soviet Union
4	Sinai	1956	Egypt, Israel, France, United Kingdom
5	Sino-Indian	1962	China, India
6	Kashmir	1965	India, Pakistan
7	Vietnam	1965	Australia, North Vietnam, South Vietnam, United States
8	Six-Day	1967	Egypt, Jordan, Israel, Syria
9	Soviet-Czech	1968	Czechoslovakia, Soviet Union
10	Football	1969	El Salvador, Honduras
11	Indo-Pakistani	1971	India, Pakistan
12	Yom Kippur	1973	Egypt, Israel, Syria
13	Cyprus	1974	Cyprus, Turkey
14	Ogaden	1977	Ethiopia, Somalia
15	Cambodian-Vietnamese	1978	Cambodia, Vietnam, China
16	Ugandan-Tanzanian	1978	Tanzania, Uganda
17	Afghanistan	1979	Afghanistan, Soviet Union
18	Persian Gulf	1980	Iran, Iraq
19	Angola	1981	Angola, South Africa
20	Falklands	1982	Argentina, United Kingdom
21	Saharan	1983	Chad, Libya
22	Lebanon	1967	France, Israel, Syria, Lebanon, United States
23	Panama	1989	Panama, United States
24	Persian Gulf	1990	Iraq, United Nations: United States and 7 other countries
25	Yugoslavia	1990	Croatia, Serbia
26	Peruvian-Ecuadorian	1995	Peru
27	Albania	1995	Albania
28	Burundi	1995	Burundi, Rwanda
29	East Timor	1995	Indonesia, New Guinea insurgency
30	Cameroon	1996	Cameroon, Nigeria
31	Northern Iraq	1996	Iraq, Kurdish insurgency
32	Eritrea	1997	Eritrea, Yemen
33	Iraq	1998	Great Britain, Iraq, United States
34	Kosovo	1999	Albania, NATO, Yugoslavia
35	War on Terrorism	2001	Afghanistan, United States

[1] "Conflict" implies at least 1,000 battle deaths.
[2] "Belligerent" implies country supplied at least 5% of the combat troops in the conflict.

The Korean War and the Vietnam War dominated the post–World War II period in terms of international military conflict. But numerous smaller conflicts have taken place, with fewer numbers of belligerents and with fewer battle and related casualties. These smaller international conflicts have been mostly territorial conflicts, reflecting the continual readjustment of political boundaries and loyalties brought about by the end of colonial empires, and the dissolution of the Soviet Union. Many of these conflicts were not wars in the more traditional sense, in which two or more countries formally declare war on one another, severing diplomatic ties, and devoting their entire national energies to the war effort. Rather, many of these conflicts were and are undeclared wars, sometimes fought between rival groups within the same country with outside support from other countries.

Map 66 International Terrorist Incidents, 2000

Americans have made a virtual mantra of the saying "The world has changed," as a consequence of the terrorist attacks on the World Trade Center and the Pentagon on September 11, 2001. As this map and the accompanying bar graphs point out, the world did not change, but the focus of a major terrorist attack shifted from Asia to North America. Many other areas of the world have lived with terrorism and terrorist activity for years. In 2000 and 2001, despite the enormous losses in the United States in the 9/11 attacks, more lives were lost in Asia and Africa as a result of terrorism than were lost in North America. The world did not change, but Americans' perception of that world and their place in it has certainly changed.

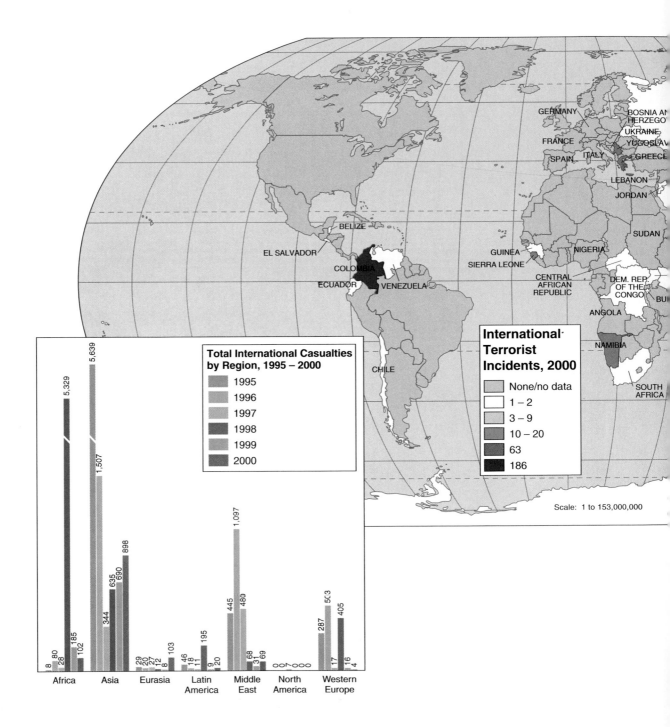

International Terrorist Incidents, 2000

	None/no data
	1 – 2
	3 – 9
	10 – 20
	63
	186

Scale: 1 to 153,000,000

Total International Casualties by Region, 1995 – 2000

- 1995
- 1996
- 1997
- 1998
- 1999
- 2000

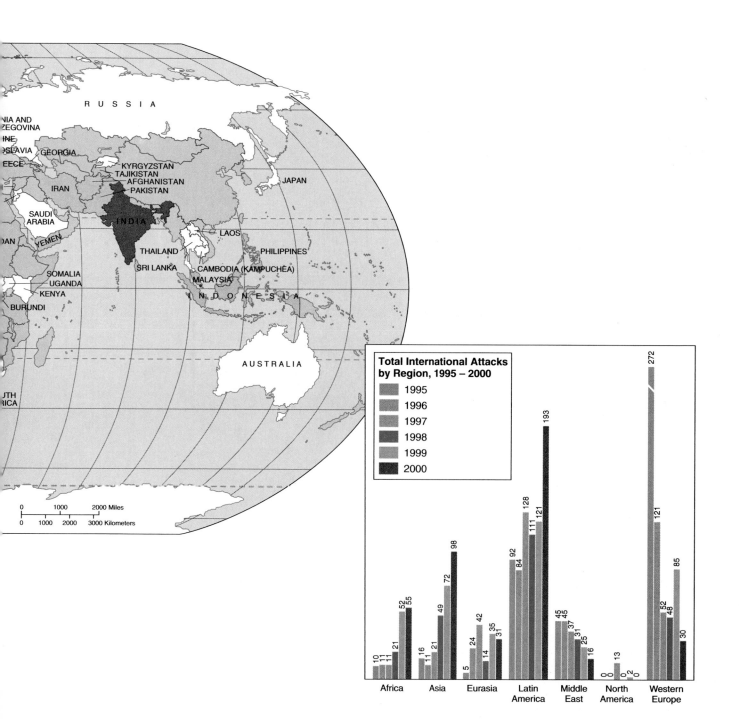

Total International Attacks by Region, 1995 – 2000

1995
1996
1997
1998
1999
2000

Africa
10
11
11
21
52
55

Asia
16
11
21
49
72
98

Eurasia
5
24
42
14
35
31

Latin America
92
84
128
111
121
193

Middle East
45
45
37
31
25
16

North America
0
0
13
0
2
0

Western Europe
272
121
52
48
85
30

Map 67 The Political Geography of a Global Religion: The Islamic World

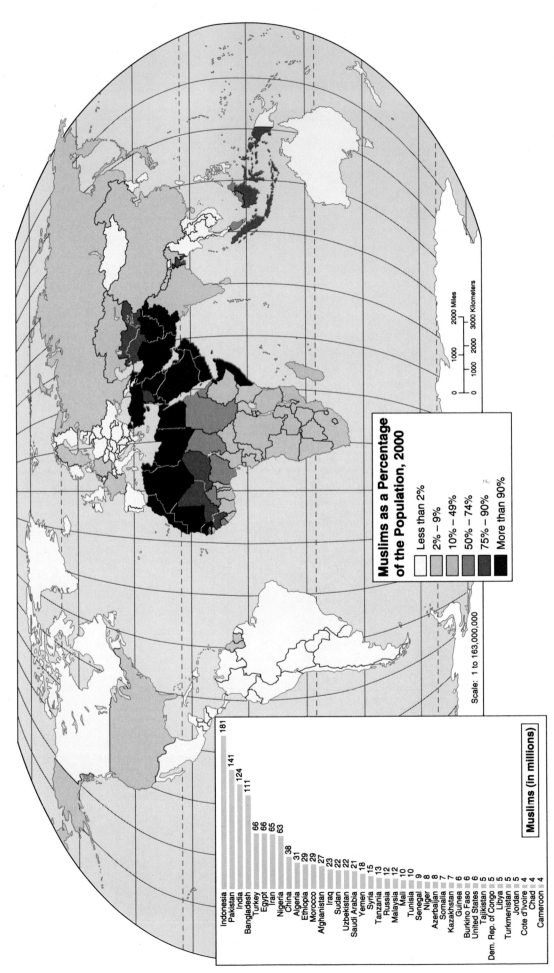

Muslims as a Percentage of the Population, 2000

- Less than 2%
- 2% – 9%
- 10% – 49%
- 50% – 74%
- 75% – 90%
- More than 90%

Scale: 1 to 163,000,000

0 1000 2000 Miles
0 1000 2000 3000 Kilometers

Muslims (in millions)

Country	Millions
Indonesia	181
Pakistan	141
India	124
Bangladesh	111
Turkey	66
Egypt	66
Iran	65
Nigeria	63
China	38
Algeria	31
Ethiopia	29
Morocco	29
Afghanistan	27
Iraq	23
Sudan	22
Uzbekistan	22
Saudi Arabia	21
Yemen	18
Syria	15
Tanzania	13
Russia	12
Malaysia	12
Mali	10
Tunisia	10
Senegal	9
Niger	8
Azerbaijan	8
Somalia	7
Kazakhstan	7
Guinea	6
Burkino Faso	6
United States	6
Tajikistan	5
Dem. Rep. of Congo	5
Libya	5
Turkmenistan	5
Jordan	5
Cote d'Ivoire	4
Chad	4
Cameroon	4

Islam, as a religion, does not promote conflict. The term *jihad*, often mistranslated to mean "holy war," in fact refers to the struggle to find God and to promote the faith. In spite of the beneficent nature of Islamic teachings, the tensions between Muslims and adherents of other faiths often flare into warfare. A comparison of this map with the map of international conflict will show a disproportionate number of wars in that portion of the world where Muslims are either majority or significant minority populations. The reasons for this are based more in the nature of government, cultures, and social

structure, than in the tenets of the faith of Islam. Nevertheless, the spatial correlations cannot be ignored. Similarly, terrorist incidents falling considerably short of open armed warfare, are spatially consistent with the distribution of Islam and even more consistent with the presence of Islamic fundamentalism or "Islamism," which tends to be less tolerant and more aggressive than the mainstream of the religion. Terrorism is also consistent with those areas where the legacy of colonialism or the persistent presence of non-Islamic cultures intrude into the Islamic world.

–82–

Map 68 Global Distribution of Minority Groups

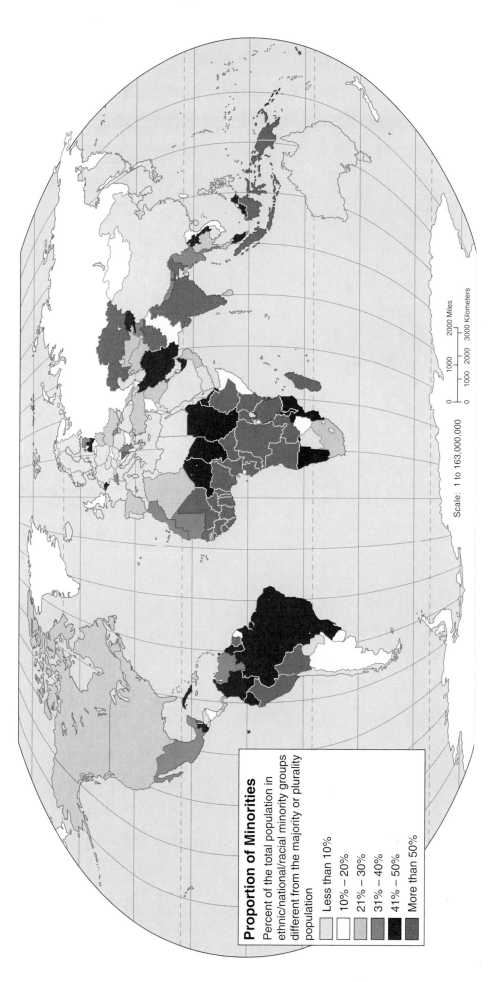

Proportion of Minorities

Percent of the total population in ethnic/national/racial minority groups different from the majority or plurality population

- Less than 10%
- 10% – 20%
- 21% – 30%
- 31% – 40%
- 41% – 50%
- More than 50%

Scale: 1 to 163,000,000

0   1000   2000 Miles

0   1000   2000   3000 Kilometers

The presence of minority ethnic, national, or racial groups within a country's population can add a vibrant and dynamic mix to the whole. Plural societies with a high degree of cultural and ethnic diversity should, according to some social theorists, be among the world's most healthy. Unfortunately, the reality of the situation is quite different from theory or expectation. The presence of significant minority populations played an important role in the disintegration of the Soviet Union; the continuing existence of minority populations within the new states formed from former Soviet republics threatens the viability and stability of those young political units. In Africa, national boundaries were drawn by colonial powers without regard for the geographical distribution of ethnic groups, and the continuing tribal conflicts that have resulted hamper both economic and political development. Even in the most highly developed regions of the world, the presence of minority ethnic populations poses significant problems: witness the separatist movement in Canada, driven by the desire of some French-Canadians to be independent of the English majority, and the continuing ethnic conflict between Flemish-speaking and Walloon-speaking Belgians. This map, by arraying states on a scale of homogeneity to heterogeneity, indicates areas of existing and potential social and political strife.

-83-

Map **69** World Refugee Population

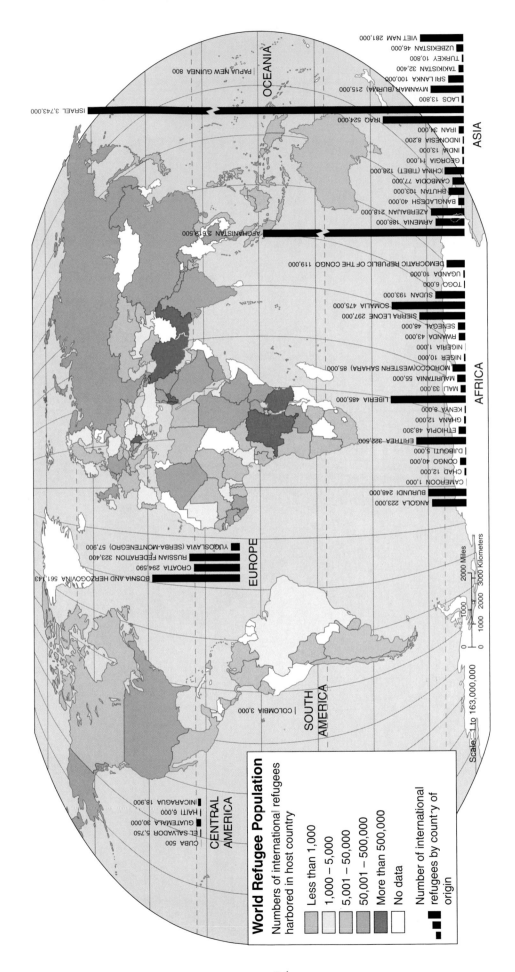

World Refugee Population

Numbers of international refugees harbored in host country

- Less than 1,000
- 1,000 – 5,000
- 5,001 – 50,000
- 50,001 – 500,000
- More than 500,000
- No data

Number of international refugees by country of origin

CENTRAL AMERICA
- CUBA 500
- HAITI 6,000
- GUATEMALA 30,000
- EL SALVADOR 5,750
- NICARAGUA 18,900

SOUTH AMERICA
- COLOMBIA 3,000

EUROPE
- BOSNIA AND HERZOGVINA 561,143
- CROATIA 294,590
- RUSSIAN FEDERATION 323,400
- YUGOSLAVIA (SERBA-MONTENEGRO) 57,900

AFRICA
- ANGOLA 223,000
- BURUNDI 248,000
- CAMEROON 1,000
- CHAD 12,000
- CONGO 40,000
- DJIBOUTI 1,500
- ERITREA 322,500
- ETHIOPIA 48,300
- GHANA 12,000
- KENYA 8,000
- LIBERIA 485,000
- MALI 33,000
- MAURITANIA 55,000
- MOROCCO(WESTERN SAHARA) 85,000
- NIGER 10,000
- NIGERIA 1,000
- RWANDA 43,000
- SENEGAL 48,000
- SIERRA LEONE 297,000
- SOMALIA 475,000
- SUDAN 193,000
- TOGO 6,000
- UGANDA 10,000
- DEMOCRATIC REPUBLIC OF THE CONGO 119,000

ASIA
- AFGHANISTAN 2,619,500
- ARMENIA 188,000
- AZERBAIJAN 218,000
- BANGLADESH 40,000
- BHUTAN 103,000
- CAMBODIA 77,000
- CHINA (TIBET) 128,000
- GEORGIA 14,000
- INDIA 13,000
- INDONESIA 8,200
- IRAN 34,000
- IRAQ 524,000
- ISRAEL 3,743,000
- LAOS 13,800
- MYANMAR (BURMA) 215,000
- SRI LANKA 100,000
- TAJIKISTAN 32,400
- TURKEY 10,800
- UZBEKISTAN 46,000
- VIET NAM 281,000

OCEANIA
- PAPUA NEW GUINEA 800

Scale—1 to 163,000,000

0   1000   2000   3000 Kilometers
0   1000   2000 Miles

Refugees are persons who have been driven from their homes, normally by armed conflict, and have sought refuge by relocating. The most numerous refugees have traditionally been international refugees, who have crossed the political boundaries of their homelands into other countries. This refugee population is recognized by international agencies, and the countries of refuge are often rewarded financially by those agencies for their willingness to take in externally displaced persons. In recent years, largely because of an increase in civil wars, there have been growing numbers of internally displaced persons—those who leave their homes but stay within their country of origin. There are no rewards for harboring such internal refugee populations.

Part VII

World Regions

Map 70 North America: Political Map

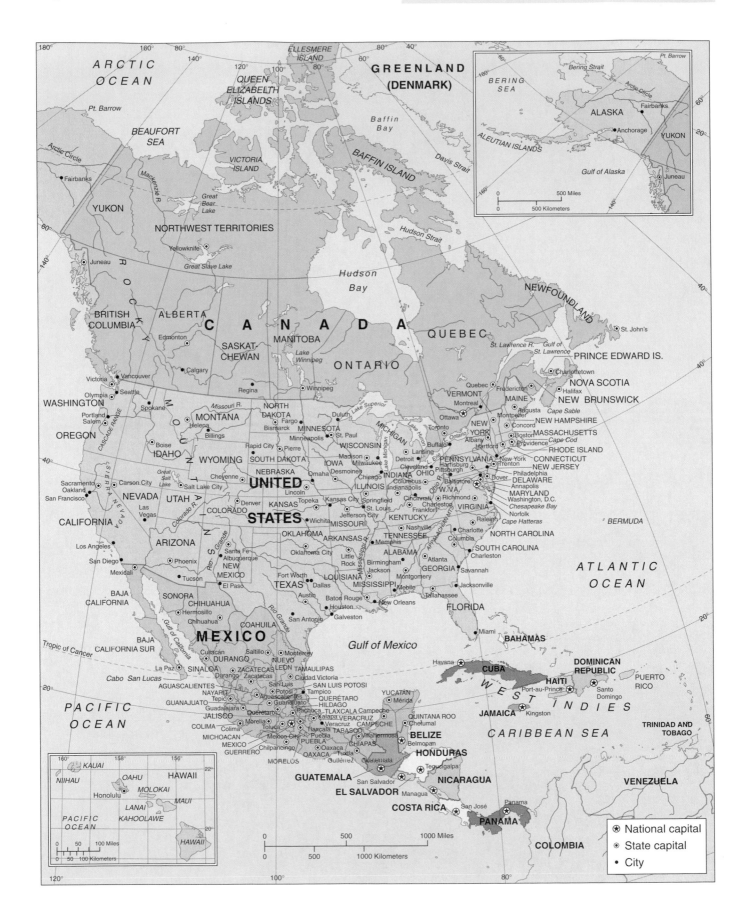

Map 71   North America: Physical Map

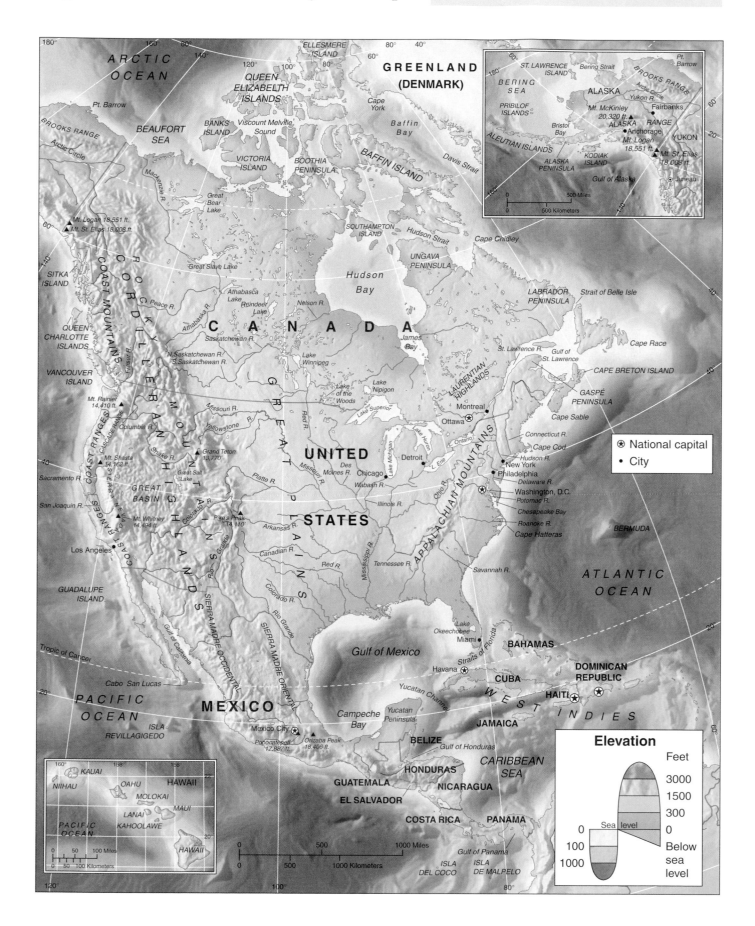

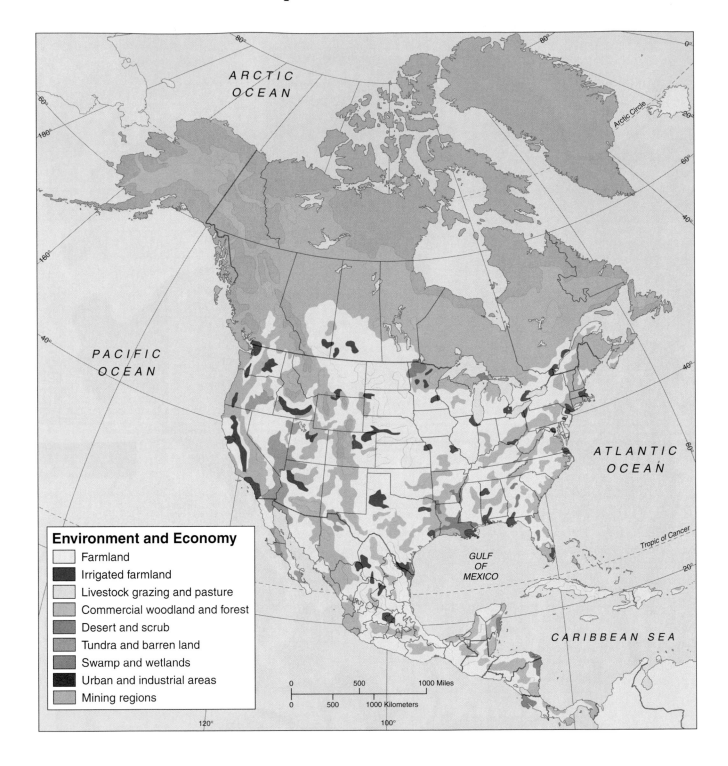

Environment and Economy
- ☐ Farmland
- ■ Irrigated farmland
- ☐ Livestock grazing and pasture
- ▨ Commercial woodland and forest
- ▨ Desert and scrub
- ▨ Tundra and barren land
- ▨ Swamp and wetlands
- ■ Urban and industrial areas
- ▨ Mining regions

Map 72a Environment and Economy: The Use of Land

The use of land in North America represents a balance between agriculture, resource extraction, and manufacturing that is unmatched. The United States, as the world's leading industrial power, is also the world's leader in commercial agricultural production. Canada, despite its small population, is a ranking producer of both agricultural and industrial products and Mexico has begun to emerge from its developing nation status to become an important industrial and agricultural

nation as well. The countries of Middle America and the Caribbean are just beginning the transition from agriculture to modern industrial economies. Part of the basis for the high levels of economic productivity in North America is environmental: a superb blend of soil, climate, and raw materials. But just as important is the cultural and social mix of the plural societies of North America, a mix that historically aided the growth of the economic diversity necessary for developed economies.

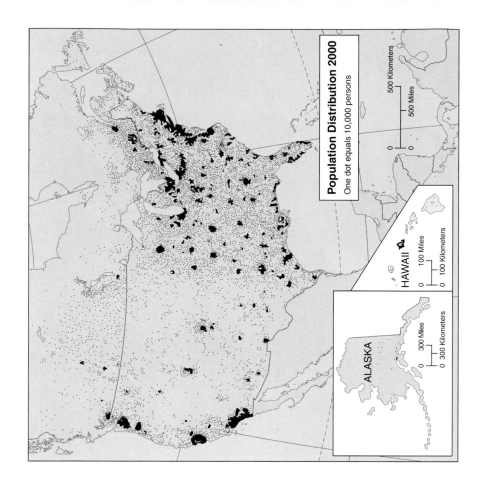

Map 72c Population Distribution: Clustering

Population Distribution 2000
One dot equals 10,000 persons

ALASKA

HAWAII

Although population clustering is characteristic of highly economically developed regions, the Anglo-American population exhibits a remarkably clustered pattern. The primary reasons for this remarkable development are agricultural technology and affluence. A highly developed agricultural technology allows a small number of farmers, using sophisticated machinery, to grow and harvest enormous quantities of agricultural produce on large farms. In the United States, the world's leader in commercial agricultural production, only 2 percent of the population are farmers and that population is thinly distributed over wide areas. The vast majority of Americans live and work in city-suburb systems in which the widespread availability of private automobiles allows people to live considerable distances from where they work, keeping overall urban population densities relatively low but allowing for extensive urbanization—with cities large enough in area to be visible as population clusters on maps at this scale.

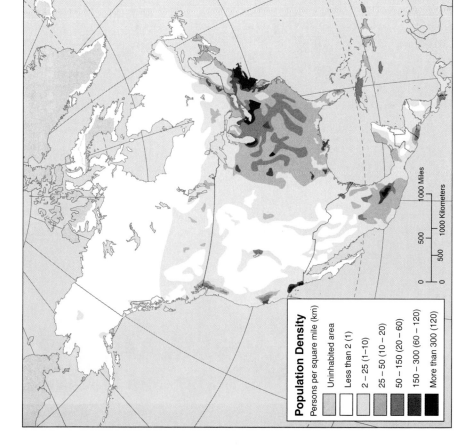

Map 72b Population Density

Population Density
Persons per square mile (km)

Uninhabited area
Less than 2 (1)
2 – 25 (1–10)
25 – 50 (10 – 20)
50 – 150 (20 – 60)
150 – 300 (60 – 120)
More than 300 (120)

North America contains nearly 500 million people and the United States, with over 250 million inhabitants, is the third most populous country in the world, after China and India. Most of the present North American population has roots in the Old World. The native populations of the Americas had little or no resistance to Old World diseases in 1500 and within a couple of centuries of first contact with Europeans, most of the native peoples had either died out or preserved their genetic heritage by mixing with disease-resistant Old World populations. This left a North American population that is largely European, but with significant minorities resulting from the slave laborers imported from Africa, and from the mixture of native Americans with Europeans and/or Africans. The density of that population is largely the consequence of environmental factors (good soil, the availability of water, the presence of other resources) and cultural/economic ones (agricultural production, urbanization, industrialization).

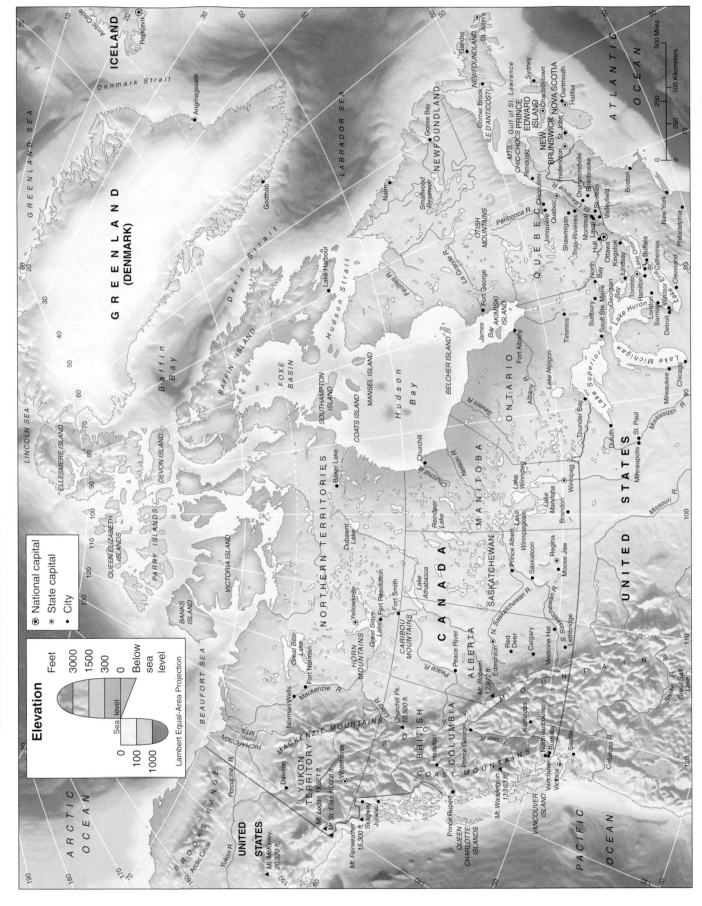

Map 73 Canada

Elevation

Feet	
3000	
1500	
300	
0	Below sea level

Sea level

0
100
1000

Lambert Equal-Area Projection

⊛ National capital
⊙ State capital
• City

500 Miles
500 Kilometers
250
250

ICELAND
Reykjavik
Arctic Circle
Denmark Strait
GREENLAND SEA
Angmagssalik
GREENLAND (DENMARK)
Godthåb
LINCOLN SEA
ELLESMERE ISLAND
QUEEN ELIZABETH ISLANDS
DEVON ISLAND
PARRY ISLANDS
Baffin Bay
Davis Strait
BAFFIN ISLAND
FOXE BASIN
Nain
LABRADOR SEA
Lake Harbour
Hudson Strait
SOUTHAMPTON ISLAND
COATS ISLAND
MANSEL ISLAND
BELCHER ISLAND
Hudson Bay
AKIMISKI ISLAND
James Bay
Fort George
Feuilles R.
La Grande R.
Péribonca R.
OTISH MOUNTAINS
Smallwood Reservoir
NEWFOUNDLAND
Goose Bay
Gander
Corner Brook
St. John's
ÎLE D'ANTICOSTI
Gulf of St. Lawrence
Sydney
PRINCE EDWARD ISLAND
Charlottetown
NEW BRUNSWICK
NOVA SCOTIA
Dartmouth
Halifax
St. John
Fredericton
Rimouski
CHIC-CHOCS MTS.
Chicoutimi
Jonquière
Shawinigan
Trois-Rivières
QUEBEC
Saguenay R.
Drummondville
Sherbrooke
Montreal
Laval
St-Jean
Valleyfield
Hull
Ottawa
Boston
New York
Philadelphia
ATLANTIC OCEAN
North Bay
Sudbury
Timmins
Fort Albany
Albany R.
Sault Ste. Marie
Lake Nipigon
Thunder Bay
Lake Superior
ONTARIO
Lake Huron
Georgian Bay
Kingston
Lindsay
Lake Ontario
Toronto
Hamilton
St. Catharines
Buffalo
London
Sarnia
Windsor
Detroit
Lake Erie
Cleveland
Lake Michigan
Chicago
Milwaukee
Duluth
Minneapolis
St. Paul
Mississippi R.
Missouri R.
UNITED STATES
Churchill
Churchill R.
Nelson R.
Severn R.
Baker Lake
Dubawnt Lake
NORTHERN TERRITORIES
Reindeer Lake
Lake Winnipeg
MANITOBA
Winnipeg
Brandon
Lake Manitoba
Lake Winnipegosis
Prince Albert
N. Saskatchewan R.
Saskatoon
SASKATCHEWAN
Regina
Moose Jaw
S. Saskatchewan R.
Lake Athabasca
Fort Smith
Fort Resolution
Great Slave Lake
Yellowknife
CARIBOU MOUNTAINS
Peace River
Peace R.
Edmonton
Red Deer
Calgary
Medicine Hat
Lethbridge
ALBERTA
Mt. Robson 12,972 ft.
Great Bear Lake
Fort Norman
Norman Wells
Mackenzie R.
Liard R.
HORN MOUNTAINS
MACKENZIE MOUNTAINS
RICHARDSON MTS.
Mackenzie
Dawson
Yukon R.
Porcupine R.
YUKON TERRITORY
Whitehorse
Mt. Logan 19,551 ft.
Mt. St. Elias 18,008 ft.
Mt. Fairweather 15,300 ft.
Skagway
Juneau
BROOKS RANGE
UNITED STATES
Mt. McKinley 20,320 ft.
Arctic Circle
ARCTIC OCEAN
BEAUFORT SEA
BANKS ISLAND
VICTORIA ISLAND
Churchill Pk. 10,500 ft.
BRITISH COLUMBIA
Hazelton
Prince Rupert
Prince George
QUEEN CHARLOTTE ISLANDS
COAST MOUNTAINS
Fraser R.
Kamloops
Mt. Waddington 13,163 ft.
Vancouver
North Vancouver
Burnaby
VANCOUVER ISLAND
Victoria
Seattle
PACIFIC OCEAN
Columbia R.
Snake R.
ROCKY MOUNTAINS
Great Salt Lake
CANADA

Map 74 United States

Map 75 Middle America

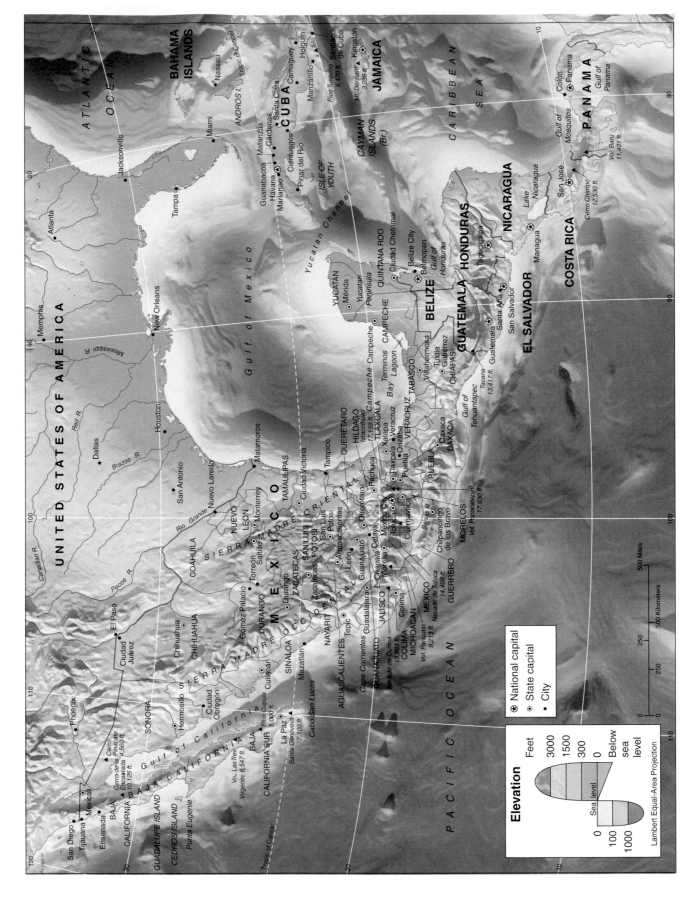

Map 76 The Caribbean

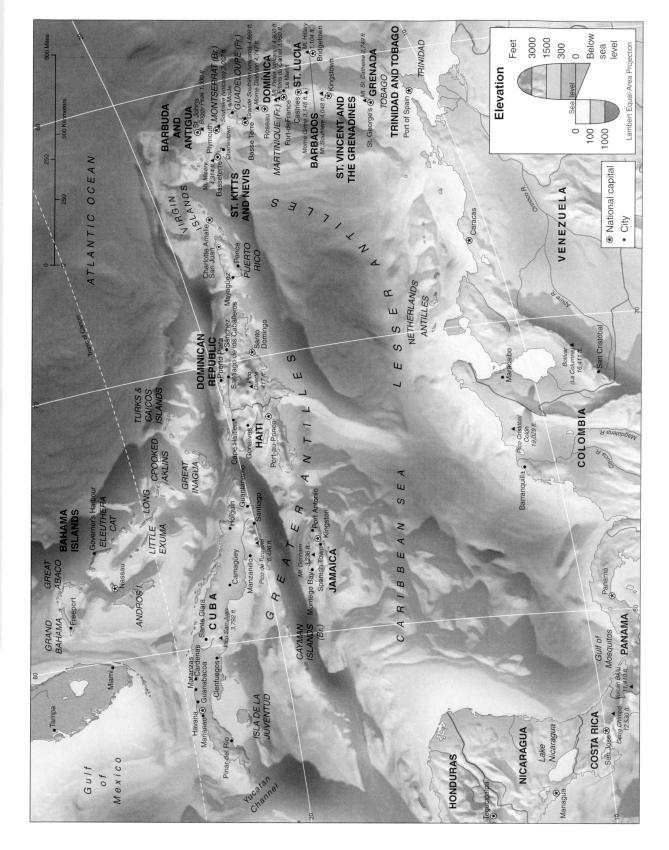

Elevation

Feet	
3000	
1500	
300	
0	Sea level
Below sea level	

0	Sea level
100	
1000	

Lambert Equal-Area Projection

⊛ National capital
• City

ATLANTIC OCEAN

Tropic of Cancer

500 Miles

500 Kilometers

BAHAMA ISLANDS
GRAND BAHAMA
Freeport
GREAT ABACO
Nassau
ANDROS I.
Governor's Harbour
ELEUTHERA
CAT
LITTLE
LONG
EXUMA
CROOKED
AKLINS
GREAT INAGUA
TURKS & CAICOS ISLANDS

Miami
Tampa

Gulf of Mexico

Havana
Marianao ⊛
Guanabacoa
Matanzas
Cárdenas
Santa Clara
Cienfuegos
Pinar del Río
ISLA DE LA JUVENTUD
Pico San Juan 3,792 ft.
CUBA
Camagüey
Manzanillo
Holguín
Guantánamo
Santiago
Cape Haitien
Gonaïves
HAITI
Port-au-Prince
Pico de Turquino 6,496 ft.
Mt. Denham 3,236 ft.
Montego Bay
Spanish Town
Kingston
Port Antonio
JAMAICA
CAYMAN ISLANDS (Br.)

Yucatan Channel

CARIBBEAN SEA

HONDURAS
Tegucigalpa ⊛
NICARAGUA
Managua ⊛
Lake Nicaragua
COSTA RICA
San José ⊛
Cerro Chirripó 12,530 ft.
Volcán Baru 11,410 ft.
PANAMA
Gulf of Mosquitos
Panama ⊛

Pico Duarte 10,417 ft.
DOMINICAN REPUBLIC
Santiago de los Caballeros
Puerto Plata
Sánchez
Santo Domingo
Mayagüez
Ponce
San Juan
PUERTO RICO
Charlotte Amalie
VIRGIN ISLANDS

BARBUDA AND ANTIGUA
St. Johns
Mt. Misery 4,314 ft.
Basseterre
ST. KITTS AND NEVIS
Charlestown
Plymouth
Boggy Peak 1,330 ft.
MONTSERRAT (Br.)
Soufrière (volcano) 3,021 ft.
GUADELOUPE (Fr.)
Le Moule
Basse Terre
Grande Soufrière (volcano) 4,869 ft.
Roseau
DOMINICA
Morne Diablotin 4,747 ft.
MARTINIQUE (Fr.)
Fort-de-France
Mt. Pelée (volcano) 4,800 ft.
Castries
ST. LUCIA
Morne Gimie 3,145 ft.
Mt. Hillary 1,104 ft.
BARBADOS
Bridgetown
Pitons du Carbet 3,960 ft.
ST. VINCENT AND THE GRENADINES
Mt. Soufrière 4,048 ft.
Kingstown
GRENADA
St. George's ⊛
Mt. St. Catherine 2,749 ft.
TOBAGO
TRINIDAD AND TOBAGO
Port of Spain ⊛
TRINIDAD

GREATER ANTILLES

LESSER ANTILLES

NETHERLANDS ANTILLES

Caracas ⊛
VENEZUELA
Maracaibo
Bolívar (La Columna) 16,411 ft.
San Cristóbal
Apure R.
Orinoco R.

Barranquilla
COLOMBIA
Pico Cristóbal Colón 19,029 ft.
Magdalena R.
Cauca R.

-93-

Map 77 South America: Political Map

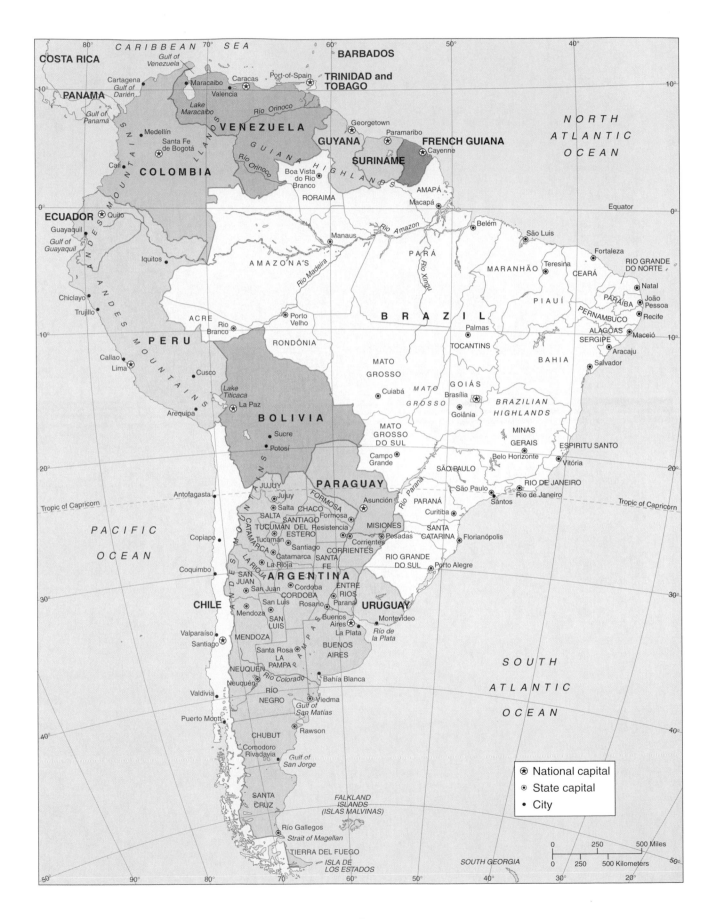

Map 78 South America: Physical Map

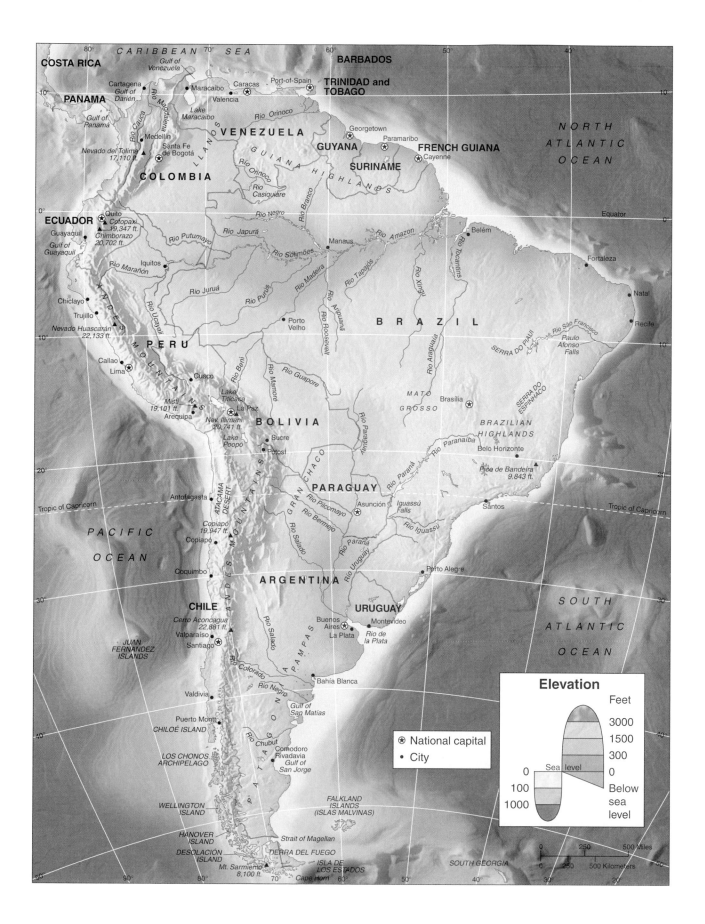

Elevation

	Feet
	3000
	1500
	300
Sea level	0
0	
100	Below
1000	sea level

⊛ National capital

• City

South America: Thematic Maps

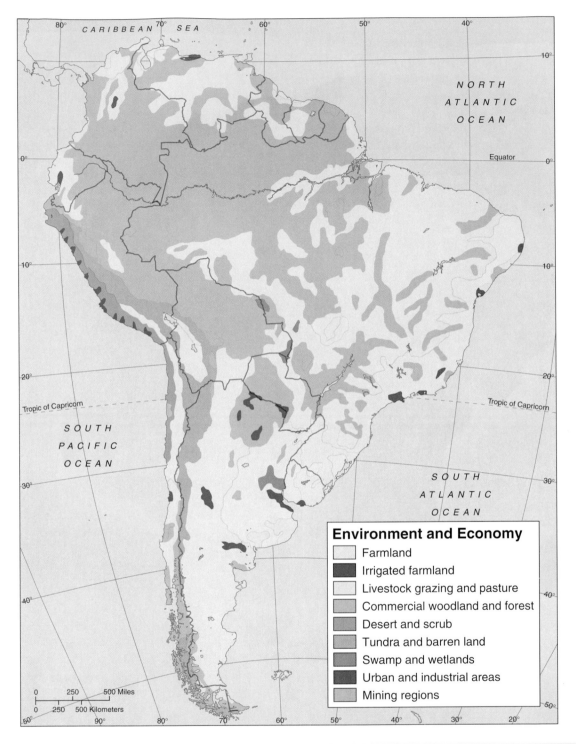

Environment and Economy

- Farmland
- Irrigated farmland
- Livestock grazing and pasture
- Commercial woodland and forest
- Desert and scrub
- Tundra and barren land
- Swamp and wetlands
- Urban and industrial areas
- Mining regions

Map 79a Environment and Economy

South America is a region just beginning to emerge from a colonial-dependency economy in which raw materials flowed from the continent to more highly developed economic regions. With the exception of Brazil, Argentina, Chile, and Uruguay, most of the continent's countries still operate under the traditional mode of exporting raw materials in exchange for capital that tends to accumulate in the pockets of a small percentage of the population. The land use patterns of the continent are, therefore, still dominated by resource extraction and agriculture. A problem posed by these patterns is that little of the continent's land area is actually suitable for either commercial forestry or commercial crop agriculture without extremely high environmental costs. Much of the agriculture, then, is based on high value tropical crops that can be grown in small areas profitably, or on extensive livestock grazing. Even within the forested areas of the Amazon Basin where forest clearance is taking place at unprecedented rates, much of the land use that replaces forest is grazing.

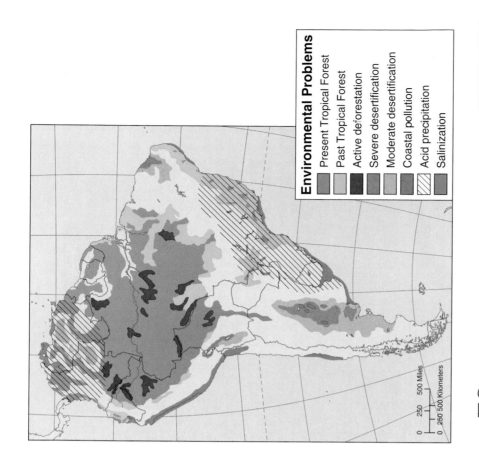

Map 79c Environmental Problems

Environmental Problems

- Present Tropical Forest
- Past Tropical Forest
- Active deforestation
- Severe desertification
- Moderate desertification
- Coastal pollution
- Acid precipitation
- Salinization

The drainage basin of the Amazon River and its tributaries, along with adjacent regions, is the world's largest remaining area of tropical forest. Much of the periphery of this vast forested region has already been cleared for farming and grazing and the ax and chainsaw and the flames are working their way steadily toward the interior. Tropical deforestation produces a loss of the biological diversity represented by the world's most biologically productive ecosystem, along with changes in soil and soil-water systems. South America has other environmental problems: *desertification* in which grassland and/or scrub vegetation is converted to desert through overgrazing or other unwise agricultural practices; soil *salinization* in which soils become increasingly salty as the consequence of the over-application of irrigation water; *coastal and estuarine pollution* resulting from unregulated or unchecked use of coastal waters for industrial, commercial, and transportation purposes; and *acid precipitation* resulting from the combination of airborne industrial wastes and automobile-truck exhausts with water vapor to produce dry or wet acidic fallout.

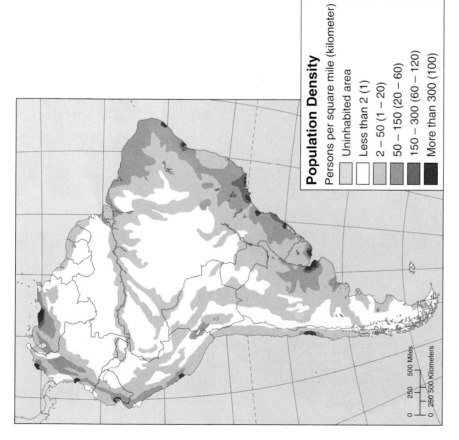

Map 79b Population Density

Population Density

Persons per square mile (kilometer)

- Uninhabited area
- Less than 2 (1)
- 2 – 50 (1 – 20)
- 50 – 150 (20 – 60)
- 150 – 300 (60 – 120)
- More than 300 (100)

Since so much of interior South America is uninhabitable (the high Andes) or only sparsely populated (the interior of the Amazon Basin), the continent's population tends to be peripheral—approximately 90 percent of the continent's nearly 375 million people live within 150 miles of the sea. This population also tends to be heavily urbanized. Over 80 percent of South America's population lives in cities and the continent has three of the world's 15 largest cities—Rio de Janeiro, Saõ Paulo, and Buenos Aires. Saõ Paulo is the world's third largest urban agglomeration after Tokyo and New York. As in North America, most of the population of South America can trace at least part of its ancestry to the Old World. Throughout the Spanish-speaking parts of the continent the population is predominantly *mestizo* or mixed European and native South American. In Portuguese-speaking Brazil, in addition to a *mestizo* population, there is a significant admixture of African blood, the result of a large slave labor force imported from Africa to work the sugar, cotton, and other plantations of the colonial period.

Map 80 Northern South America

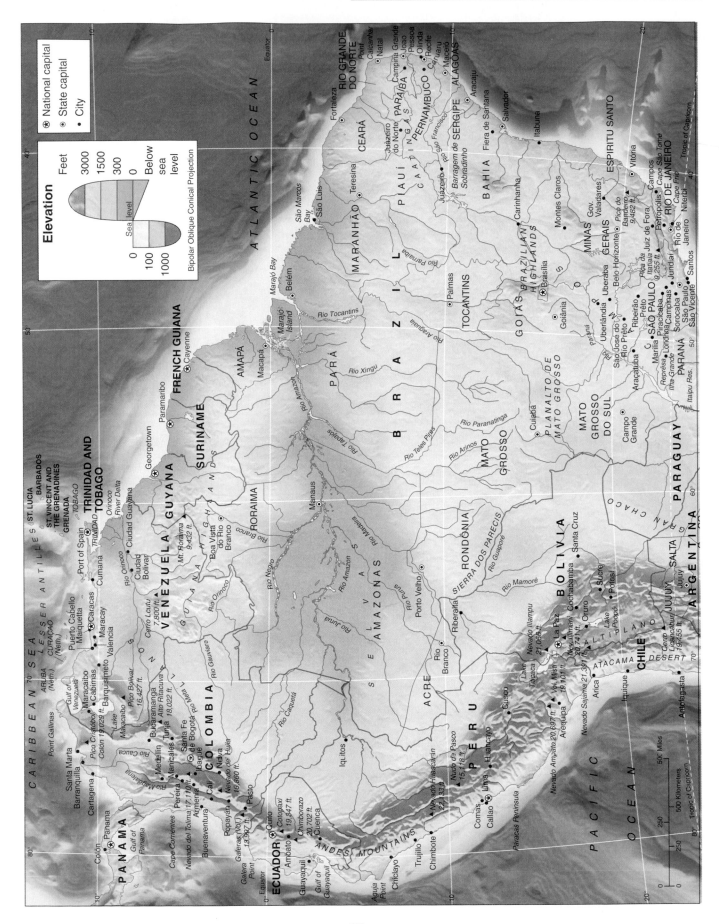

Map 81 Southern South America

Map 82 Europe: Political Map

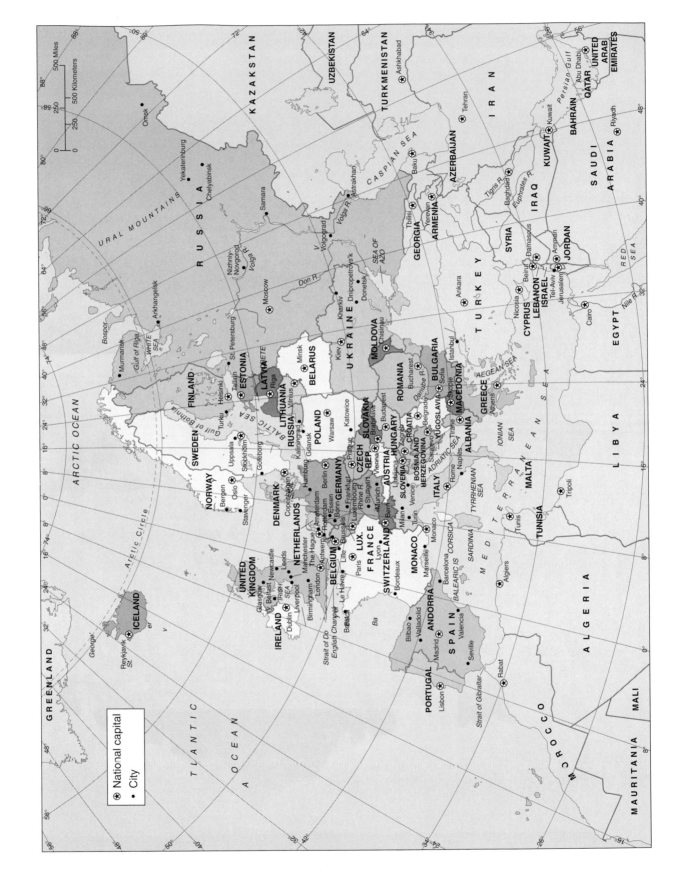

Map **83** Europe: Physical Map

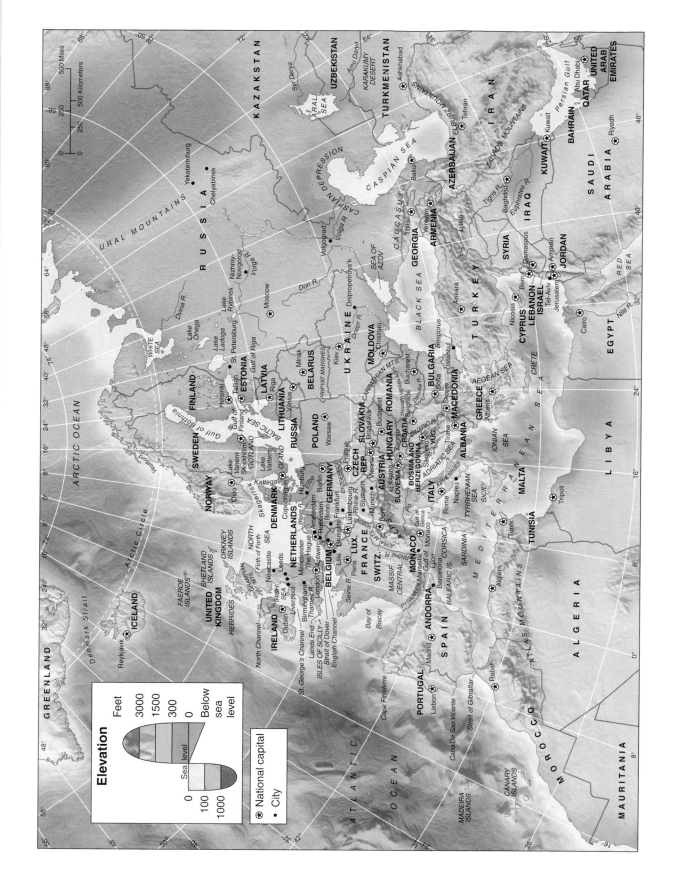

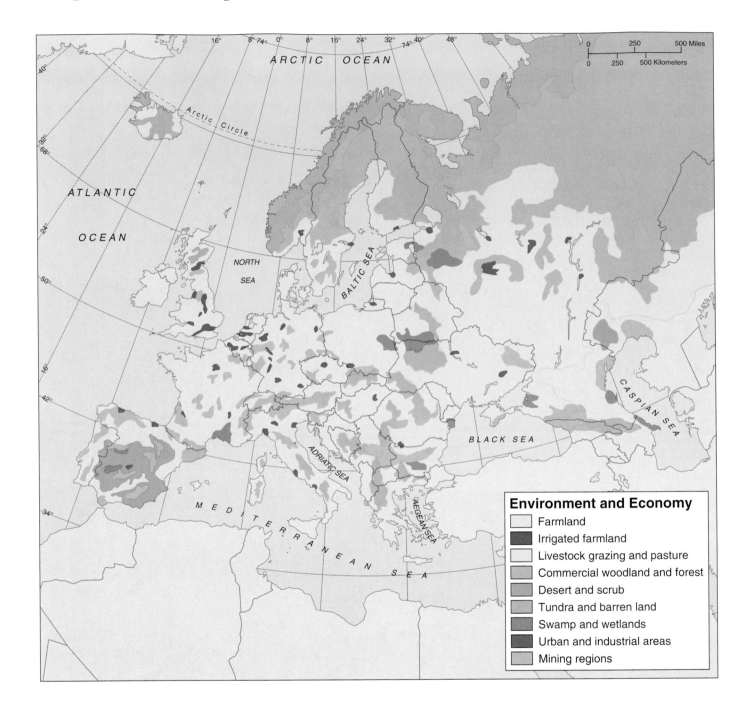

Environment and Economy

- Farmland
- Irrigated farmland
- Livestock grazing and pasture
- Commercial woodland and forest
- Desert and scrub
- Tundra and barren land
- Swamp and wetlands
- Urban and industrial areas
- Mining regions

Map 84a Environment and Economy: The Use of Land

More than any other continent, Europe bears the imprint of human activity—mining, forestry, agriculture, industry, and urbanization. Virtually all of western and central Europe's natural forest vegetation is gone, lost to clearing for agriculture beginning in prehistory, to lumbering that began in earnest during the Middle Ages, or more recently, to disease and destruction brought about by acid precipitation. Only in the far north and the east do some natural stands remain. The region is the world's most heavily industrialized and the industrial areas on the map represent only the largest and most significant. Not shown are the industries that are found in virtually every small town and village and smaller city throughout the industrial countries for Europe. Europe also possesses abundant raw materials and a very productive agricultural base. The mineral resources have long been in a state of active exploitation and the mining regions shown on the map are, for the most part, old regions in upland areas that are somewhat less significant now than they may have been in the past. Agriculturally, the northern European plain is one of the world's great agricultural regions but most of Europe contains decent land for agriculture.

Map 84b Population Density

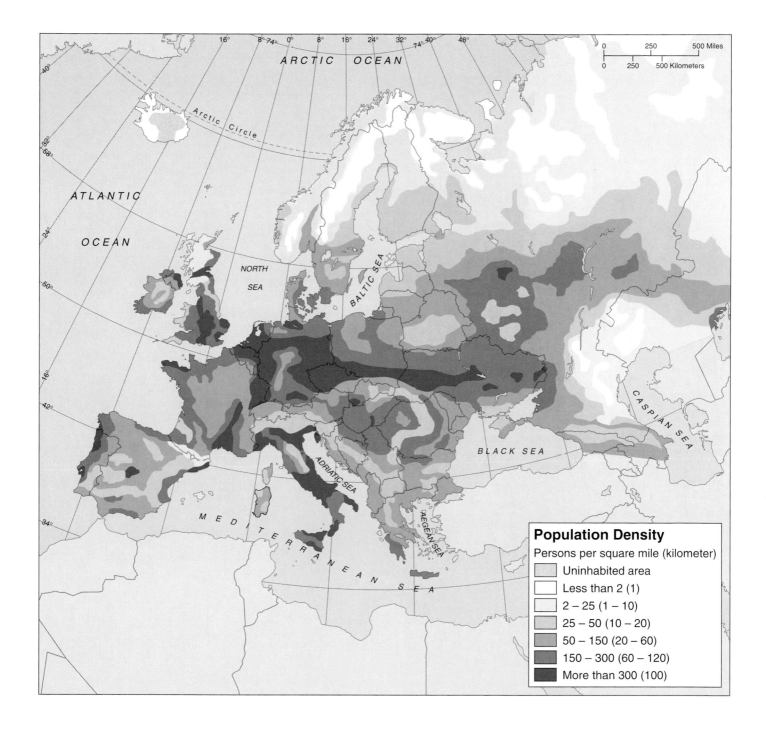

Population Density

Persons per square mile (kilometer)

	Uninhabited area
	Less than 2 (1)
	2 – 25 (1 – 10)
	25 – 50 (10 – 20)
	50 – 150 (20 – 60)
	150 – 300 (60 – 120)
	More than 300 (100)

Europe is one of the most densely settled regions of the world with an overall population density nearing 200 persons per square mile (80 per square kilometer), the consequence of a high level of urbanization and an economic system that is heavily industrialized. Even in agricultural regions, the population density is high. Beyond high density, the two chief identifying marks of the European population are remarkable diversity and unusual dynamics. For a small part of the world, Europe has cultural and ethnic diversity that is rarely matched elsewhere; more than 60 languages are spoken in an area not much larger than the United States. The population dynamics of Europe show a mature population that has passed through the "Demographic Transition"—a remarkable increase and then decline in growth rates resulting from the rise of an urban-industrial society. Only in other heavily industrialized regions of the world are found the very small overall growth rates characteristic of Europe.

Map 84c Political Changes in the 1990s

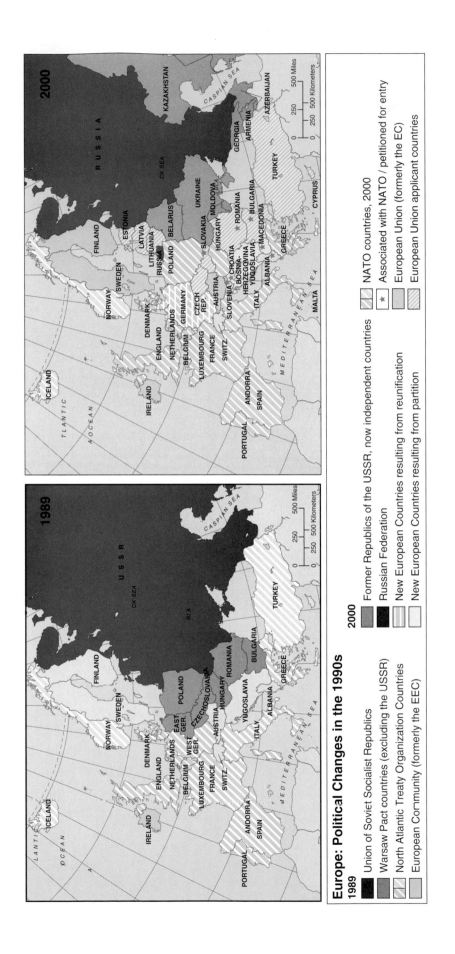

Europe: Political Changes in the 1990s

1989

■	Union of Soviet Socialist Republics
▨	Warsaw Pact countries (excluding the USSR)
▨	North Atlantic Treaty Organization Countries
▨	European Community (formerly the EEC)

2000

▨	Former Republics of the USSR, now independent countries
■	Russian Federation
▨	New European Countries resulting from reunification
▨	New European Countries resulting from partition
▨	NATO countries, 2000
★	Associated with NATO / petitioned for entry
▨	European Union (formerly the EC)
▨	European Union applicant countries

During the last decade of the twentieth century, one of the most remarkable series of political geographic changes of the last 500 years took place. The bipolar "East-West" structure that had characterized Europe's political geography since the end of World War II altered in the space of a very few years. In the mid-1980s, as Soviet influence over eastern and central Europe weakened, those countries began to turn to the capitalist West. Between 1989, when the country of Hungary was the first Soviet satellite to open its borders to travel, and 1991, when the Soviet Union dissolved into 15 independent countries, abrupt change in political systems occurred. The result is a new map of Europe that includes a number of countries not present on the map of 1989. These countries have emerged as the result of reunification, separation, or independence from the old Soviet Union. The new political structure has been accompanied by growing economic cooperation.

-104-

Map 85 Western Europe

Map **86** Eastern Europe

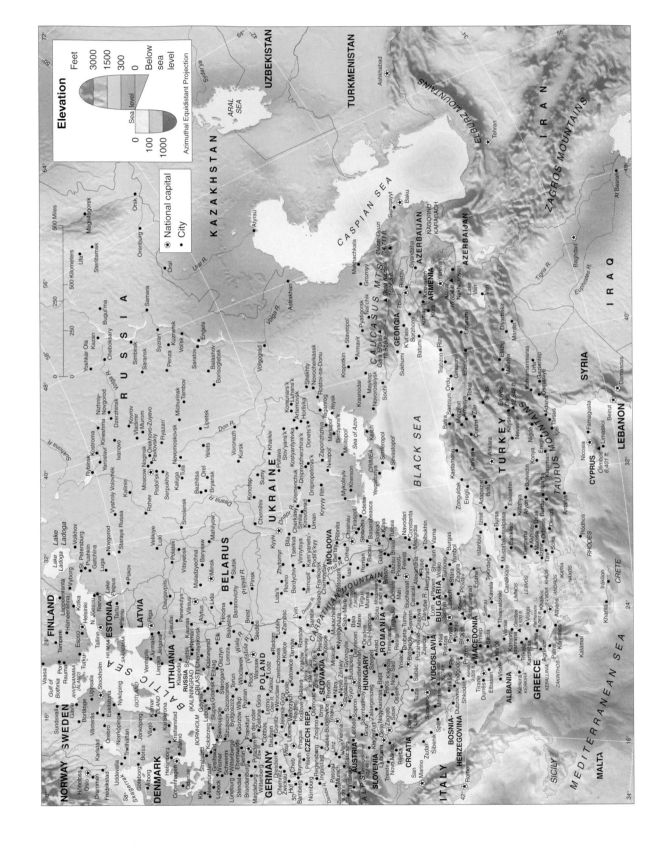

Map **87** Northern Europe

Map 88 Africa: Political Map

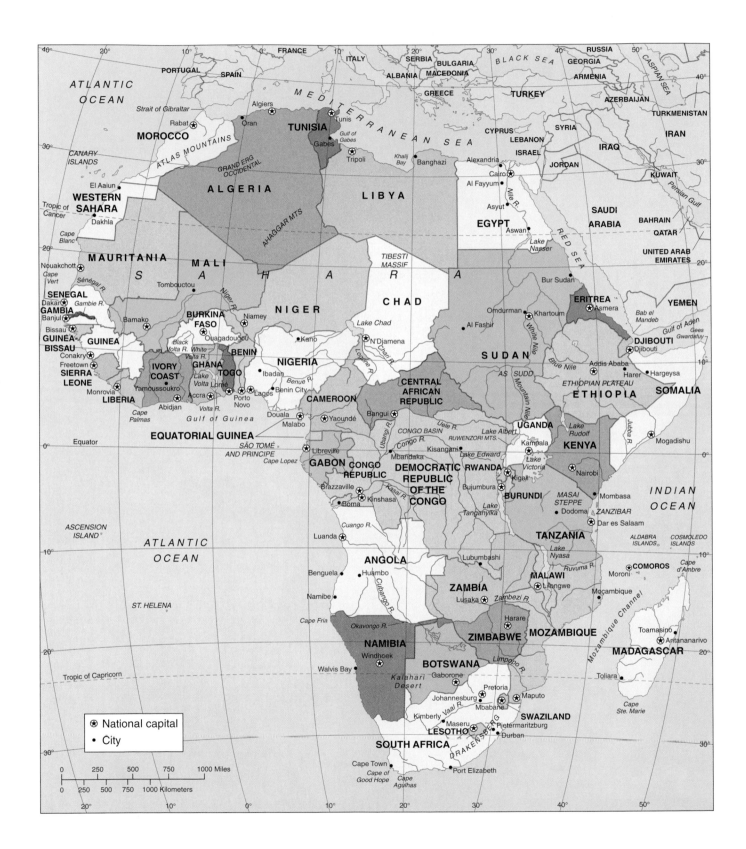

Map 89 Africa: Physical Map

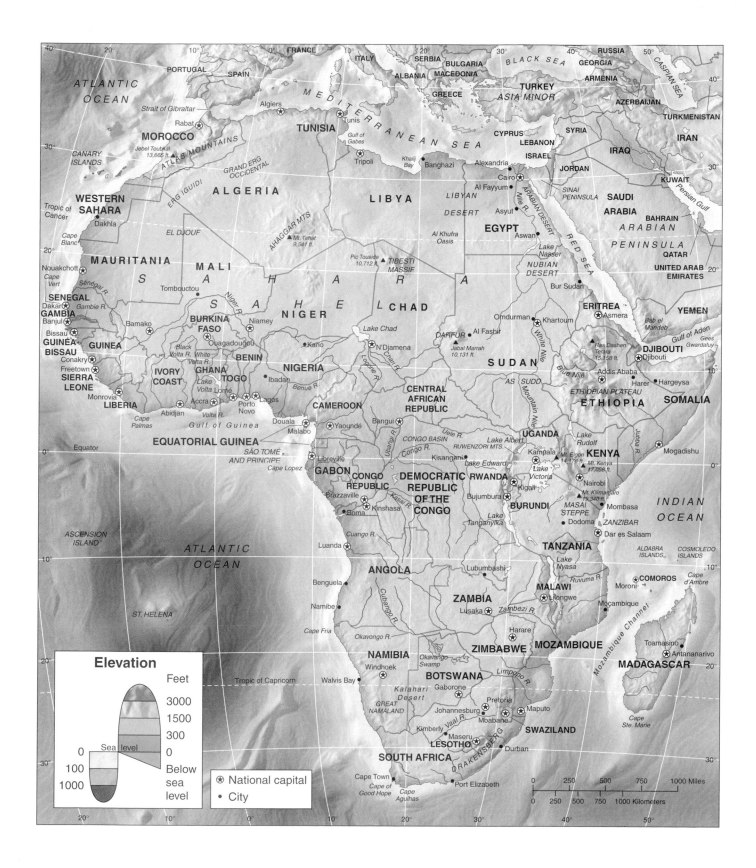

Africa: Thematic Maps

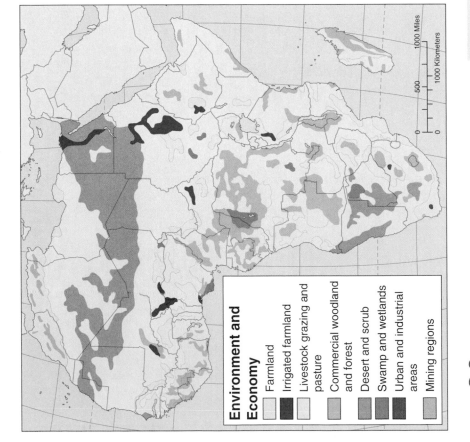

Environment and Economy

Farmland
Irrigated farmland
Livestock grazing and pasture
Commercial woodland and forest
Desert and scrub
Swamp and wetlands
Urban and industrial areas
Mining regions

Map 90a Environment and Economy

Africa's economic landscape is dominated by subsistence, or marginally-commercial agricultural activities and raw material extraction, engaging three-fourths of Africa's workers. Much of this grazing land is very poor desert scrub and bunch grass that is easily impacted by cattle, sheep, and goats. Growing human and livestock populations place enormous stress on this fragile support capacity and the result is desertification: the conversion of even the most minimal of grazing environments to a small quantity of land suitable for crop farming. Although the continent has approximately 20 percent of the world's total land area, the proportion of Africa's arable land is small. The agricultural environment is also uncertain; unpredictable precipitation and poor soils hamper crop agriculture.

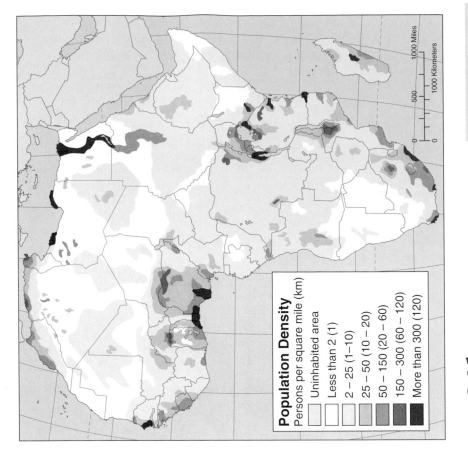

Population Density
Persons per square mile (km)

Uninhabited area
Less than 2 (1)
2 – 25 (1–10)
25 – 50 (10 – 20)
50 – 150 (20 – 60)
150 – 300 (60 – 120)
More than 300 (120)

1000 Miles
0   500   1000 Kilometers

Map 90b Population Density

Nearly 800 million people occupy the African continent, approximately one-eighth of the world's population. In general, this population has two chief characteristics: a low level of quality of life and growth rates that are among the world's highest. On a continent beset by poverty, recurrent internal civil and tribal war, and a host of environmental problems, the populations of many African countries are nevertheless increasing at a rate above 3 percent per year. The bulk of the population is concentrated in relatively small areas of the Mediterranean coastal regions of the north, the bulge of West Africa, and the eastern coastal and highland zone stretching from South Africa to Kenya. Where populations in other parts of the world tend to avoid highland locations, in Africa highlands often tend to be the most densely settled regions: moister with better soils, freer of insect pests, and somewhat cooler.

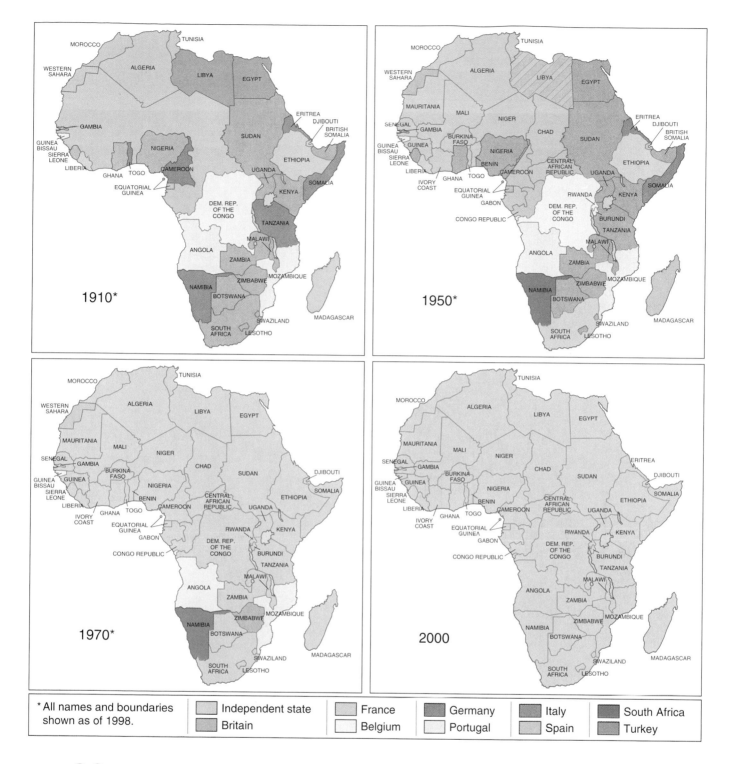

1910*

1950*

1970*

2000

*All names and boundaries shown as of 1998.	Independent state	France	Germany	Italy	South Africa	
	Britain	Belgium	Portugal	Spain	Turkey	

Map 90c Colonial Patterns

In few parts of the world has the transition from colonialism to independence been as abrupt as on the African continent. Most African states did not become colonies until the nineteenth century and did not become independent until the twentieth, nearly all of them after World War II. Much of the colonial power in Africa is social and economic. The African colony provided the mother country with raw materials in exchange for marginal economic returns, and many African countries still exist in this colonial dependency relationship. An even more important component of the colonial legacy of Europe in Africa is geopolitical. When the world's colonial powers joined at the Conference of Berlin in 1884, they divided up Africa to fit their own needs, drawing boundary lines on maps without regard for terrain or drainage features, or for tribal/ethnic linguistic, cultural, economic, or political borders. Traditional Africa was enormously disrupted by this process. After independence, African countries retained boundaries that are legacies of the colonial past and African countries today are beset by internal problems related to tribal and ethnic conflicts, the disruption of traditional migration patterns, and inefficient spatial structures of market and supply.

Map **91** Northern Africa

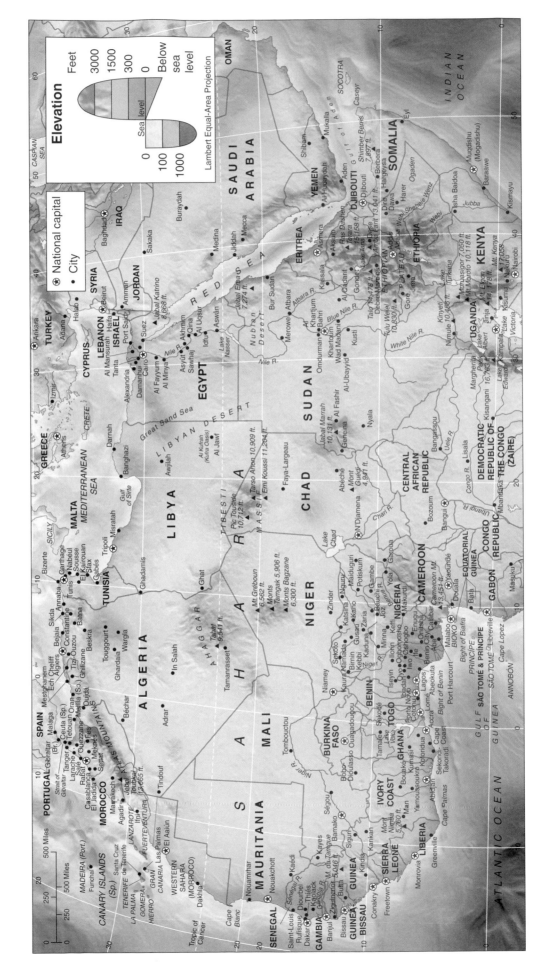

Map 92 Southern Africa

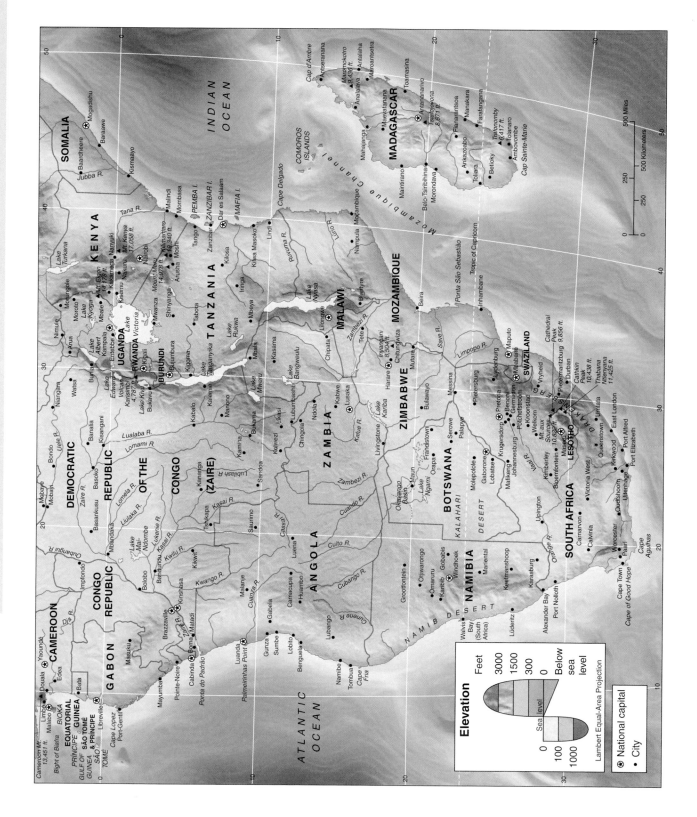

Elevation

Feet	
3000	
1500	
300	
0	Sea level
Below sea level	

| 0 | 100 | 1000 |

Lambert Equal-Area Projection

⊛ National capital
• City

500 Miles
500 Kilometers
250
250
0
0

Map 93 Asia: Political Map

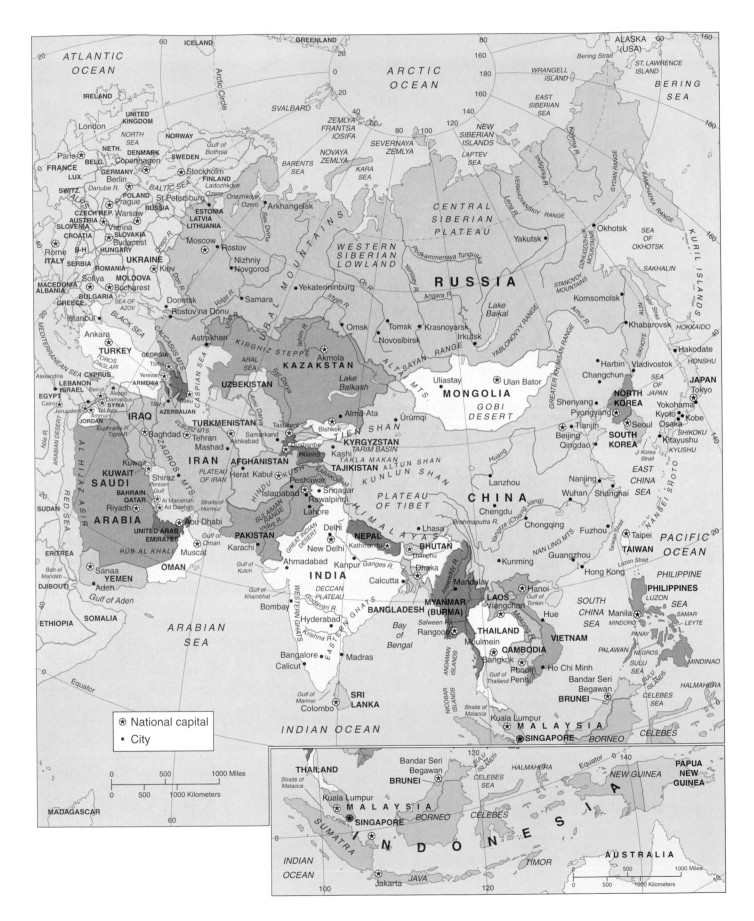

National capital

• City

Map 94 Asia: Physical Map

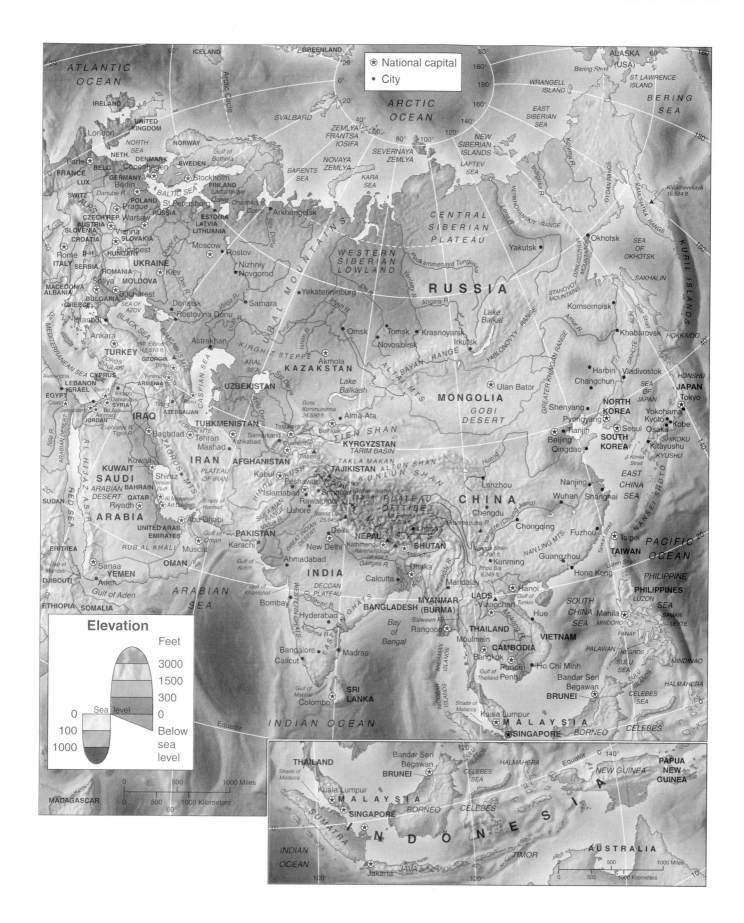

Asia: Thematic Maps

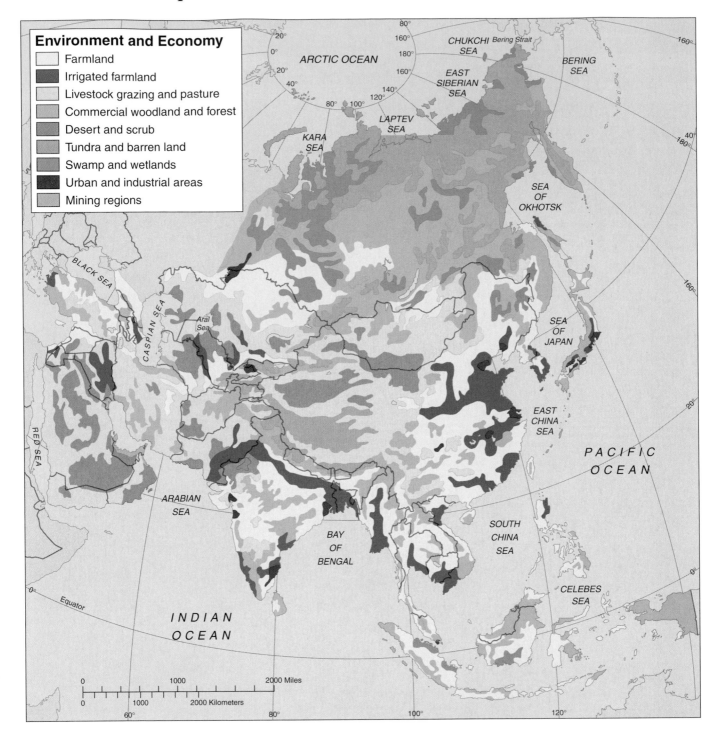

Environment and Economy

- Farmland
- Irrigated farmland
- Livestock grazing and pasture
- Commercial woodland and forest
- Desert and scrub
- Tundra and barren land
- Swamp and wetlands
- Urban and industrial areas
- Mining regions

Map 95a Environment and Economy

Asia is a land of extremes of land use with some of the world's most heavily industrialized regions, barren and empty areas, and productive and densely populated farm regions. Asia is a region of rapid industrial growth. Yet Asia remains an agricultural region with three out of every four workers engaged in agriculture. Asian commercial agriculture and intensive subsistence agriculture is characterized by irrigation. Some of Asia's irrigated lands are desert requiring additional water. But most of the Asian irrigated regions have sufficient precipitation for crop agriculture and irrigation is a way of coping with seasonal drought—the wet-and-dry cycle of the monsoon—often gaining more than one crop per year on irrigated farms. Agricultural yields per unit area in many areas of Asia are among the world's highest. Because the Asian population is so large and the demands for agricultural land so great, Asia is undergoing rapid deforestation and some areas of the continent have only small remnants of a once-abundant forest reserve.

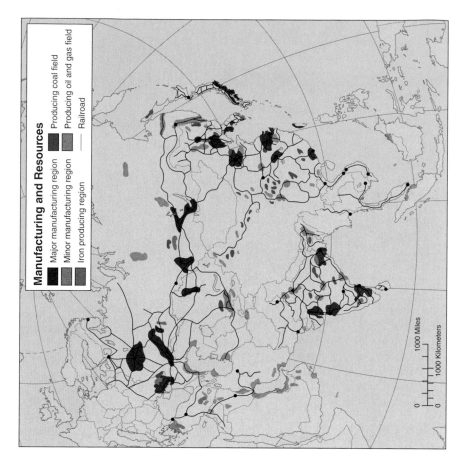

Population Density

Persons per square mile (km)

- Uninhabited area
- Less than 2 (1)
- 2 – 25 (1–10)
- 25 – 50 (10 – 20)
- 50 – 150 (20 – 60)
- 150 – 300 (60 – 120)
- More than 300 (120)

1000 Miles

1000 Kilometers

Map 95b Population Density

With one-third of the world's land area and nearly two-thirds of the world's population, Asia is more densely settled than any other region of the world. In some of the continent's farming regions, agricultural population density exceeds 2,000 persons per square mile. In some portions of the continent, particularly the Islamic areas of Central and Southwest Asia, this already large population is growing very rapidly with some countries having population growth rates above 3 percent per year and doubling times between 20 and 25 years. The populations of neither China nor India are growing particularly rapidly but since both countries have population bases that are enormous, the absolute number of Indians and Chinese added to the world's population each year is staggering. In spite of these massive populations, Asia also contains areas that are either completely uninhabited or have population densities that are as low as any on earth.

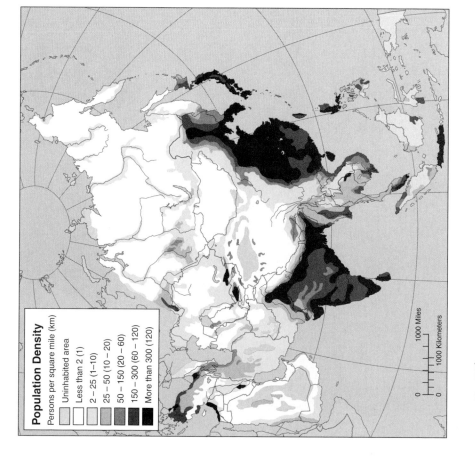

Manufacturing and Resources

- Major manufacturing region
- Minor manufacturing region
- Iron producing region
- Producing coal field
- Producing oil and gas field
- — Railroad

1000 Miles

1000 Kilometers

Map 95c Industrialization: Manufacturing and Resources

International economists have predicted that the next century will be one of Asian economic dominance, as other Asian nations approach the economic levels of industrial giant Japan. Asian countries have begun to industrialize rapidly and have increased industrial production more than 100 percent in the last decade. But nothing guarantees industrial output, it must master the great distances separating critical raw material and production sites from the locations of the markets for them. Such a mastery is achieved only by the development of efficient transportation systems. Usually that means water travel and here Asia is remarkably deficient when compared with Europe and North America. Nevertheless, the mixed prospects for Asian economic growth into the next century are a great deal better than would have been predicted a decade ago.

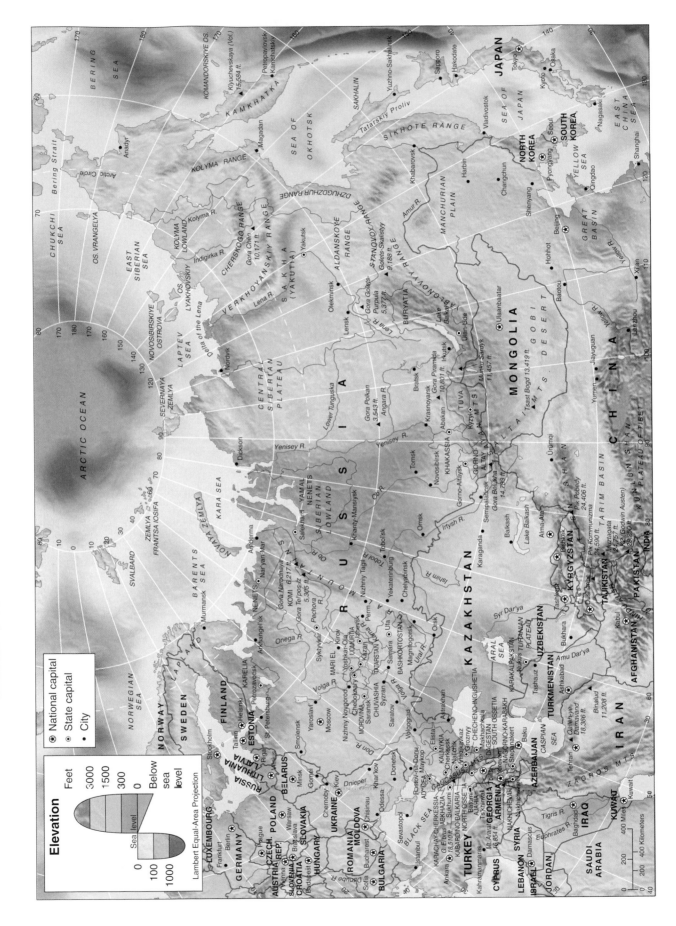

Map 96 Northern Asia

Elevation

Feet
3000
1500
300
0
Below sea level

Sea level

0
100
1000

Lambert Equal-Area Projection

⊛ National capital
⊙ State capital
• City

Map 97 Southwestern Asia

Elevation

	Feet
	3000
	1500
	300
0 — Sea level	0
100	Below
1000	sea level

Simple Conic Projection

⊛ National capital
• City

Map **98** Southern Asia

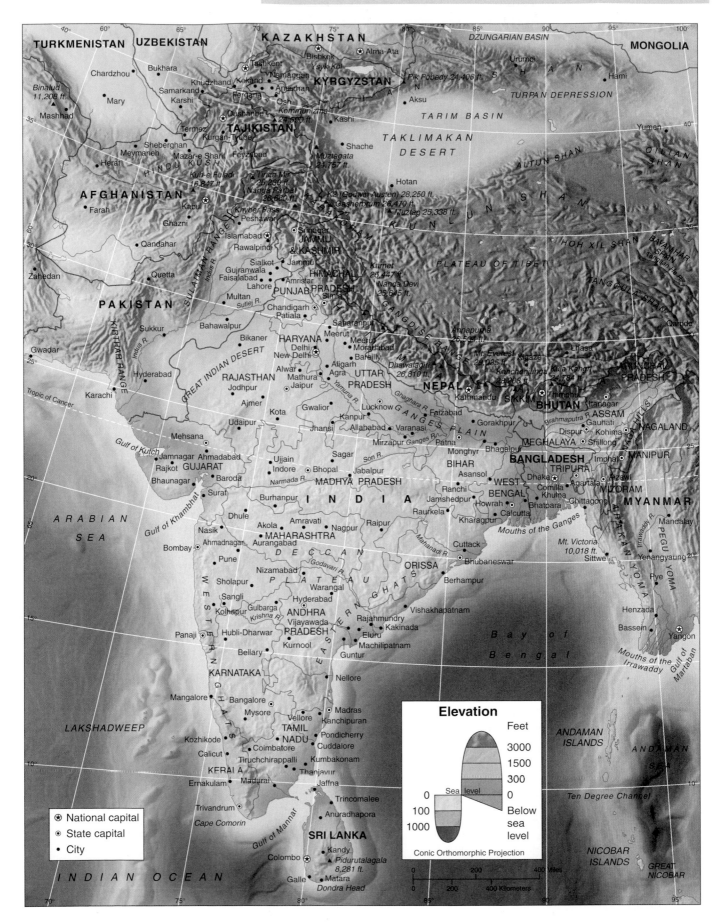

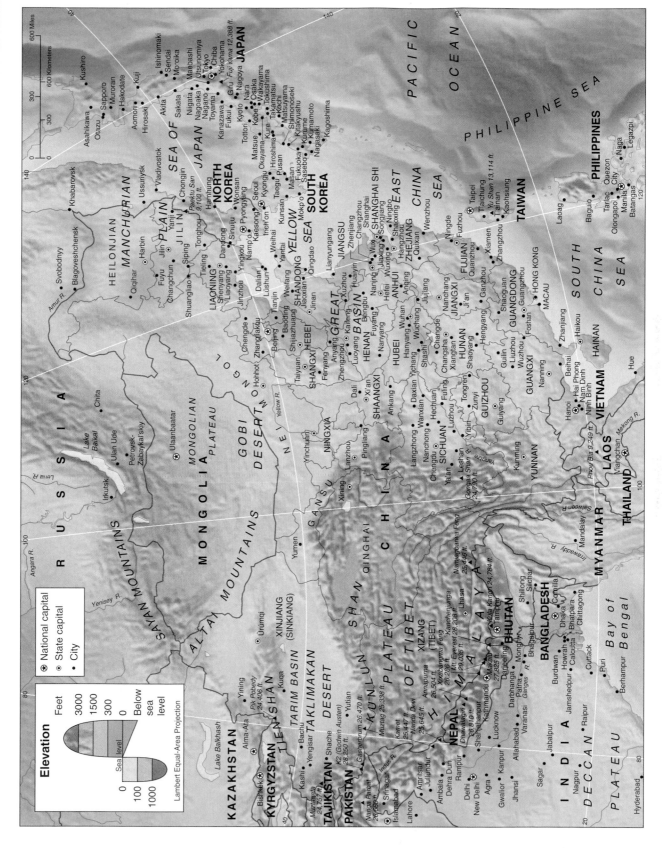

Map **99** East Asia

Elevation

Feet
3000
1500
300
0
Below
sea
level

Sea level

0
100
1000

Lambert Equal-Area Projection

⊛ National capital
◉ State capital
• City

600 Miles
600 Kilometers
300
300
0
0

Map 100 Southeast Asia

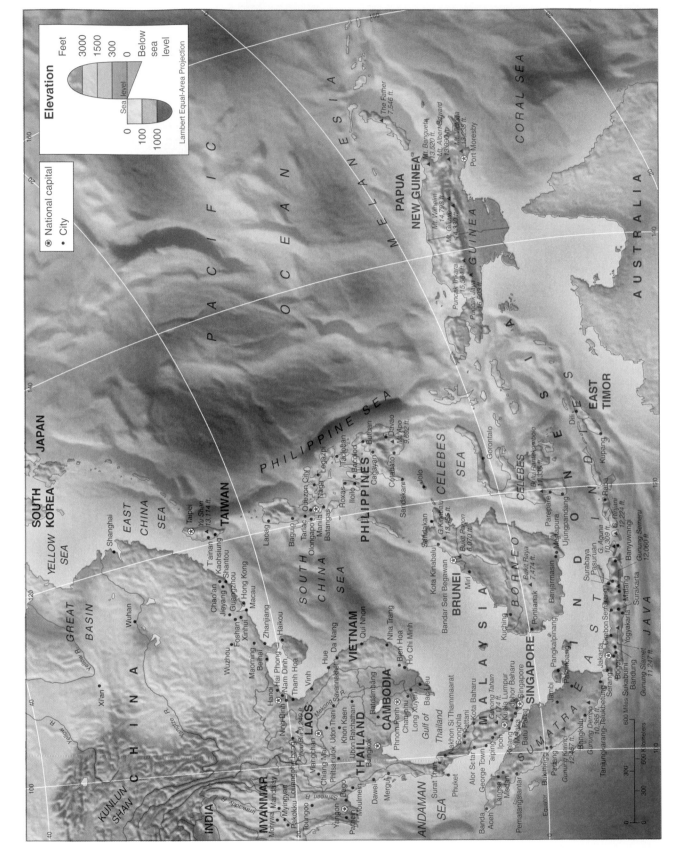

Elevation

Feet
3000
1500
300
0
Below sea level

Sea level
0
100
1000

Lambert Equal-Area Projection

⊛ National capital
• City

JAPAN

SOUTH KOREA

CHINA

GREAT BASIN

Yellow R.

Xi'an

Wuhan

Shanghai

YELLOW SEA

EAST CHINA SEA

PACIFIC OCEAN

Chao'an
Jieyang
Guangzhou
Shantou
Foshan
Xinhui
Macau
Hong Kong
Zhanjiang
Maoming
Beihai
Haikou

TAIWAN
Taipei
Yu Shan 13,114 ft.
T'ainan
Kaohsiung

SOUTH CHINA SEA

PHILIPPINE SEA

Laoag
Baguio
Tarlac
Quezon City
Olongapo
Manila
Batangas
Naga
Legazpi
Roxas
Iloilo
Bacolod
Butuan
Cagayan
Davao
Mt. Apo 9,692 ft.
Cotabato
Jolo
Sandakan

PHILIPPINES

CELEBES SEA

MELANESIA

PAPUA NEW GUINEA
The Father 7,546 ft.
Mt. Banguela 13,520 ft.
Mt. Wilhelm 14,793 ft.
Mt. Albert Edward 13,090 ft.
Mt. Giluwe 14,330 ft.
Port Moresby

GUINEA

CORAL SEA

Puncak Trikora 15,584 ft.
Puncak Jaya 16,503 ft.

NEW GUINEA

EAST INDIES

CELEBES SEA

Gorontalo
Parepare
Ujungpandang
G. Rantekombola 11,335 ft.

Dili
Kupang

EAST TIMOR

AUSTRALIA

BORNEO

Bandar Seri Begawan
BRUNEI
Miri
Kota Kinabalu
G. Kinabalu 13,455 ft.
Bukit Raya 7,474 ft.
Bukit Payan 6,070 ft.
Pontianak
Banjarmasin

Kuching

MALAYSIA

SINGAPORE
Batu Pahat
Johor Baharu
Kelang
Kuala Lumpur
Melaka
Gunung Tahan 7,174 ft.
Ipoh

SUMATRA

Pangkalpinang
Palembang
Jambi

Pekanbaru
Padang
Bukittinggi
Gunung Kerinci 12,467 ft.
Gunung Demo 12,467 ft.
Bengkulu
Tanjungarang-Telukbetung
Gunung Demo 10,364 ft.

Jakarta
Bogor
Bandung
Cirebon
Semarang
Tegal
Surakarta
Yogyakarta
Gunung Slamet 11,247 ft.
Gunung Semeru 12,060 ft.

JAVA

Surabaya
Pasuruan
Banyuwangi
Malang
G. Agung 10,309 ft.
G. Rinjani 12,224 ft.
Raba

INDIA

KUNLUN SHAN

MYANMAR
Monywa
Mandalay
Myingyan
Pakokku
Loikaw
Prome
Toungoo
Yangon
Bago
Pathein
Dawei
Mergui

ANDAMAN SEA

Irrawaddy R.

Salween

Mekong

LAOS
Louangphrabang
Vangchan
Viangchan

THAILAND
Chiang Mai
Phitsanulok
Udon Thani
Khon Kaen
Ubon Ratchathani
Nakhon Ratchasima
Bangkok
Gulf of Thailand
Nakhon Si Thammarat
Surat Thani
Songkhla
Phuket
Pattani
Hat Yai
Alor Setar
George Town
Taiping
Langsa
Medan
Banda Aceh
Pematangsiantar
Pematang

Equator

CAMBODIA
Phnom Penh
Battambang
Long Xuyen
Chau Phu
Bac Lieu

VIETNAM
Hanoi
Ninh Binh
Nam Dinh
Thanh Hoa
Vinh
Hue
Da Nang
Savannakhet
Qui Nhon
Nha Trang
Bien Hoa
Ho Chi Minh

Hai Phong

Phou Bia 9,249 ft.

600 Miles
500 Kilometers

300
300
0
0

-122-

Map 101 Australasia: Political Map

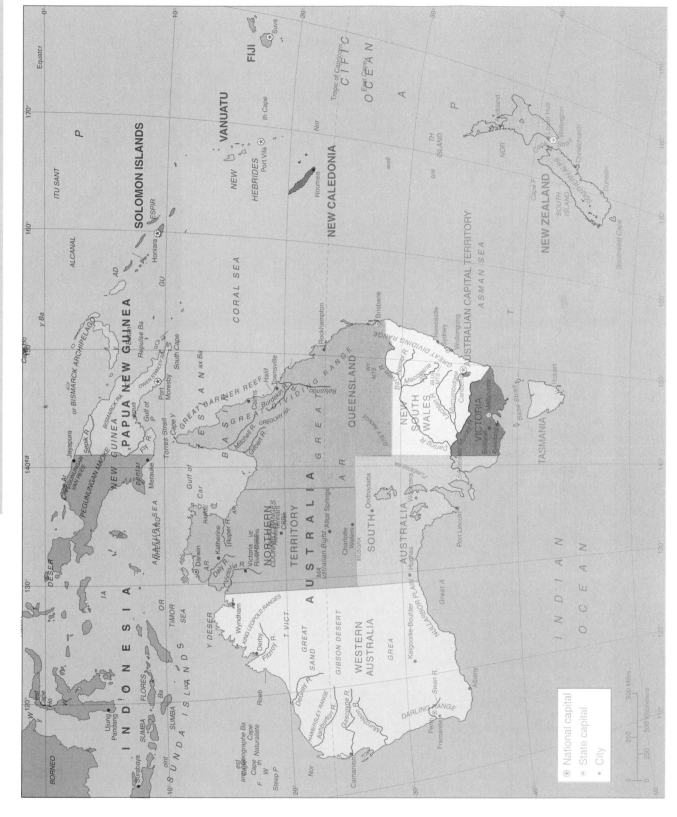

National capital ⊛
State capital ⊙
City •

Map **102** Australasia: Physical Map

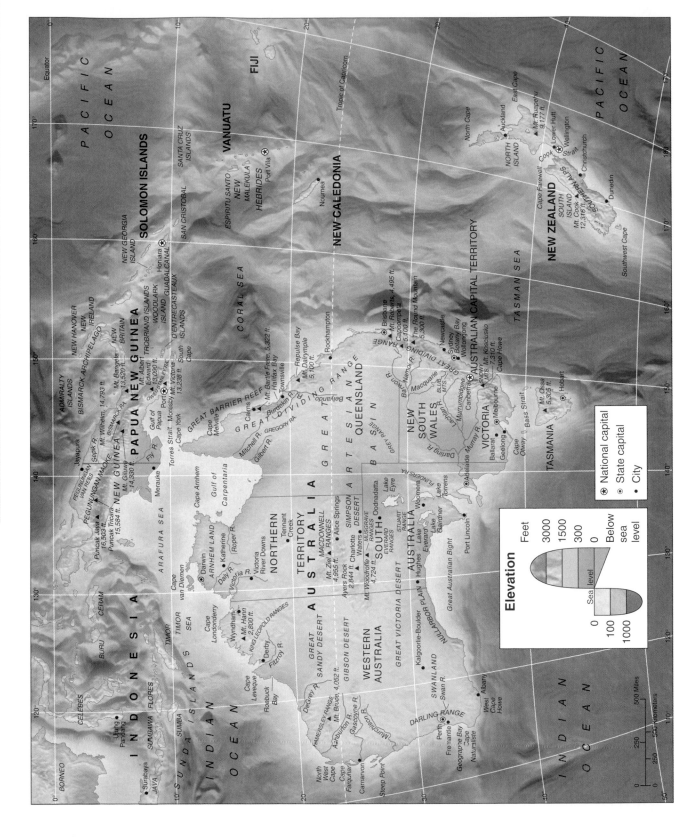

Elevation

Feet	
3000	
1500	
300	
0	
Below sea level	

Sea level	
0	
100	
1000	

⊛ National capital
◉ State capital
• City

500 Miles
250
0
0 250 500 Kilometers

Australasia: Thematic Maps

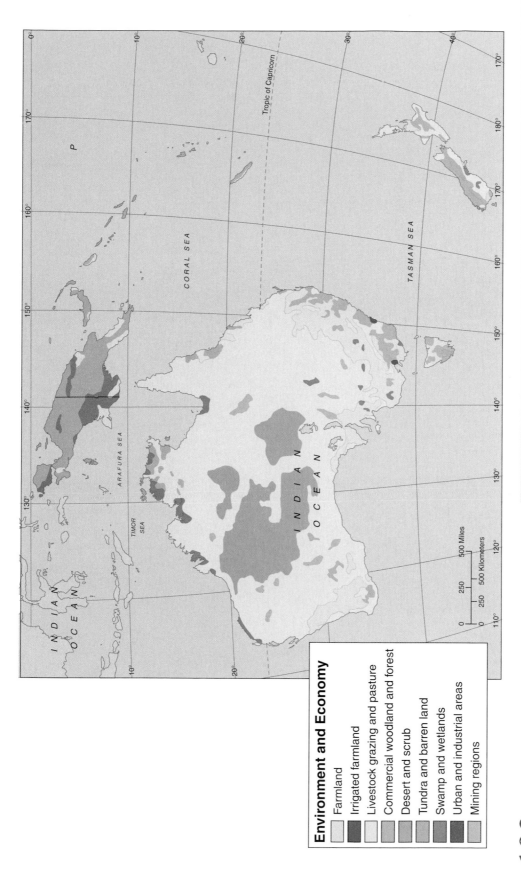

Environment and Economy

- Farmland
- Irrigated farmland
- Livestock grazing and pasture
- Commercial woodland and forest
- Desert and scrub
- Tundra and barren land
- Swamp and wetlands
- Urban and industrial areas
- Mining regions

Map 103a Environment and Economy

Australasia is dominated by the world's smallest and most uniform continent. Flat, dry, and mostly hot, Australia has the simplest of land use patterns: where rainfall exists so does agricultural activity. Two agricultural patterns dominate the map: livestock grazing, primarily sheep, and wheat farming, although some sugar cane production exists in the north and some cotton is grown elsewhere. Only about 6 percent of the continent consists of arable land so the areas of wheat farming, dominant as they may be in the context of Australian agriculture, are small. Australia also supports a healthy mineral resource economy, with iron and copper and precious metals making up the bulk of the extraction. Elsewhere in the region, tropical forests dominate Papua New Guinea, with some subsistence agriculture and livestock. New Zealand's temperate climate with abundant precipitation supports a productive livestock industry and little else besides tourism—which is an important economic element throughout the remainder of the region as well.

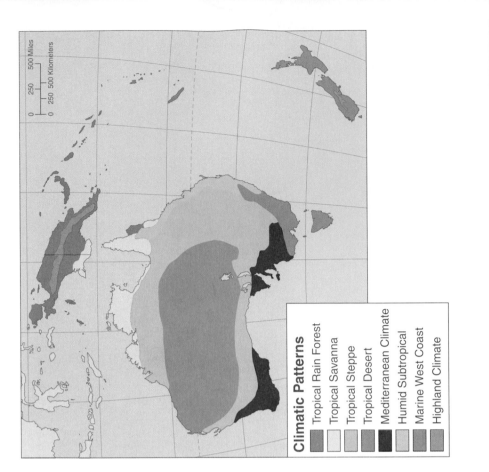

Map 103c Climate Patterns

Because of its nearly uniform surface, with only a few low and scattered uplands, Australian climate is a consequence only of the two great climate controls—latitudinal position and location relative to continental margin and interior. The continent bestrides the 30th parallel of latitude and its climatic pattern is dominated by the subtropical high pressure system with dry air masses that are responsible for the existence of great deserts. Toward the equator, the desert grades into steppe, savanna, and tropical forest as the subtropical high gives way to equatorial low pressure and abundant precipitation. Toward the pole, arid land fades into more well-watered steppe grasslands, the Mediterranean type climate of the southern margins of the continent, and the marine west coast climate of the Australian southeast and New Zealand. This latter climate is where most of the region's people live.

Climatic Patterns

- Tropical Rain Forest
- Tropical Savanna
- Tropical Steppe
- Tropical Desert
- Mediterranean Climate
- Humid Subtropical
- Marine West Coast
- Highland Climate

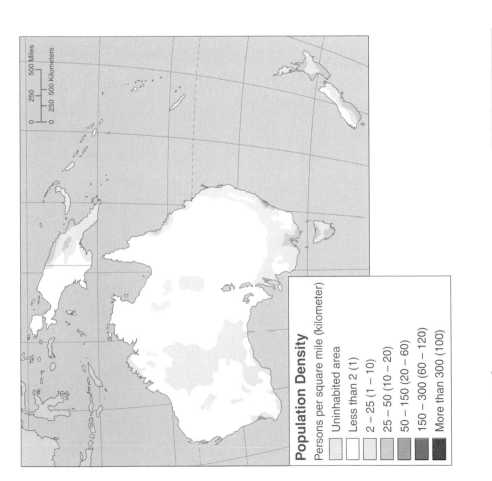

Map 103b Population Density

The region's small population is remarkably diverse. To the north are the Melanesian-New Guinea peoples, while Europeans dominate the populations of Australia and New Zealand, both of which have significant indigenous populations. Throughout most of the smaller island groups, the bulk of the population is Melanesian but with a scattering of Europeans. The distribution of population is extremely uneven with New Guinea, southeastern Australia, anc New Zealand supporting the bulk of the region's people while the remainder of the region—meaning nearly all of Australia—is either sparsely populated or devoid of population altogether. Nowhere do population densities reach the levels they do in other major regions of the world and densities of 50 persons per square mile are the highest to be found. Population location is dependent on precipitation and population growth patterns are culturally variable.

Population Density

Persons per square mile (kilometer)

- Uninhabited area
- Less than 2 (1)
- 2 – 25 (1 – 10)
- 25 – 50 (10 – 20)
- 50 – 150 (20 – 60)
- 150 – 300 (60 – 120)
- More than 300 (100)

Map 104 The Pacific Rim

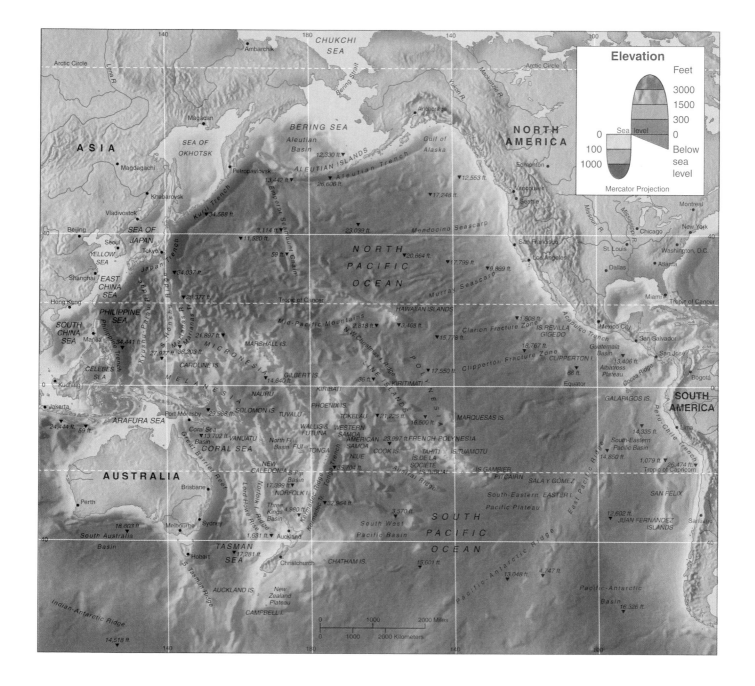

Map **105** The Atlantic Ocean

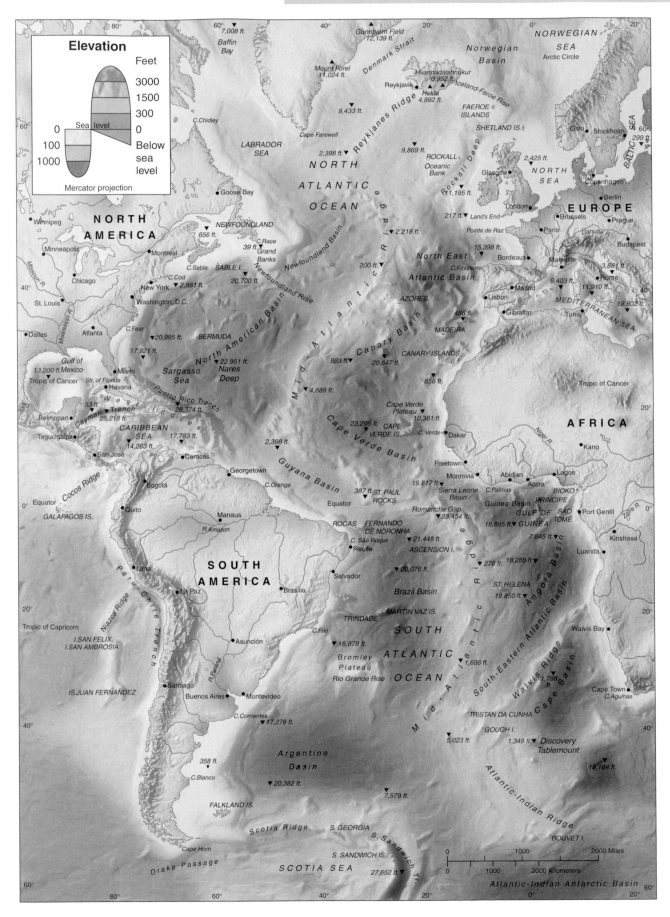

Elevation

Feet

- 3000
- 1500
- 300
- Sea level — 0
- 0 — Below sea level
- 100
- 1000

Mercator projection

80° 60° 40° 20° 0° 20°

7,008 ft.
Baffin Bay
Gunnbjørn Fjeld 12,139 ft.
NORWEGIAN SEA
Arctic Circle
Mount Forel 11,024 ft.
Denmark Strait
Norwegian Basin
Hvannadalshnukur 6,952 ft.
Reykjavik
Hekla 4,892 ft.
Iceland-Faroe Rise
9,433 ft.
FAEROE ISLANDS
C.Chidley
Reykjanes Ridge
SHETLAND IS.
Oslo
Stockholm
Cape Farewell
LABRADOR SEA
2,398 ft.
9,869 ft.
ROCKALL Oceanic Bank
Rockall Deep
2,425 ft.
Glasgow
NORTH SEA
Copenhagen
299 ft.
BALTIC SEA
60°
Goose Bay
NORTH ATLANTIC OCEAN
11,195 ft.
London
Berlin
EUROPE
Winnipeg
NORTH AMERICA
217 ft. Land's End
Brussels
Prague
Minneapolis
NEWFOUNDLAND
656 ft.
2,218 ft.
Pointe de Raz
Paris
Danube R.
Budapest
Montreal
C.Race 39 ft. Grand Banks
Newfoundland Basin
200 ft.
North East
15,398 ft.
C.Finisterre
Bordeaux
Marseille
3,881 ft.
Chicago
C.Sable
SABLE I.
Newfoundland Rise
Atlantic Basin
Madrid
9,403 ft.
Rome
11,910 ft.
St. Louis
C.Cod
20,700 ft.
AZORES
Lisbon
MEDITERRANEAN SEA
16,802 ft.
40°
New York
2,881 ft.
Washington, D.C.
North American Basin
486 ft.
Gibraltar
Tunis
Dallas
C.Fear
20,995 ft.
BERMUDA
Canary Basin
MADEIRA
Atlanta
17,921 ft.
22,951 ft. Nares Deep
883 ft.
CANARY ISLANDS
20,647 ft.
Tropic of Cancer
Gulf of Mexico 13,200 ft.
Miami
Sargasso Sea
4,689 ft.
856 ft.
Tropic of Cancer
Str. of Florida
Havana
West Indies
Puerto Rico Trench 28,374 ft.
Cape Verde Plateau
AFRICA
53 ft.
Cayman Trench
10,381 ft.
Kano
Belmopan
25,218 ft.
CARIBBEAN SEA
17,783 ft.
23,295 ft. CAPE VERDE IS.
C. Verde
Dakar
20°
Tegucigalpa
14,263 ft.
Cape Verde Basin
San José
Caracas
Freetown
Georgetown
Guyana Basin
Monrovia
19,817 ft.
Abidjan
Accra
Lagos
Cocos Ridge
Bogotá
C.Orange
2,398 ft.
Sierra Leone Basin
C.Palmas
BIOKO
PRINCIPE
Equator
Quito
387 ft. ST. PAUL ROCKS
Equator
Guinea Basin
GULF OF GUINEA
SÃO TOMÉ
Port Gentil
Equator
GALAPAGOS IS.
Romanche Gap 25,454 ft.
18,895 ft.
7,645 ft.
Manaus
ROCAS
FERNANDO DE NORONHA
Kinshasa
R.Amazon
C. São Roque
21,448 ft.
Luanda
Lima
Recife
ASCENSION I.
276 ft.
18,288 ft.
SOUTH AMERICA
Salvador
20,076 ft.
ST. HELENA
Angola Basin
La Paz
Brasília
Brazil Basin
19,850 ft.
Peru-Chile Trench
20°
MARTIN VAZ IS.
Walvis Bay
TRINDADE
SOUTH
Tropic of Capricorn
Nazca Ridge
C.Frio
18,879 ft.
ATLANTIC
South-Eastern Atlantic Basin
Walvis Ridge
I.SAN FELIX
I.SAN AMBROSIA
Asunción
Bromley Plateau
OCEAN
1,686 ft.
Cape Basin
1,788 ft.
Cape Town
Rio Grande Rise
C.Agulhas
R.Paraná
ST. HELENA
IS.JUAN FERNANDEZ
Santiago
South-Eastern Atlantic Basin
Buenos Aires
Montevideo
TRISTAN DA CUNHA
C.Corrientes
17,278 ft.
GOUGH I.
40°
5,023 ft.
1,349 ft. Discovery Tablemount
358 ft.
Argentine Basin
18,164 ft.
C.Blanco
20,382 ft.
FALKLAND IS.
7,579 ft.
Atlantic-Indian Ridge
Scotia Ridge
S. GEORGIA
BOUVET I.
Cape Horn
S. SANDWICH IS.
Drake Passage
SCOTIA SEA
27,652 ft.
Atlantic-Indian Antarctic Basin
60°

0   1000   2000 Miles
0   1000   2000 Kilometers

Map **106** The Indian Ocean

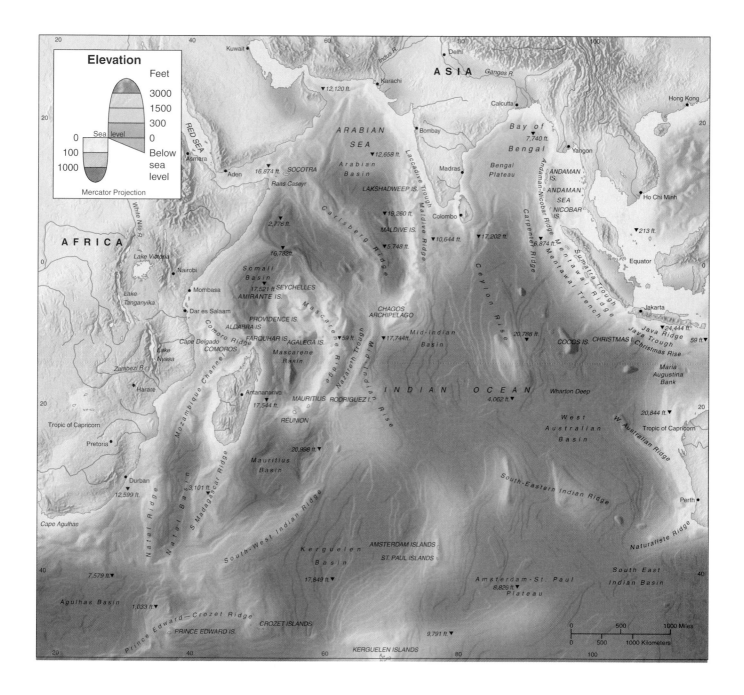

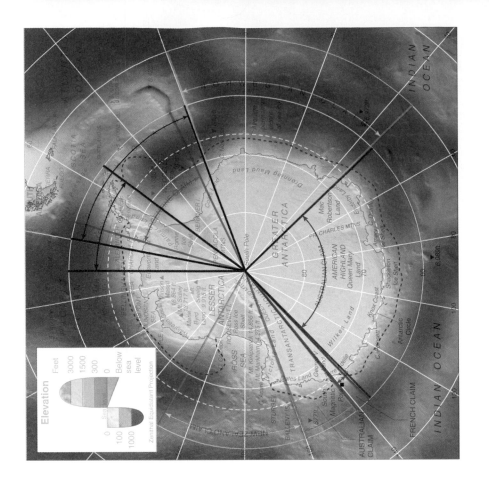

Map 108 Antarctica

Map 107 The Arctic

Part VIII

Tables

Table A
World Countries: Area, Population, and Population Density, 2000

COUNTRY	AREA		POPULATION	DENSITY	
	(Mi^2)	(Km^2)	(2000)[a]	(Pop/Mi^2)	(Pop/Km^2)
Afghanistan	251,826	652,229	21,764,955	86	33
Albania	11,100	28,749	3,134,222	282	109
Algeria	919,595	2,381,750	30,291,341	33	13
Andorra	175	453	66,824	382	148
Angola	481,354	1,246,706	13,134,452	27	11
Antigua and Barbuda	171	443	66,422	388	150
Argentina	1,073,400	2,780,104	37,031,814	34	13
Armenia	11,506	29,801	3,786,997	329	127
Australia	2,966,155	7,682,337	19,137,645	6	2
Austria	32,377	83,856	8,079,902	250	96
Azerbaijan	33,436	86,599	8,041,278	240	93
Bahamas	5,382	13,939	304,233	57	22
Bahrain	267	692	639,753	2,396	924
Bangladesh	55,598	143,999	137,439,261	2,472	954
Barbados	166	430	267,498	1,611	622
Belarus	80,155	207,601	10,186,996	127	49
Belgium	11,783	30,518	10,249,370	870	336
Belize	8,866	22,963	226,325	26	10
Benin	43,475	112,600	6,271,732	144	56
Bhutan	18,200	47,138	2,085,136	115	44
Bolivia	424,165	1,098,587	8,328,685	20	8
Bosnia-Herzegovina	19,776	51,233	3,977,106	201	78
Botswana	231,803	600,362	1,541,256	7	3
Brazil	3,286,488	8,511,999	170,406,280	52	20
Brunei	2,228	5,770	328,305	147	57
Bulgaria	42,823	110,912	7,949,331	186	72
Burkina Faso	105,869	274,201	11,535,072	109	42
Burundi	10,745	27,830	6,356,252	592	228
Cambodia	69,898	181,036	13,104,030	187	72
Cameroon	183,569	475,443	14,875,513	81	31
Canada	3,849,674	9,970,650	30,756,698	8	3
Cape Verde	1,557	4,033	426,958	274	106
Central African Republic	240,535	622,985	3,717,293	15	6
Chad	495,755	1,284,005	7,885,299	16	6
Chile	292,259	756,950	15,211,308	52	20
China	3,705,392	9,596,960	1,275,132,866	344	133
Colombia	439,734	1,138,910	42,104,701	96	37
Comoros	838	2,170	705,929	842	325
Congo Republic	132,047	342,002	3,018,426	23	9
Costa Rica	19,730	51,101	4,023,502	204	79
Côte d'Ivoire	124,502	322,460	16,013,139	129	50
Croatia	21,824	56,538	4,653,749	213	82
Cuba	42,804	110,862	11,199,176	262	101
Cyprus	3,571	9,250	783,827	219	85
Czech Republic	30,387	78,703	10,271,830	338	131

Table A (Continued)
World Countries: Area, Population, and Population Density, 2000

COUNTRY	AREA		POPULATION	DENSITY	
	(Mi^2)	(Km^2)	(2000)[a]	(Pop/Mi^2)	(Pop/Km^2)
Democratic Republic of the Congo (formerly Zaire)	905,564	2,345,410	50,948,236	56	22
Denmark	16,629	43,070	5,320,065	320	124
Djibouti	8,494	22,000	632,096	74	29
Dominica	290	750	71,540	247	95
Dominican Republic	18,815	48,730	8,372,695	445	172
Ecuador	109,484	283,563	12,645,796	116	45
Egypt	386,662	1,001,454	67,884,476	176	68
El Salvador	8,124	21,041	6,277,897	773	298
Equatorial Guinea	10,831	28,052	456,703	42	16
Eritrea	46,842	121,320	3,658,777	78	30
Estonia	17,413	45,100	1,393,470	80	31
Ethiopia	435,184	1,127,127	62,907,788	145	56
Fiji	7,054	18,270	813,607	115	45
Finland	130,127	337,030	5,171,806	40	15
France	176,460	547,030	59,237,668	336	108
Gabon	103,347	267,669	1,230,088	12	5
The Gambia	4,363	11,300	1,302,723	299	115
Georgia	26,911	69,699	5,262,050	196	75
Germany	137,803	356,910	82,016,767	595	230
Ghana	92,098	238,534	19,305,633	210	81
Greece	50,942	131,940	10,609,962	208	80
Grenada	131	340	89,018	680	262
Guatemala	42,042	108,889	1,138,532	27	10
Guinea	94,926	245,858	8,154,267	86	33
Guinea-Bissau	13,948	36,125	1,199,198	86	33
Guyana	83,000	214,970	760,513	9	4
Haiti	10,714	27,749	8,142,471	760	293
Honduras	43,277	112,087	6,416,635	148	57
Hungary	35,920	93,033	9,967,555	277	107
Iceland	39,768	103,000	279,291	7	3
India	1,269,340	3,287,590	1,008,937,356	795	307
Indonesia	741,097	1,919,440	212,092,024	286	110
Iran	636,294	1,648,000	70,330,053	111	43
Iraq	168,754	437,072	22,946,245	136	52
Ireland	27,137	70,285	3,803,085	140	54
Israel[b]	8,019	20,769	6,040,427	753	291
Italy	116,305	301,230	57,529,998	495	191
Jamaica	4,244	10,992	2,576,085	607	234
Japan	145,882	377,835	127,096,314	871	336
Jordan	35,445	89,213	4,913,115	139	55
Kazakhstan	1,049,156	2,717,313	16,172,470	15	6
Kenya	224,961	582,649	30,668,697	136	53
Kiribati	277	717	91,985	332	128
Korea, North	46,540	120,539	22,268,377	478	185

COUNTRY	AREA		POPULATION	DENSITY	
	(Mi^2)	(Km^2)	(2000)[a]	(Pop/Mi^2)	(Pop/Km^2)
Korea, South	38,023	98,480	46,740,141	1,229	475
Kuwait	6,880	17,819	1,914,404	278	107
Kyrgyzstan	76,641	198,500	4,920,847	64	25
Laos	91,429	236,801	5,278,563	58	22
Latvia	24,749	64,100	2,420,546	98	38
Lebanon	4,015	10,399	3,496,489	871	336
Lesotho	11,720	30,355	2,034,667	174	67
Liberia	43,000	111,370	2,913,064	68	26
Libya	679,362	1,759,547	5,289,730	8	3
Liechtenstein	62	161	32,207	519	200
Lithuania	25,174	65,201	3,696,093	147	57
Luxembourg	998	2.585	436,818	438	168,982
Macedonia	9,781	25,333	2,033,975	208	80
Madagascar	226,658	587,044	15,970,364	70	27
Malawi	45,747	118,485	11,308,352	247	95
Malaysia	127,317	329,750	22,218,485	175	67
Maldives	115	298	290,959	2,530	976
Mali	478,767	1,240,006	11,350,798	24	9
Malta	124	320	389,941	3,145	1,219
Marshall Islands	70	181	68,126	973	376
Mauritania	397,954	1,030,700	2,664,528	7	3
Mauritius	718	1,860	1,161,371	1,618	624
Mexico	761,603	1,972,550	98,872,230	130	50
Micronesia	271	702	133,144	491	190
Moldova	13,012	33,701	4,295,453	330	127
Monaco	1.21	1.95	31,693	26,193	16,253
Mongolia	604,427	1,565,000	2,533,299	4	2
Morocco	172,413	446,550	29,878,403	173	67
Mozambique	309,494	801,590	18,292,382	59	23
Myanmar (Burma)	261,969	678,500	47,748,939	182	70
Namibia	318,259	824,290	1,756,597	6	2
Nauru	8	21	11,845	1,481	564
Nepal	54,363	140,800	23,042,704	424	164
Netherlands	14,413	37,330	15,863,747	1,101	425
New Zealand	103,738	268,680	3,778,004	36	14
Nicaragua	49,998	129,494	5,071,423	101	39
Niger	489,191	1,267,004	10,831,545	22	9
Nigeria	356,669	923,772	113,061,753	319	123
Norway	125,182	324,220	4,469,030	36	14
Oman	82,030	212,458	2,538,161	31	12
Pakistan	310,402	803,940	141,256,186	455	176
Palau	177	458	18,766	106	41
Panama	30,193	78,200	2,855,694	95	37
Papua New Guinea	178,259	461,690	4,809,215	27	10
Paraguay	157,048	406,754	5,496,467	35	14

COUNTRY	AREA		POPULATION	DENSITY	
	(Mi^2)	(Km^2)	(2000)[a]	(Pop/Mi^2)	(Pop/Km^2)
Peru	496,225	1,285,222	25,661,679	52	20
Philippines	115,831	300,002	75,653,257	653	252
Poland	120,728	312,685	38,605,447	320	123
Portugal	35,552	92,080	10,015,505	282	109
Qatar	4,247	11,000	565,439	133	51
Romania	91,699	237,500	2,243,771	24	9
Russia	6,592,745	17,075,200	145,491,166	22	9
Rwanda	10,169	26,338	7,608,928	748	289
St. Kitts and Nevis	104	269	38,819	373	144
St. Lucia	238	616	147,783	621	240
St. Vincent/Grenadines	131	340	115,461	881	340
Samoa	1,104	2,860	158,633	144	55
San Marino	23	60	26,937	1,171	449
Soã Tomé and Principe	372	963	159,883	430	166
Saudi Arabia	756,982	1,960,582	20,346,231	27	10
Senegal	75,749	196,190	9,420,518	124	48
Seychelles	175	453	79,326	453	175
Sierra Leone	27,699	71,740	4,404,740	159	61
Singapore	244	633	4,018,114	16,468	6,348
Slovakia	18,859	48,845	5,398,693	286	111
Slovenia	7,836	20,296	1,987,682	254	98
Solomon Islands	10,985	28,450	447,428	41	16
Somalia	246,201	637,660	8,777,879	36	14
South Africa	471,444	1,221,040	43,309,197	92	35
Spain	194,885	504,752	39,910,274	205	79
Sri Lanka	25,332	65,610	18,923,749	747	288
Sudan	967,500	2,505,824	31,095,162	32	12
Suriname	63,039	163,270	417,159	7	3
Swaziland	6,704	17,363	924,786	138	53
Sweden	173,732	449,966	8,842,094	51	20
Switzerland	15,943	41,292	7,170,407	450	174
Syria	71,498	185,180	16,188,760	226	87
Taiwan	13,892	35,980	22,191,087	1,597	617
Tajikistan	55,251	143,100	6,086,983	110	43
Tanzania	364,900	945,090	35,119,255	96	37
Thailand	198,456	514,000	62,805,574	316	122
Togo	21,925	56,786	4,526,972	206	80
Tonga	290	751	102,321	353	136
Trinidad and Tobago	1,980	5,128	1,294,368	654	252
Tunisia	63,170	163,610	9,458,661	150	58
Turkey	301,382	780,580	66,667,997	221	85
Turkmenistan	188,456	488,101	4,737,256	25	10
Tuvalu	10	26	10,838	1,084	417
Uganda	93,135	236,040	23,300,162	250	99
Ukraine	233,090	603,703	49,568,167	213	82

World Countries: Area, Population, and Population Density, 2000

COUNTRY	AREA		POPULATION	DENSITY	
	(Mi^2)	(Km^2)	$(2000)^a$	(Pop/Mi^2)	(Pop/Km^2)
United Arab Emirates	31,969	82,880	2,605,958	82	31
United Kingdom	94,525	244,820	59,414,643	629	243
United States	3,717,797	9,629,091	283,230,243	76	29
Uruguay	68,039	176,220	3,337,080	49	19
Uzbekistan	172,742	447,402	24,880,545	144	56
Vanuatu	5,699	14,760	196,803	35	13
Venezuela	352,145	912,055	24,169,742	69	27
Vietnam	127,243	329,560	78,136,913	614	237
Yemen	203,850	527,970	18,348,746	90	35
Yugoslavia (Serbia-Montenegro)	39,517	102,350	10,552,420	267	103
Zambia	290,586	752,617	10,421,339	36	14
Zimbabwe	150,803	390,580	12,627,277	84	32

[a]Primary source for population figures: United Nations Population Division.

[b]The figures for Israel do not include the West Bank and Gaza. These territories combined have a population of 3,190,943 (mid-2000), and an area of approximately 2,400 square miles. The West Bank has an estimated population density of 780 persons per square mile and Gaza an estimated population density of 8,700 persons per square mile.

Sources: *World Development Indicators 2001* (The World Bank, 2001); *The New York Times 2001 Almanac* (Penguin Putnam, Inc., New York, 2000); *World Population Prospects: The 2000 Revision* (United Nations Population Information Network, 2001).

Table B
World Countries: Form of Government, Capital City, Major Languages

Notes: Unless indicated otherwise, republics are multiparty. "Theocratic" normally refers to fundamentalist Islamic rule. "Transitional" governments are those still in the process of change from a previous form (e.g., single-party communist state to multiparty republic).

COUNTRY	GOVERNMENT	CAPITAL	MAJOR LANGUAGES
Afghanistan	Theocratic republic/transitional	Kabul	Dari, Pashtu, Uzbek, Turkmen
Albania	Multiparty democracy	Tiranë	Albanian, Greek
Algeria	Military-transitional	Algiers	Arabic, Berber, dialects, French
Andorra	Parliamentary democracy	Andorra	Catalán, French, Spanish
Angola	Multiparty democracy/transitional	Luanda	Portugese; Bantu and other African
Antigua and Barbuda	Parliamentary democracy	St. John's	English, local dialects
Argentina	Federal republic	Buenos Aires	Spanish, English, Italian, German
Armenia	Republic	Yerevan	Armenian, Azerbaijani, Russian
Australia	Federal parliamentary democracy	Canberra	English, indigenous
Austria	Federal republic	Vienna	German
Azerbaijan	Republic	Baku	Azerbaijani, Russian, Armenian
Bahamas	Parliamentary democracy; independent commonwealth	Nassau	English
Bahrain	Traditional monarchy	Al Manamah	Arabic, English, Farsi, Urdu
Bangladesh	Republic	Dhaka	Bangla, English
Barbados	Parliamentary democracy	Bridgetown	English
Belarus	Republic	Minsk	Byelorussian, Russian
Belgium	Constitutional monarchy	Brussels	Dutch (Flemish), French, German
Belize	Parliamentary state	Belmopan	English, Spanish, Garifuna, Mayan
Benin	Multiparty republic	Porto-Novo	French, Fon, Yoruba
Bhutan	Monarchy; special treaty relationship with India	Timphu	Dzongha, Tibetan, Nepalese
Bolivia	Republic	La Paz	Spanish, Quechua, Aymara
Bosnia-Herzegovina	Republic/transitional	Sarajevo	Croatian, Serbian, Bosnian
Botswana	Parliamentary republic	Gaborone	English, Setswana
Brazil	Federal republic	Brasília	Portugese, Spanish, English, French
Brunei	Constitutional monarchy	Bandar Seri Begawan	Malay, English, Chinese
Bulgaria	Parliamentary republic	Sofia	Bulgarian
Burkina Faso	Provisional military	Ouagadougou	French, indigenous
Burundi	Republic/transitional	Bujumbura	French, Kirundi, Swahili
Cambodia	Multiparty democracy/transitional (under UN supervision)	Phnom Penh	Khmer, French
Cameroon	Multiparty republic	Yaoundé	English, French, indigenous
Canada	Federal parliamentary	Ottawa	English, French
Cape Verde	Republic	Cidade de Praia	Portugese, Crioulu
Central African Republic	Republic	Bangui	French, Sangho, Arabic, Hunsa, Swahili
Chad	Republic	N'Djamena	French, Arabic, Sara, Sango, other indigenous
Chile	Republic	Santiago	Spanish
China	Single-party communist state	Beijing	Various Chinese dialects
Colombia	Republic	Bogotá	Spanish
Comoros	Republic	Moroni	Arabic, French, Comoran
Congo Republic	Multiparty republic	Brazzaville	French, Lingala, Kikongo
Costa Rica	Republic	San José	Spanish
Côte d'Ivoire	Multiparty republic	Abidjan	French, indigenous

Table B *(Continued)*
World Countries: Form of Government, Capital City, Major Languages

Notes: Unless indicated otherwise, republics are multiparty. "Theocratic" normally refers to fundamentalist Islamic rule. "Transitional" governments are those still in the process of change from a previous form (e.g., single-party communist state to multiparty republic).

COUNTRY	GOVERNMENT	CAPITAL	MAJOR LANGUAGES
Croatia	Parliamentary democracy	Zaghreb	Serbo-Croatian
Cuba	Single-party communist state	Havana	Spanish
Cyprus	Republic	Nicosia	Greek, Turkish, English
Czech Republic	Federal republic	Prague	Czech, Slovak, Hungarian
Democratic Republic of the Congo (formerly Zaire)	Republic/transitional from military dictatorship	Kinshasa	French, Lingala, Kingwana, Kikongo, Tshuliba
Denmark	Constitutional monarchy	Copenhagen	Danish, Faroese, German
Djibouti	Republic	Djibouti	French, Somali, Afar, Arabic
Dominica	Republic	Roseau	English, French
Dominican Republic	Republic	Santo Domingo	Spanish
Ecuador	Republic	Quito	Spanish, Quechua, indigenous
Egypt	Republic	Cairo	Arabic
El Salvador	Republic	San Salvador	Spanish, Nahua
Equatorial Guinea	Republic	Malabo	Spanish, indigenous, English
Eritrea	Transitional government	Asmara	Afar, Amharic, Arabic, Tigre, other indigenous
Estonia	Parliamentary democracy	Tallinn	Estonian, Russian, Ukranian, Finnish
Ethiopia	Federal republic	Addis Ababa	Amharic, Tigrinya, Orominga, Somali, Arabic, English
Fiji	Republic	Suva	English, Fijian, Hindustani
Finland	Republic	Helsinki	Finnish, Swedish
France	Republic	Paris	French
Gabon	Multiparty presidential republic	Libreville	French, Fang, indigenous
The Gambia	Multiparty democratic republic	Banjul	English, Mandinka, Wolof, Fula
Georgia	Republic	Tbilisi	Georgian, Russian, Armenian
Germany	Federal republic	Berlin	German
Ghana	Parliamentary democracy	Accra	English, Akan, indigenous
Greece	Parliamentary democratic republic	Athens	Greek
Grenada	Parliamentary state	St. George's	English, French
Guatemala	Republic	Guatemala City	Spanish, Quiche, Cakchiquel, other indigenous
Guinea	Republic	Conakry	French, indigenous
Guinea-Bissau	Multiparty republic	Bissau	Portuguese, Crioulo, indigenous
Guyana	Republic (within U.K. Commonwealth)	Georgetown	English, indigenous
Haiti	Republic	Port-au-Prince	Creole, French
Honduras	Republic	Tegucigalpa	Spanish, indigenous
Hungary	Republic	Budapest	Hungarian
Iceland	Republic	Reykjavík	Icelandic
India	Federal republic	New Delhi	English, Hindi
Indonesia	Republic	Jakarta	Bahasa Indonesian, English, Dutch, Javanese
Iran	Theocratic republic	Tehran	Farsi, Turkish, Kurdish, Arabic
Iraq	Single-party republic/ dictatorship	Baghdad	Arabic, Kurdish, Assyrian, Armenian
Ireland	Republic	Dublin	English, Irish Gaelic
Israel	Parliamentary democracy	Jerusalem	Hebrew, Arabic, Yiddish
Italy	Republic	Rome	Italian
Jamaica	Parliamentary state	Kingston	English, Creole

Notes: Unless indicated otherwise, republics are multiparty. "Theocratic" normally refers to fundamentalist Islamic rule. "Transitional" governments are those still in the process of change from a previous form (e.g., single-party communist state to multiparty republic).

COUNTRY	GOVERNMENT	CAPITAL	MAJOR LANGUAGES
Japan	Constitutional monarchy	Tokyo	Japanese
Jordan	Constitutional monarchy	Amman	Arabic
Kazakhstan	Republic/transitional	Alma-Ata	Kazakh, Russian, German, Ukranian
Kenya	Republic	Nairobi	English, Swahili, indigenous
Kiribati	Republic	Tarawa	English, Gilbertese
Korea, North	Single-party communist state	Pyongyang	Korean
Korea, South	Republic	Seoul	Korean
Kuwait	Constitutional monarchy	Kuwait	Arabic, English
Kyrgyzstan	Republic	Bishkek	Kirghiz, Russian, Uzbek, Ukranian
Laos	Single-party communist state	Vientiane	Lao, French, Thai, indigenous
Latvia	Multiparty republic	Riga	Lettish, Lithuanian, Russian
Lebanon	Republic	Beirut	Arabic, French, Armenian, English
Lesotho	Constitutional monarchy	Maseru	English, Sesotho, Zulu, Xhosa
Liberia	Republic/transitional	Monrovia	English, indigenous
Libya	Single party/military dictatorship	Tripoli	Arabic
Liechtenstein	Constitutional monarchy	Vaduz	German
Lithuania	Republic	Vilnius	Lithuanian, Russian, Polish
Luxembourg	Constitutional monarchy	Luxembourg	French, Luxembourgian, German
Macedonia	Republic/transitional	Skopje	Macedonian, Albanian, Turkish, Serbo-Croatian
Madagascar	Republic	Antananarivo	Malagasy, French
Malawi	Multiparty republic	Lilongwe	Chichewa, English, Tombuka
Malaysia	Constitutional monarchy	Kuala Lumpur	Malay, Chinese, English
Maldives	Republic	Male	Divehi
Mali	Single-party republic	Bamako	French, Bambara, indigenous
Malta	Parliamentary democracy	Valletta	English, Maltese
Marshall Islands	Constitutional government (free association with U.S.)	Majuro	English, Polynesian dialects, Japanese
Mauritania	Republic/transitional	Nouakchott	Arabic, Wolof, Pular, French
Mauritius	Parliamentary state	Port Louis	English, Creole, French, Hindi, Urdu, Bojpoori
Mexico	Federal republic	Mexico City	Spanish, indigenous
Micronesia	Constitutional government (free association with U.S.)	Palikir	English, Trukese, Pohnpeian, Yapese
Moldova	Republic	Kishinev	Moldavian, Russian, Gagauz
Monaco	Constitutional monarchy	Monaco	French, English, Italian, Monegasque
Mongolia	Republic	Ulan Bator	Khalkha Mongol, Turkic, Russian, Chinese
Morocco	Constitutional monarchy	Rabat	Arabic, Berber dialects, French
Mozambique	Republic	Maputo	Portuguese, indigenous
Myanmar (Burma)	Provisional military	Rangoon	Burmese, indigenous
Namibia	Republic	Windhoek	Afrikaans, English, German, indigenous
Nauru	Republic	Yaren district	Nauruan, English
Nepal	Parliamentary democracy	Kathmandu	Nepali, Maithali, Bhojpuri, indigenous
Netherlands	Constitutional monarchy	Amsterdam	Dutch
New Zealand	Parliamentary democracy	Wellington	English, Maori
Nicaragua	Republic	Managua	Spanish, English, indigenous
Niger	Provisional military	Niamey	French, Hausa, Djerma, indigenous
Nigeria	Military/transitional	Abuja	English, Hausa, Fulani, Yorbua, Ibo

Notes: Unless indicated otherwise, republics are multiparty. "Theocratic" normally refers to fundamentalist Islamic rule. "Transitional" governments are those still in the process of change from a previous form (e.g., single-party communist state to multiparty republic).

COUNTRY	GOVERNMENT	CAPITAL	MAJOR LANGUAGES
Norway	Constitutional monarchy	Oslo	Norwegian, Lapp
Oman	Monarchy	Muscat	Arabic, English, Baluchi, Urdu
Pakistan	Federal Islamic republic	Islamabad	Punjabi, Sindhi, Siraiki, Pashtu, Urbu, English, others
Palau	Constitutional government (free association with U.S.)	Koror	English, Palauan
Panama	Democratic republic	Panama	Spanish, English, indigenous
Papua New Guinea	Parliamentary democracy	Port Moresby	Various indigenous, English, Motu
Paraguay	Republic/transitional	Asunción	Spanish, Guarani
Peru	Republic	Lima	Quechua, Spanish, Aymara
Philippines	Republic	Manila	English, Pilipino, Tagalog
Poland	Republic	Warsaw	Polish
Portugal	Republic	Lisbon	Portuguese
Qatar	Traditional monarchy	Doha	Arabic, English
Romania	Republic	Bucharest	Romanian, Hungarian, German
Russia	Federational republic	Moscow	Russian, numerous other
Rwanda	Republic/transitional	Kigali	French, Kinyarwanda, English, Kiswahili
St. Kitts and Nevis	Constitutional monarchy	Basseterre	English
St. Lucia	Parliamentary democracy	Castries	English, French
St. Vincent/Grenadines	Parliamentary monarchy	Kingstown	English, French
Samoa	Constitutional monarchy	Apia	Samoan, English
San Marino	Republic	San Marino	Italian
São Tome and Príncipe	Republic	São Tome	Portuguese, Fang
Saudi Arabia	Islamic monarchy	Riyadh	Arabic
Senegal	Republic	Dakar	French, Wolof, indigenous
Seychelles	Republic	Victoria	English, French, Creole
Sierra Leone	Transitional government	Freetown	English, Krio, indigenous
Singapore	Republic	Singapore	Mandarin Chinese, English, Malay
Slovakia	Parliamentary democracy	Bratislava	Slovak, Hungarian, Polish
Slovenia	Republic	Ljubljana	Slovenian, Serbo-Croatian
Solomon Islands	Parliamentary state	Honiara	Mostly indigenous, English
Somalia	Transitional	Mogadishu	Arabic, Somali, English, Italian
South Africa	Republic	Pretoria	Afrikaans, English, Zulu, Xhosa, other
Spain	Parliamentary monarchy	Madrid	Castilian Spanish, Catalan, Galician, Basque
Sri Lanka	Republic	Colombo	English, Sinhala, Tamil
Sudan	Provisional military	Khartoum	Arabic, Nubian, various others
Suriname	Constitutional democracy	Paramaribo	Dutch, Sranang Tongo, English
Swaziland	Monarchy	Mbabane	English, siSwati
Sweden	Constitutional monarchy	Stockholm	Swedish
Switzerland	Federal republic	Bern	German, French, Italian, Romansch
Syria	Republic (authoritarian)	Damascus	Arabic, Kurdish, Armenian, Aramaic
Taiwan	Multiparty democracy	Taipei	Chinese dialects
Tajikistan	Republic	Dushanbe	Tajik, Uzbek, Russian
Tanzania	Republic	Dar es Salaam	Kiswahili, Swahili, English, indigenous
Thailand	Constitutional monarchy	Bangkok	Thai, English
Togo	Republic/transitional	Lomé	French, indigenous

Table B *(Continued)*
World Countries: Form of Government, Capital City, Major Languages

Notes: Unless indicated otherwise, republics are multiparty. "Theocratic" normally refers to fundamentalist Islamic rule. "Transitional" governments are those still in the process of change from a previous form (e.g., single-party communist state to multiparty republic).

COUNTRY	GOVERNMENT	CAPITAL	MAJOR LANGUAGES
Tonga	Constitutional monarchy	Nuku'alofa	Tongan, English
Trinidad and Tobago	Parliamentary democracy	Port of Spain	English, Hindi, French, Spanish
Tunisia	Republic	Tunis	Arabic, French
Turkey	Parliamentary republic	Ankara	Turkish, Kurdish, Arabic
Turkmenistan	Republic	Ashkhabad	Turkmen, Russian, Uzbek, Kazakh
Tuvalu	Constitutional monarchy	Funafuti	Tuvaluan, English
Uganda	Republic	Kampala	English, Luganda, Swahili, indigenous
Ukraine	Republic	Kiev	Ukranian, Russian
United Arab Emirates	Federated monarchy	Abu Dhabi	Arabic, English, Farsi, Hindi, Urdu
United Kingdom	Constitutional monarchy	London	English, Welsh, Gaelic
United States	Federal republic	Washington, D.C.	English, Spanish
Uruguay	Republic	Montevideo	Spanish, Portunol/Brazilero
Uzbekistan	Republic	Tashkent	Uzbek, Russian, Kazakh, Tajik, Tatar
Vanatu	Republic	Port Vila	English, French, Bislama (pidgin)
Venezuela	Federal republic	Caracas	Spanish, indigenous
Vietnam	Single-party communist state	Hanoi	Vietnamese, French, Chinese, English
Yemen	Republic	San'aa	Arabic
Yugoslavia (Serbia-Montenegro)	Republic	Belgrade	Serb, Albanian, Hungarian
Zambia	Single-party republic	Lusaka	English, Tonga, Lozi, other indigenous
Zimbabwe	Republic	Harare	English, Shona, Sindebele

Sources: *The New York Times 2001 Almanac* (Penguin Putnam, New York, 2000); *The World Factbook 2000* (CIA, Washington, DC, 2000).

Table C
World Countries: Basic Economic Indicators, 1999

COUNTRY	GROSS NATIONAL INCOME (GNI) 1999 [a, b]		PURCHASING POWER PARITY GNI 1999			AVERAGE ANNUAL % GROWTH IN GDP		STRUCTURE OF ECONOMIC OUTPUT (GDP) 1999 (value added in % of GDP)			
	Total ($US billions)	Per Capita ($US)	Total ($US billions)	Per Capita ($US)	Rank	1980-1990	1990-1999	Agriculture	Industry	Manufacturing	Services
Albania	3.1	930	11	3,240	137	1.5	3.2	53	26	12	21
Algeria	46.5	1,550	145[c]	4,840[c]	105	2.7	1.6	11	51	10	38
Angola	3.3	270	14[c]	1,100[c]	183	3.7	0.4	7	77	4	16
Argentina	276.1	7,550	437	11,940	57	−0.4	4.9	5	28	18	67
Armenia	1.9	490	9	2,360	–	–	−3.2	29	33	23	39
Australia	397.3	20,950	452	23,850	20	3.4	4.1	3	25	13	72
Austria	205.7	25,430	199	24,600	16	2.2	1.9	2	29	19	69
Azerbaijan	3.7	460	20	2,450	145	–	−9.6	23	35	5	41
Bangladesh	47.1	370	196	1,530	167	4.3	4.7	25	24	15	50
Belarus	26.3	2,620	69	6,880	83	–	−3.0	13	42	35	45
Belgium	252.1	24,650	263	25,710	12	2.0	1.7	1	25	18	73
Benin	2.3	380	6	920	189	2.9	4.7	38	14	8	48
Bolivia	8.1	990	19	2,300	150	−0.2	4.2	18	18	15	64
Bosnia-Herzegovina	4.7	1,210	–	–	–	–	35.2	15	27	21	58
Botswana	5.1	3,240	10	6,540	85	10.3	4.3	4	45	5	51
Brazil	730.4	4,350	1,148	6,840	84	2.7	3.0	9	31	23	61
Bulgaria	11.6	1,410	42	5,070	102	3.4	−2.7	15	23	15	62
Burkina Faso	2.6	240	11[c]	960[c]	187	3.6	3.8	31	28	22	40
Burundi	0.8	120	4[c]	570[c]	203	4.4	−2.9	52	17	9	30
Cambodia	3.0	260	16	1,350	174	–	4.8	51	15	6	35
Cameroon	8.8	600	22	1,490	168	3.4	1.3	44	19	10	38
Canada	614.0	20,140	776	25,440	14	3.3	2.7	–	–	–	–
Central African Republic	1.0	290	4[c]	1150[c]	181	1.4	1.8	55	20	9	25
Chad	1.6	210	6[c]	840[c]	191	3.7	2.1	36	15	12	49
Chile	69.6	4,630	126	8,410	72	4.2	7.2	8	34	16	57
China	979.9	780	4,452	3,550	127	10.2	10.7	18	49	38	33
Colombia	90.0	2,170	232	5,580	93	3.6	3.3	13	26	14	61
Congo Republic	1.6	550	2	540	205	3.3	−0.5	10	49	7	41
Costa Rica	12.8	3,570	28	7,880	76	3.0	5.1	11	37	30	53
Côte d'Ivoire	10.4	670	24	1,540	166	0.7	3.7	26	26	21	48
Croatia	20.2	4,530	32	7,260	80	–	0.2	9	32	20	59
Cuba	–	_[g]	–	–	–	–	–	–	–	–	–
Czech Republic	51.6	5,020	132	12,840	54	1.7	0.8	4	43	–	53
Democratic Republic of the Congo (formerly Zaire)	–	_[d]	–	–	–	1.6	−5.1	58	17	–	25
Denmark	170.7	32,050	136	25,600	13	2.3	2.4	2	21	14	76
Dominican Republic	16.1	1,920	44	5,210	100	3.1	5.8	11	34	17	54
Ecuador	16.8	1,360	35	2,820	140	2.0	2.2	12	37	21	50
Egypt	86.5	1,380	217	3,460	128	5.4	4.4	17	32	20	51
El Salvador	11.8	1,920	26	4,260	117	0.2	5.0	10	29	23	00
Eritrea	0.8	200	4	1,040	185	–	5.0	17	29	15	54
Estonia	4.9	3,400	12	8,190	74	2.2	−1.3	6	26	15	69
Ethiopia	6.5	100	39	620	201	2.3	4.6	52	11	7	37
Finland	127.8	24,730	117	22,600	25	3.3	2.4	3	28	21	68
France	1,453.2	24,170	1,349	23,020	23	2.3	1.5	3	23	–	74
Gabon	4.0	3,300	6	5,280	98	0.9	3.2	8	41	5	51
The Gambia	0.4	330	2[c]	1550[c]	164	3.6	2.8	31	13	6	56
Georgia	3.4	620	14	2,540	144	0.4	–	36	13	8	51

Table C *(Continued)*
World Countries: Basic Economic Indicators, 1999

COUNTRY	GROSS NATIONAL INCOME (GNI) 1999 [a, b]		PURCHASING POWER PARITY GNI 1999			AVERAGE ANNUAL % GROWTH IN GDP		STRUCTURE OF ECONOMIC OUTPUT (GDP) 1999 (value added in % of GDP)			
	Total ($US billions)	Per Capita ($US)	Total ($US billions)	Per Capita ($US)	Rank	1980-1990	1990-1999	Agriculture	Industry	Manufacturing	Services
Germany	2,103.8	25,620	1,930	23,510	21	2.2	1.3	1	28	21	71
Ghana	7.5	400	35[c]	1850[c]	161	3.0	4.3	36	25	9	39
Greece	127.6	12,110	166	15,800	48	1.8	2.2	7	20	11	72
Guatemala	18.6	1,680	40	3,630	126	0.8	4.2	23	20	13	57
Guinea	3.6	490	14	1,870	158	–	4.2	24	37	4	39
Guinea-Bissau	0.2	160	1	630	200	4.0	0.3	62	12	10	26
Haiti	3.6	460	11[c]	1470[c]	169	–0.2	–1.3	29	22	7	48
Honduras	4.8	760	14	2,270	151	2.7	3.3	16	32	20	52
Hong Kong, China	165.1	24,570	152	22,570	26	6.9	3.9	0	15	6	85
Hungary	46.8	4,640	111	11,050	60	1.3	1.0	6	34	25	61
India	441.8	440	2,226	2,230	153	5.8	6.0	28	26	16	46
Indonesia	125.0	600	550	2,660	143	6.1	4.7	19	43	25	37
Iran	113.7	1,810	347	5,520	95	1.7	3.6	21	31	17	48
Iraq	–	_g	–	–	–	–6.8	–	–	–	–	–
Ireland	80.6	21,470	84	22,460	27	3.2	6.9	5	34	–	62
Israel	99.6	16,310	110	18,070	40	3.5	5.2	–	–	–	–
Italy	1,162.9	20,170	1,268	22,000	32	2.4	1.4	3	26	19	71
Jamaica	6.3	2,430	9	3,390	130	2.0	0.3	7	32	14	61
Japan	4,054.5	32,030	3,186	25,170	15	4.0	1.3	2	36	24	62
Jordan	7.7	1,630	18	3,880	122	2.5	5.3	2	26	16	72
Kazakhstan	18.7	1,250	71	4,790	106	–	–5.9	11	32	–	57
Kenya	10.7	360	30	1,010	186	4.2	2.2	23	16	11	61
Korea, North	–	_d	–	–	–	–	–	–	–	–	–
Korea, South	397.9	8,490	728	15,530	49	9.4	5.7	5	44	32	51
Kuwait	–	_h	–	–	–	1.3	–	–	–	–	–
Kyrgyzstan	1.5	300	12	2,420	147	–	–5.4	38	27	12	36
Laos	1.5	290	7[c]	1430[c]	170	–	6.6	53	22	17	25
Latvia	5.9	2,430	15	6,220	89	3.5	–4.8	4	28	15	68
Lebanon	15.8	3,700	–	–	–	–	7.7	12	27	17	61
Lesotho	1.2	550	5[c]	2350[c]	149	4.4	4.4	18	38	–	44
Libya	–	_i	–	–	–	–5.7	–	–	–	–	–
Lithuania	9.8	2,640	24	6,490	86	–	–4.0	9	32	18	59
Macedonia	3.3	1,660	9	4,590	109	–	–0.8	12	35	–	53
Madagascar	3.7	250	12	790	193	1.1	1.7	30	14	11	56
Malawi	2.0	180	6	570	203	2.5	3.6	38	18	14	45
Malaysia	76.9	3,390	173	7,640	78	5.3	7.3	11	46	32	43
Mali	2.6	240	8	740	195	2.8	3.6	47	17	4	37
Mauritania	1.0	390	4	1,550	164	1.8	4.2	25	29	10	46
Mauritius	4.2	3,540	11	8,950	68	6.2	5.1	6	33	25	61
Mexico	428.9	4,440	780	8,070	75	0.7	2.7	5	28	21	67
Moldova	1.5	410	9	2,100	155	3.0	–11.0	25	22	15	53
Mongolia	0.9	390	4	1,610	163	5.4	0.7	32	30	–	39
Morocco	33.7	1,190	94	3,320	134	4.2	2.3	15	33	17	53
Mozambique	3.8	220	14[c]	810[c]	192	–0.1	6.2	33	25	13	42
Myanmar (Burma)	–	_d	–	–	–	0.6	6.3	60	9	7	31
Namibia	3.2	1,890	9[c]	5580[c]	93	0.9	3.4	13	33	15	55
Nepal	5.2	220	30	1,280	176	4.6	4.9	42	21	9	37
Netherlands	397.4	25,140	386	24,410	17	2.3	2.7	3	24	16	74
New Zealand	53.3	13,990	67	17,630	43	1.8	3.1	–	–	–	–

Table C *(Continued)*
World Countries: Basic Economic Indicators, 1999

COUNTRY	GROSS NATIONAL INCOME (GNI) 1999 [a, b]		PURCHASING POWER PARITY GNI 1999			AVERAGE ANNUAL % GROWTH IN GDP		STRUCTURE OF ECONOMIC OUTPUT (GDP) 1999 (value added in % of GDP)			
	Total ($US billions)	Per Capita ($US)	Total ($US billions)	Per Capita ($US)	Rank	1980-1990	1990-1999	Agriculture	Industry	Manufacturing	Services
Nicaragua	2.0	410	10c	2060c	156	−2.0	3.2	32	23	14	46
Niger	2.0	190	8c	740c	195	−0.1	2.4	41	17	6	42
Nigeria	31.6	260	95	770	194	1.6	2.4	39	33	5	28
Norway	149.3	33,470	126	28,140	8	2.8	3.8	2	31	–	67
Oman	_i	–	–	8.4	5.9	–	–	–	–		
Pakistan	62.9	470	250	1,860	159	6.3	3.8	27	23	16	49
Panama	8.7	3,080	15c	5450c	96	0.5	4.2	7	17	8	76
Papua New Guinea	3.8	810	11c	2260c	152	1.9	4.7	30	46	8	24
Paraguay	8.4	1,560	23c	4380c	113	2.5	2.4	29	26	14	45
Peru	53.7	2,130	113	4,480	111	−0.3	5.0	7	38	24	55
Philippines	78.0	1,050	296	3,990	120	1.0	3.2	18	30	21	52
Poland	157.4	4,070	324	8,390	73	1.8	4.5	3	31	18	65
Portugal	110.2	11,030	158	15,860	47	3.1	2.5	4	27	–	69
Puerto Rico	–	_i	–	–	–	4.0	3.1	–	–	–	–
Romania	33.0	1,470	134	5,970	90	0.5	−0.8	16	31	22	53
Russia	329.0	2,250	1,022	6,990	82	–	−6.1	7	38	–	56
Rwanda	2.0	250	7	880	190	2.2	−1.5	46	20	12	34
Saudi Arabia	139.4	6,900	223	11,050	60	0.0	1.6	7	48	10	45
Senegal	4.7	500	13	1,400	172	3.1	3.3	18	26	17	56
Sierra Leone	0.7	130	2	440	207	0.3	−4.7	43	27	4	31
Singapore	95.4	24,150	88	22,310	28	6.6	8.0	0	36	26	64
Slovakia	20.3	3,770	56	10,430	64	2.0	1.8	4	32	22	64
Slovenia	19.9	10,000	32	16,050	46	–	2.4	4	38	28	58
South Africa	133.6	3,170	367c	8710c	70	1.2	1.9	4	32	19	64
Spain	583.1	14,800	704	17,850	42	3.0	2.2	4	28	–	69
Sri Lanka	15.6	820	61	3,230	138	4.0	5.3	21	27	16	52
Sudan	9.4	330	–	–	–	0.4	8.2	40	18	9	42
Sweden	236.9	26,750	196	22,150	31	2.3	1.6	–	–	–	–
Switzerland	273.9	38,380	205	28,760	7	2.0	0.6	–	–	–	–
Syria	15.2	970	54	3,450	129	1.5	5.7	–	–	–	–
Tajikistan	1.7	280	–	–	–	–	–	19	25	21	57
Tanzania[j]	8.5j	260j	16	500	206	j	2.8	45	15	7	40
Thailand	121.1	2,010	358	5,950	91	7.6	4.7	10	40	32	50
Togo	1.4	310	6	1,380	173	1.7	2.4	41	21	9	38
Trinidad and Tobago	6.1	4,750	10	7,690	77	−0.8	2.7	2	40	8	58
Tunisia	19.8	2,090	54	5,700	92	3.3	4.6	13	28	18	59
Turkey	186.5	2,900	415	6,440	87	5.4	3.8	16	24	15	60
Turkmenistan	3.2	670	16	3,340	132	–	−6.8	27	45	34	28
Uganda	6.8	320	25c	1160c	180	2.9	7.2	44	18	9	38
Ukraine	42.0	840	168	3,360	131	–	−10.7	13	38	33	49
United Arab Emirates	–	–	–	–	–	−3.5	2.9	–	–	–	–
United Kingdom	1,403.8	23,590	1,322	22,220	29	3.2	2.5	1	25	–	74
United States	8,879.5	31,910	8,878	31,910	4	3.0	3.3	–	–	–	–
Uruguay	20.6	6,220	29	8,750	69	0.4	3.8	6	27	17	67
Uzbekistan	17.6	720	54	2,230	153	–	−1.2	33	24	11	43
Venezuela	87.3	3,680	129	5,420	97	1.1	1.7	5	36	14	59
Vietnam	28.7	370	144	1,860	159	4.6	8.1	25	34	18	40
West Bank and Gaza	5.1	1,780	–	–	–		3.7	9	29	16	62

Table C *(Continued)*
World Countries: Basic Economic Indicators, 1999

COUNTRY	GROSS NATIONAL INCOME (GNI) 1999 [a, b]		PURCHASING POWER PARITY GNI 1999			AVERAGE ANNUAL % GROWTH IN GDP		STRUCTURE OF ECONOMIC OUTPUT (GDP) 1999 (value added in % of GDP)			
	Total ($US billions)	Per Capita ($US)	Total ($US billions)	Per Capita ($US)	Rank	1980-1990	1990-1999	Agriculture	Industry	Manufacturing	Services
Yemen	6.1	360	12	730	197	–	3.2	17	40	11	42
Yugoslavia (Serbia-Montenegro)	_g	–	–	–	–	–	–	–	–	–	
Zambia	3.2	330	7	720	199	1.0	0.2	25	24	12	51
Zimbabwe	6.3	530	32	2,690	141	3.6	2.8	20	25	17	55

a. Calculated using the World Bank Atlas method.
b. Gross National Income (GNI) has replaced GNP in the World Bank Atlas Method's estimate of national income.
c. The estimate is based on regression; others are extrapolated from the latest International Comparison Programme benchmark estimates.
d. Estimated to be low income ($755 or less).
f. GNP data refer to GDP.
g. Estimated to be lower middle income ($756 to $2,995).
h. Estimated to be high income ($9266 or more).
i. Estimated to be upper middle income ($2,996-9,265 to $9,655).
j. Data refer to mainland Tanzania only.

Source: World Development Indicators (World Bank, 2001)

Table D
World Countries: Population Growth, 1950–2025

COUNTRY	POPULATION (thousands)			AVERAGE ANNUAL POPULATION CHANGE (percent)		AVERAGE ANNUAL INCREMENT TO THE POPULATION (mid-year population, in thousands)		
	1950	2000[a]	2025[a]	1975-1980	1995-00[a]	1985-90	1995-2000	2005-2010
WORLD	2,521,495.0	6,055,049.0	7,823,703.0	1.7	1.3	85,831.0		
AFRICA						359.8		
Algeria	8,753	31,471	46,611	3.1	2.3	631.8	566.0	539.3
Angola	4,131	12,878	25,107	2.7	3.2	130.2	187.2	267.5
Benin	2,046	6,097	11,109	2.5	2.7	135.6	184.8	204.6
Botswana	389	1,622	2,242	3.5	1.9	43.3	21.5	−15.4
Burkina Faso	3,654	11,937	23,321	2.5	2.7	233.8	307.0	360.3
Burundi	2,456	6,695	11,569	2.3	1.7	95.2	123.0	166.5
Cameroon	4,466	15,085	26,484	2.8	2.7	326.5	370.9	374.8
Central African Republic	1,314	3,615	5,704	2.3	1.9	57.5	62.0	60.0
Chad	2,658	7,651	13,908	2.1	2.6	171.2	259.8	340.4
Congo (Zaire)	12,184	51,654	104,788	3.0	2.6	1,146.2	1,234.4	1,907.8
Congo Republic	808	2,943	5,689	2.9	2.8	56.3	62.4	67.2
Côte d'Ivoire	2,776	14,786	23,345	3.9	1.8	411.1	350.2	394.2
Egypt	21,834	68,470	95,615	2.4	1.9	1,318.5	1,199.9	1,125.0
Equatorial Guinea	226	453	795	−0.7	2.5	8.7	11.1	13.6
Eritrea	1,140	3,850	6,681	2.6	3.8	35.4	134.9	153.2
Ethiopia	18,434	62,565	115,382	2.4	2.4	1,530.7	1,670.9	1,848.4
Gabon	469	1,226	1,981	3.1	2.6	11.6	13.9	8.6
The Gambia	294	1,305	2,151	3.1	3.2	33.0	42.3	48.0
Ghana	4,900	20,212	36,876	1.9	2.7	435.4	380.2	285.2
Guinea	2,550	7,430	12,497	1.5	0.8	173.7	63.1	193.9
Guinea-Bissau	505	1,213	1,946	4.7	2.2	22.1	28.4	35.1
Kenya	6,265	30,080	41,756	3.8	2.0	723.5	604.9	206.8
Lesotho	734	2,153	3,506	2.5	2.2	41.5	39.6	11.5
Liberia	824	3,154	6,618	3.1	8.2	−2.9	236.3	107.3
Libya	1,029	5,605	8,647	4.4	2.4	92.8	92.2	136.3
Madagascar	4,230	15,942	28,964	2.5	3.0	308.1	433.2	590.5
Malawi	2,881	10,925	19,958	3.3	2.4	416.6	169.9	103.8
Mali	3,520	11,234	21,295	2.1	2.4	164.0	305.1	390.7
Mauritania	825	2,670	4,766	2.5	2.7	47.5	65.2	94.9
Morocco	8,953	28,351	38,670	2.3	1.8	565.7	535.1	515.0
Mozambique	6,198	19,680	30,612	2.8	2.5	78.7	359.2	75.1
Namibia	511	1,726	2,338	2.7	2.2	58.6	33.5	7.5
Niger	2,400	10,730	21,495	3.2	3.2	207.6	259.7	322.0
Nigeria	30,703	111,506	183,041	2.8	2.4	2,530.7	3,225.5	3,161.6
Rwanda	2,120	7,733	12,427	3.3	7.7	187.9	311.1	49.1
Senegal	2,500	9,481	16,743	2.8	2.6	191.6	278.0	336.2
Sierra Leone	1,944	4,854	8,085	2.0	3.0	106.3	143.8	162.5
Somalia	2,264	10,097	21,211	7.0	4.2	45.8	192.4	266.1
South Africa	13,683	40,377	46,015	2.2	1.5	943.4	383.4	−419.6
Sudan	9,190	29,490	46,264	3.1	2.1	634.6	902.5	1,059.5
Tanzania	7,886	33,517	57,918	3.1	2.3	799.7	832.1	970.9
Togo	1,329	4,629	8,482	2.7	2.6	122.0	160.9	115.3
Tunisia	3,530	9,586	12,843	2.6	1.4	168.9	124.3	104.8
Uganda	4,762	21,778	44,435	3.2	2.8	590.9	598.7	877.5

Table D (Continued)
World Countries: Population Growth, 1950–2025

COUNTRY	POPULATION (thousands)			AVERAGE ANNUAL POPULATION CHANGE (percent)		AVERAGE ANNUAL INCREMENT TO THE POPULATION (mid-year population, in thousands)		
	1950	2000[a]	2025[a]	1975-1980	1995-00[a]	1985-90	1995-2000	2005-2010
Zambia	2,440	9,169	15,616	3.4	2.2	210.9	174.1	190.9
Zimbabwe	2,730	11,669	15,092	3.0	1.4	308.9	78.1	−58.6
NORTH AND CENTRAL AMERICA						353.9		
Belize	69	241	370	1.7	2.4	5.0	6.3	7.2
Canada	13,737	31,147	37,896	1.2	1.0	369.8	331.8	289.5
Costa Rica	862	4,023	5,929	3.0	2.5	76.6	65.4	58.0
Cuba	5,850	11,201	11,798	0.9	0.4	93.2	48.4	37.3
Dominican Republic	2,353	8,495	11,164	2.4	1.6	141.1	136.4	147.1
El Salvador	1,951	6,276	9,062	2.1	2.0	87.1	110.8	117.7
Guatemala	2,969	11,385	19,816	2.5	2.6	255.9	317.6	366.3
Haiti	3,261	8,222	11,988	2.1	1.7	111.8	89.1	114.6
Honduras	1,380	6,485	10,656	3.4	2.7	117.2	151.1	134.9
Jamaica	1,403	2,583	3,245	1.2	0.9	18.3	16.8	23.9
Mexico	27,737	98,881	130,196	2.7	1.6	1,594.3	1,572.4	1,425.0
Nicaragua	1,134	5,074	8,696	3.1	2.7	91.0	107.6	100.9
Panama	860	2,856	3,779	2.5	1.6	44.8	39.8	32.5
Trinidad and Tobago	636	1,295	1,493	1.3	0.5	6.5	−4.9	−6.1
United States	157,813	278,357	325,573	0.9	0.8	2,296.3	2,503.8	2,429.2
SOUTH AMERICA								
Argentina	17,150	37,032	47,160	1.5	1.3	445.4	427.4	401.1
Bolivia	2,714	8,329	13,131	2.4	2.3	127.7	155.2	128.3
Brazil	53,975	170,115	217,930	2.4	1.3	2,756.3	1,975.6	1,285.4
Chile	6,082	15,211	19,548	1.5	1.4	212.2	189.7	148.3
Colombia	12,568	42,321	59,758	2.3	1.9	636.0	681.0	630.9
Ecuador	3,387	12,646	17,796	2.8	2.0	262.0	264.1	257.2
Guyana	423	861	1,045	0.7	0.7	−3.2	−3.5	4.1
Paraguay	1,488	5,496	9,355	3.2	2.6	113.5	141.6	162.8
Peru	7,632	25,662	35,518	2.7	1.7	472.9	491.4	432.4
Suriname	215	417	525	−0.5	0.4	3.9	3.3	1.4
Uruguay	2,239	3,337	3,907	0.6	0.7	19.4	23.7	26.7
Venezuela	5,094	24,170	34,775	3.4	2.0	465.5	397.3	351.7
ASIA								
Afghanistan	8,958	22,720	44,934	0.9	2.9	170.3	879.9	724.7
Armenia	1,354	3,520	3,946	1.8	−0.3	−0.7	−13.8	7.6
Azerbaijan	2,896	7,734	9,403	1.6	0.4	103.6	23.6	61.8
Bangladesh	41,783	129,155	178,751	2.8	1.7	2,028.8	2,001.0	2,119.6
Bhutan	734	2,124	3,904	2.3	2.8	34.7	42.4	48.8
Cambodia	4,346	11,168	16,526	−1.8	2.2	313.1	271.6	314.6
China	554,760	1,277,558	1,480,412	1.5	0.9	16,833.4	11,408.3	8,726.8
Georgia	3,527	4,968	5,178	0.7	−1.1	49.9	−53.5	−14.4
India	357,561	1,013,662	1,330,449	2.1	1.6	16,448.0	16,317.4	15,140.5
Indonesia	79,538	212,107	273,442	2.1	1.4	3,283.4	3,702.8	3,388.7
Iran	16,913	67,702	94,463	3.3	1.7	1,632.8	818.3	1,000.3
Iraq	5,158	23,115	41,014	3.3	2.8	488.2	623.7	719.5

Table D (Continued)
World Countries: Population Growth, 1950–2025

COUNTRY	POPULATION (thousands)			AVERAGE ANNUAL POPULATION CHANGE (percent)		AVERAGE ANNUAL INCREMENT TO THE POPULATION (mid-year population, in thousands)		
	1950	2000[a]	2025[a]	1975-1980	1995-00[a]	1985-90	1995-2000	2005-2010
Israel	1,258	6,217	8,277	2.3	2.2	87.4	107.5	73.6
Japan	83,625	126,714	121,150	0.9	0.2	556.6	252.5	−30.4
Jordan	1,237	6,669	12,063	2.3	3.0	126.9	159.4	145.2
Kazakhstan	6,703	16,223	17,698	1.1	−0.3	148.5	−42.0	85.9
Korea, North	9,488	24,039	29,388	1.6	1.6	307.4	27.2	4,294.4
Korea, South	20,357	46,844	52,533	1.6	0.8	943.5	459.2	322.0
Kuwait	152	1,972	2,974	6.2	3.1	81.8	70.6	90.4
Kyrgyzstan	1,740	4,699	6,096	1.9	0.6	76.8	30.0	80.9
Laos	1,755	5,433	9,653	1.2	2.6	110.7	130.3	155.3
Lebanon	1,443	3,282	4,400	−0.7	1.7	11.8	48.7	46.0
Malaysia	6,110	22,244	30,968	2.3	2.0	391.7	436.4	438.2
Mongolia	761	2,662	3,709	2.8	1.6	62.1	37.4	44.4
Myanmar (Burma)	17,832	45,611	58,120	2.1	1.2	452.8	317.4	162.3
Nepal	7,862	23,930	38,010	2.5	2.4	457.5	559.0	616.3
Oman	456	2,542	5,352	5.0	3.3	58.3	80.5	104.3
Pakistan	39,513	156,483	263,000	2.6	2.8	2,984.4	2,984.8	2,936.8
Philippines	20,988	75,967	108,251	2.3	2.1	1,450.5	1,658.1	1,666.1
Saudi Arabia	3,201	21,607	39,965	5.6	3.4	527.8	678.3	922.2
Singapore	1,022	3,567	4,168	1.3	1.4	56.1	134.2	169.1
Sri Lanka	7,678	18,827	23,547	1.7	1.0	234.4	186.9	153.5
Syria	3,495	16,125	26,292	3.1	2.5	391.1	399.2	431.5
Tajikistan	1,532	6,188	8,857	2.8	1.5	149.0	115.3	168.7
Thailand	20,010	61,399	72,717	2.4	0.9	755.5	598.8	468.5
Turkey	20,809	66,591	87,869	2.1	1.7	1,083.1	895.5	732.4
Turkmenistan	1,211	4,459	6,287	2.5	1.8	85.4	83.3	95.8
United Arab Emirates	70	2,441	3,284	14.0	2.0	76.1	38.6	40.0
Uzbekistan	6,314	24,318	33,355	2.6	1.6	473.1	381.7	485.8
Vietnam	29,954	79,832	108,037	2.2	1.6	1,321.7	1,178.3	1,123.4
Yemen	4,316	18,112	38,985	3.2	3.7	436.3	524.0	782.1
EUROPE								
Albania	1,230	3,113	3,820	1.9	−0.4	60.3	50.7	34.4
Austria	6,935	8,211	8,186	−0.1	0.5	32.1	17.8	11.3
Belarus	7,745	10,236	9,496	0.6	−0.3	46.7	−7.5	−1.4
Belgium	8,639	10,161	9,918	0.1	0.1	22.2	20.9	5.3
Bosnia-Herzegovina	2,661	3,972	4,324	0.9	3.0	29.7	96.0	15.5
Bulgaria	7,251	8,225	7,023	0.3	−0.7	−9.9	−95.1	−74.3
Croatia	3,850	4,473	4,193	0.5	−0.1	10.1	−34.6	11.2
Czech Republic	8,925	10,244	9,512	0.6	−0.2	−0.1	−10.6	−14.7
Denmark	4,271	5,293	5,238	0.2	0.3	5.5	20.8	12.0
Estonia	1,101	1,396	1,131	0.6	−1.2	7.0	−10.5	−4.9
Finland	4,009	5,176	5,254	0.3	0.3	16.9	12.3	4.8
France	41,829	59,080	61,662	0.4	0.4	312.8	236.0	142.8
Germany	68,376	82,220	80,238	−0.1	0.1	339.1	229.8	152.4
Greece	7,566	10,645	9,863	1.3	0.3	44.5	22.4	11.0
Hungary	9,338	10,036	8,900	0.3	−0.4	−55.4	−31.4	−31.1
Iceland	143	281	328	0.9	0.9	2.7	1.8	1.1

COUNTRY	POPULATION (thousands)			AVERAGE ANNUAL POPULATION CHANGE (percent)		AVERAGE ANNUAL INCREMENT TO THE POPULATION (mid-year population, in thousands)		
	1950	2000[a]	2025[a]	1975-1980	1995-00[a]	1985-90	1995-2000	2005-2010
Ireland	2,969	3,730	4,404	1.4	0.7	−6.4	37.2	31.9
Italy	47,104	57,298	51,270	0.4	0.0	4.0	74.2	−67.5
Latvia	1,949	2,357	1,936	0.4	−1.5	12.3	−23.5	−12.9
Lithuania	2,567	3,670	3,399	0.7	−0.3	22.2	−10.4	−3.6
Macedonia	1,230	2,024	2,258	1.4	0.6	6.9	11.0	7.2
Moldova	2,341	4,380	4,547	0.9	0.0	49.9	−5.8	16.0
Netherlands	10,114	15,786	15,782	0.7	0.4	92.0	86.6	62.5
Norway	3,265	4,465	4,817	0.4	0.5	17.9	24.4	18.2
Poland	24,824	38,765	39,069	0.9	0.1	178.7	8.5	11.2
Portugal	8,405	9,875	9,348	1.4	0.0	5.1	15.9	9.6
Romania	16,311	22,327	19,945	0.9	−0.4	69.0	−56.3	−49.8
Russia	102,192	146,934	137,933	0.6	−0.2	820.8	−422.7	−281.7
Slovakia	3,463	5,387	5,393	1.0	0.1	23.6	9.3	6.1
Slovenia	1,473	1,986	1,818	1.0	0.0	4.6	3.6	1.2
Spain	28,009	39,630	36,658	1.1	0.0	163.2	49.1	−2.8
Sweden	7,014	8,910	9,097	0.3	0.2	40.5	9.5	0.5
Switzerland	4,694	7,386	7,587	−0.1	0.7	54.8	19.2	7.9
Ukraine	36,906	50,456	45,688	0.4	−0.4	142.7	−432.6	−264.4
United Kingdom	50,616	58,830	59,961	0.0	0.2	189.0	178.9	94.6
Yugoslavia (Serbia-Montenegro)	7,131	10,640	10,844	0.9	0.1	21.0	−5.2	−4.6
OCEANIA								
Australia	8,219	18,886	23,098	0.9	1.0	246.8	209.7	167.0
Fiji	289	817	1,104	1.9	1.2	7.8	11.3	12.8
New Zealand	1,908	3,862	4,695	0.2	1.0	12.3	50.8	38.5
Papua New Guinea	1,613	4,807	7,460	2.5	2.2	89.3	116.4	125.1
Solomon Islands	90	444	817	3.5	3.1	11.0	13.8	14.3
DEVELOPING COUNTRIES	1,667,848	4,746,022	6,459,163	2.1	1.6			
DEVELOPED COUNTRIES	852,572	1,306,083	1,359,258	0.6	0.3			

a. Data include projections based on 1990 base year population data.

Sources: United Nations Population Division and International Labour Organisation; *World Resources 2000–2001* (World Resources Institute, Washington, D.C.); U.S. Bureau of the Census International Data Base (2000).

Table E
World Countries: Basic Demographic Data, 1975–2000

COUNTRY	CRUDE BIRTHRATE (births per 1,000 population)		LIFE EXPECTANCY AT BIRTH (years)		LIFE EXPECTANCY OF FEMALES AS A PERCENTAGE OF MALES (years)		TOTAL FERTILITY RATE		PERCENTAGE OF POPULATION IN SPECIFIC AGE GROUPS					
									1980			2000		
	1975–80	1995-00	1975-80	1995-00	1975-80	1995-00	1975-80	1995-00	<15	15-65	65	<15	15-65	>65
WORLD	28.3	21.6	59.7	63.7	106.0	107.9	3.9	2.8	35.2	58.9	5.9	30.0	63.2	7.0
AFRICA	46.0	37.6	47.9	50.5	106.7	105.6	6.5	5.3	44.7	52.2	3.1			
Algeria	45.0	23.1	57.5	69.7	103.5	102.9	7.2	3.8	46.5	49.6	3.9	37.0	60.0	4.0
Angola	50.2	46.9	40.0	38.3	108.1	106.6	6.8	6.7	44.7	52.4	2.9	48.0	50.0	3.0
Benin	51.4	44.8	47.0	50.2	108.2	105.7	7.1	5.8	45.1	50.8	4.1	46.5	50.7	2.8
Botswana	46.6	29.6	56.5	39.3	106.6	104.3	6.4	4.5	48.7	49.3	2.0	41.9	55.6	2.4
Burkina Faso	50.8	45.3	42.9	46.7	106.0	102.2	7.8	6.6	47.4	49.8	2.8	47.0	50.3	2.7
Burundi	44.7	40.5	46.0	46.2	107.2	107.3	6.8	6.3	44.7	51.8	3.4	45.0	52.3	2.7
Cameroon	45.5	36.6	48.5	54.8	106.4	105.6	6.5	5.3	44.4	52.0	3.6	43.5	52.9	3.6
Central African Republic	44.1	37.5	44.5	44.0	111.9	109.3	5.9	5.0	41.7	54.4	4.0	41.6	54.4	4.0
Chad	44.1	48.8	41.0	50.5	108.1	106.5	5.9	5.5	41.9	54.5	3.6	43.0	53.4	3.6
Congo Republic	45.8	38.6	48.7	47.4	111.3	110.8	6.3	5.9	45.1	51.5	3.4	45.7	51.1	3.3
Cote d'Ivoire	50.7	40.8	47.9	45.1	107.1	102.0	7.4	5.1	46.6	51.0	2.5	43.0	54.0	3.0
Dem. Republic of the Congo (formerly Zaire)	47.8	46.4	48.0	48.8	107.1	106.1	6.5	6.2	46.0	51.1	2.8	48.0	49.1	2.9
Egypt	38.9	25.4	54.1	63.3	104.5	104.6	5.3	3.4	39.5	56.5	4.0	35.0	60.5	4.5
Equatorial Guinea	42.7	38.1	42.0	53.6	107.9	108.3	5.7	5.5	41.0	54.8	4.1	43.1	52.9	4.0
Eritrea	45.1	42.7	45.3	55.8	106.8	106.1	6.1	5.3	44.2	53.1	2.6	43.6	53.3	3.1
Ethiopia	49.0	45.1	42.0	45.2	107.9	104.7	6.8	7.0	46.1	51.2	2.7	47.1	50.2	2.8
Gabon	32.9	27.6	47.0	50.1	107.3	105.8	4.4	5.4	34.4	59.5	6.1	39.8	54.6	5.7
The Gambia	48.8	42.3	39.0	53.2	108.3	108.8	6.5	5.2	42.6	54.4	2.8	41.2	55.5	3.1
Ghana	45.1	29.8	51.0	57.4	106.9	106.9	6.5	5.3	44.9	52.3	2.8	43.4	53.6	3.0
Guinea	51.6	40.1	38.8	45.6	102.6	102.1	7.0	6.6	45.8	51.6	2.6	47.0	50.4	2.6
Guinea-Bissau	42.4	39.6	37.5	49.0	108.6	106.9	5.6	5.4	39.0	57.0	4.0	41.7	54.2	4.2
Kenya	53.6	29.4	53.4	48.0	107.8	103.9	8.1	4.9	50.1	46.5	3.4	43.4	53.7	2.9
Lesotho	41.9	31.7	51.8	50.8	107.0	103.6	5.7	4.9	41.9	53.9	4.2	41.0	55.0	4.1
Liberia	47.4	47.2	49.5	51.0	106.3	106.5	6.8	6.3	44.3	52.0	3.7	43.7	52.7	3.6
Libya	47.3	27.7	55.8	75.5	106.3	105.8	7.4	5.9	46.7	51.1	2.2	44.7	52.4	2.9
Madagascar	46.9	42.9	49.5	55.0	106.3	105.3	6.6	5.7	45.9	51.4	2.7	45.8	51.6	2.6
Malawi	57.2	38.5	43.1	37.6	103.3	102.5	7.6	6.7	47.5	50.3	2.3	46.4	50.9	2.7
Mali	50.7	49.2	40.0	46.7	108.1	105.8	7.1	6.6	46.8	50.7	2.5	47.2	50.2	2.5
Mauritania	44.7	43.4	45.5	50.8	107.3	105.8	6.5	5.0	43.7	53.3	3.0	41.5	55.2	3.3
Mauritius	26.7	16.7	64.9	71.0	108.3	..	3.1	2.3	35.6	60.8	3.6	26.6	67.3	6.0
Morocco	39.4	24.6	55.8	69.1	106.3	106.1	5.9	3.1	43.2	52.7	4.1	33.8	61.9	4.3
Mozambique	45.4	38.0	43.5	37.5	107.6	106.8	6.5	6.1	43.4	53.4	3.2	44.7	52.1	3.2
Namibia	41.9	35.2	51.3	42.5	105.0	101.9	6.0	4.9	43.1	53.4	3.5	41.6	54.6	3.8
Niger	59.7	51.5	40.5	41.3	108.2	106.4	8.1	7.1	46.8	50.8	2.5	48.6	49.0	2.4
Nigeria	46.6	40.2	45.0	51.6	107.4	106.1	6.5	6.0	44.3	3.1	2.6	45.0	52.1	2.9
Rwanda	52.8	34.8	45.0	39.3	107.4	107.7	8.5	6.0	48.8	48.8	2.4	44.7	53.0	2.4
Senegal	49.3	37.9	42.8	62.2	104.8	105.8	7.0	5.6	45.3	51.8	2.8	43.6	53.4	3.0
Sierra Leone	48.8	45.6	35.2	45.2	108.9	108.3	6.5	6.1	43.0	53.9	3.1	43.9	53.1	3.0
Somalia	50.4	47.7	42.0	46.2	107.9	108.8	7.0	7.0	46.0	51.0	3.0	48.0	49.5	2.6
South Africa	37.3	21.6	55.9	51.1	111.3	111.5	5.1	3.8	40.3	55.9	3.9	36.2	59.3	4.5
Sudan	47.1	38.6	46.7	56.5	106.2	103.7	6.7	4.6	44.9	52.4	2.7	38.7	58.1	3.2
Swaziland	46.2	40.6	49.9	40.4	109.9	..	6.5	4.5	45.9	51.3	2.9	41.7	55.6	2.7
Tanzania	47.5	40.2	49.0	52.3	107.2	104.2	6.8	5.5	47.6	50.1	2.3	45.1	52.3	2.6
Togo	45.2	38.0	48.0	54.7	107.1	104.1	6.6	6.1	44.6	52.3	3.2	45.6	51.3	3.1
Tunisia	36.4	17.4	60.0	73.7	101.7	104.4	5.7	2.9	41.6	54.6	3.8	32.2	62.9	4.9

COUNTRY	CRUDE BIRTHRATE (births per 1,000 population)		LIFE EXPECTANCY AT BIRTH (years)		LIFE EXPECTANCY OF FEMALES AS A PERCENTAGE OF MALES (years)		TOTAL FERTILITY RATE		PERCENTAGE OF POPULATION IN SPECIFIC AGE GROUPS					
									1980			2000		
	1975–80	1995-00	1975-80	1995-00	1975-80	1995-00	1975-80	1995-00	<15	15-65	65	<15	15-65	>65
Uganda	50.3	48.0	47.0	42.9	107.0	102.5	6.9	7.1	47.8	49.7	2.5	49.1	48.6	2.2
Zambia	51.6	41.9	49.3	37.2	106.9	102.5	7.2	5.5	49.4	48.2	2.4	46.3	51.4	2.3
Zimbabwe	44.2	25.0	53.8	37.8	106.9	102.3	6.6	4.7	47.9	49.5	2.6	43.6	53.7	2.7
NORTH AMERICA	15.1	12.8	73.3	78.3	111.2	108.7	1.8	1.9	22.5	66.4	11.0	21.2	66.4	12.4
Canada	15.4	11.3	74.2	79.4	110.8	107.8	1.8	1.6	22.7	67.9	9.4	19.3	68.1	12.6
United States	15.1	14.2	73.2	77.1	111.2	109.6	1.8	2.0	22.5	66.3	11.2	21.4	66.2	12.4
CENTRAL AMERICA	38.2	24.8	63.7	70.0	110.2	106.7	5.4	3.0	45.1	51.2	3.6	34.8	60.7	4.5
Belize	40.9	32.3	69.7	70.9	102.5	104.1	6.2	3.7	47.2	48.6	4.8	39.7	55.8	4.1
Costa Rica	31.7	20.7	71.0	75.8	106.4	106.7	3.9	3.0	38.9	57.5	3.6	33.1	61.8	5.1
Cuba	17.2	12.7	73.0	76.2	104.8	105.4	2.1	1.6	32.9	60.5	7.6	21.2	69.2	9.6
Dominican Republic	34.9	25.1	62.0	73.2	106.1	105.8	4.7	2.8	42.3	54.6	3.1	33.1	62.5	4.5
El Salvador	41.5	29.0	57.2	69.7	119.5	108.9	5.7	3.1	46.1	50.8	3.1	35.6	59.7	4.7
Guatemala	44.3	35.0	56.4	66.2	107.2	109.8	6.4	4.9	45.9	51.3	2.8	42.9	53.3	3.7
Haiti	36.8	32.0	50.7	49.2	106.3	109.8	5.4	4.6	40.7	54.8	4.4	40.0	56.2	3.8
Honduras	44.9	32.6	57.7	69.9	107.7	105.9	6.6	4.3	47.2	50.1	2.7	41.6	54.9	3.4
Jamaica	28.8	18.5	70.1	75.2	106.3	105.5	4.0	2.4	40.2	53.0	6.8	30.2	53.4	6.4
Mexico	37.1	23.1	65.3	71.5	110.3	107.1	5.3	2.8	45.1	51.1	3.8	33.1	62.1	4.7
Nicaragua	45.7	28.3	57.6	68.7	108.5	107.6	6.4	3.9	47.7	49.8	2.5	40.8	56.0	3.2
Panama	31.0	19.5	69.0	75.5	105.8	105.5	4.1	2.6	40.5	55.0	4.5	31.3	63.7	5.5
Trinidad and Tobago	29.3	13.8	68.3	68.0	107.6	105.5	3.4	2.1	34.2	60.2	5.5	26.1	67.4	6.5
SOUTH AMERICA	32.0	22.1	62.9	70.5	108.8	109.0	4.3	2.5	37.8	57.6	4.6	30.2	64.2	5.6
Argentina	25.7	18.6	68.7	75.0	110.4	110.0	3.4	2.6	30.5	61.4	8.1	27.7	62.6	9.7
Bolivia	41.0	28.1	50.1	63.7	108.8	105.0	5.8	4.4	42.6	53.9	3.5	30.5	56.4	4.0
Brazil	32.6	18.8	61.8	62.9	108.1	112.7	4.3	2.2	38.1	57.8	4.2	28.4	66.4	5.2
Chile	24.0	17.2	67.2	75.7	110.5	108.3	3.0	2.4	33.5	60.9	5.6	28.5	64.4	7.2
Colombia	31.7	22.9	64.0	70.3	107.3	110.4	4.1	2.7	40.0	56.2	3.7	32.5	62.8	4.6
Ecuador	38.2	26.5	61.4	71.1	105.9	107.5	5.4	3.1	42.8	53.2	4.0	33.8	61.5	4.7
Guyana	31.5	17.9	60.7	64.0	108.4	111.4	3.9	2.3	40.8	55.2	4.0	29.9	65.9	4.2
Paraguay	35.9	31.3	66.6	73.7	106.7	107.4	5.2	4.2	42.2	53.3	4.5	39.5	57.0	3.5
Peru	38.0	24.5	58.5	70.0	106.7	107.6	5.4	3.0	41.9	54.5	3.6	33.4	61.8	4.8
Suriname	29.5	21.1	65.1	71.4	107.8	107.3	4.2	2.4	39.7	55.8	4.5	32.3	62.4	5.5
Uruguay	20.3	17.4	69.6	75.2	110.2	111.4	2.9	2.3	26.9	62.5	10.5	23.9	63.5	12.7
Venezuela	34.2	21.1	67.6	73.1	109.1	108.6	4.5	3.0	40.7	56.1	3.2	34.0	61.5	4.4
ASIA	29.6	24.7	58.5	66.4	102.6	106.2	4.2	2.7	37.6	58.0	4.4	30.1	64.1	5.8
Afghanistan	50.8	41.8	40.0	45.9	100.0	102.2	7.2	6.9	43.0	54.5	2.5	41.7	55.6	2.7
Armenia	20.9	11.0	72.3	66.4	109.0	110.0	2.5	1.7	30.4	63.6	6.0	24.5	66.8	8.7
Azerbaijan	24.7	18.1	68.5	62.9	112.1	110.4	3.6	2.3	34.5	60.0	5.4	29.5	63.6	6.9
Bangladesh	47.2	25.4	46.6	60.2	97.9	100.0	6.7	3.1	46.0	50.5	3.4	35.6	61.2	3.3
Bhutan	42.8	36.2	42.6	52.4	104.8	103.3	5.9	5.9	41.3	55.6	3.1	43.0	53.8	3.2
Cambodia	30.0	33.5	31.2	56.5	108.3	107.8	4.1	4.5	39.2	57.9	2.9	40.4	56.6	3.0
China	21.5	16.1	65.3	71.4	103.0	105.8	3.3	1.8	35.5	59.7	4.7	24.9	68.4	6.7
Georgia	18.1	10.9	70.7	64.5	111.9	111.6	2.4	1.9	25.7	65.1	9.1	22.0	65.4	12.6
India	34.7	24.8	52.9	62.5	96.3	101.6	4.8	3.1	38.5	57.4	4.0	32.7	62.3	5.0
Indonesia	35.4	22.6	52.8	68.0	104.9	106.3	4.7	2.6	41.0	55.6	3.3	30.8	64.6	4.7

Table E *(Continued)*
World Countries: Basic Demographic Data, 1975–2000

COUNTRY	CRUDE BIRTHRATE (births per 1,000 population)		LIFE EXPECTANCY AT BIRTH (years)		LIFE EXPECTANCY OF FEMALES AS A PERCENTAGE OF MALES (years)		TOTAL FERTILITY RATE		PERCENTAGE OF POPULATION IN SPECIFIC AGE GROUPS					
									1980			2000		
	1975–80	1995-00	1975-80	1995-00	1975-80	1995-00	1975-80	1995-00	<15	15-65	65	<15	15-65	>65
Iran	44.7	18.3	58.6	69.7	101.4	101.4	6.5	4.8	44.9	51.8	3.3	43.1	52.8	4.1
Iraq	41.9	35.0	61.4	66.5	103.0	104.9	6.6	5.3	46.0	51.3	2.7	41.4	55.5	3.1
Israel	26.0	19.3	73.1	78.6	104.9	105.2	3.4	2.8	33.2	58.2	8.6	28.1	62.4	9.5
Japan	15.2	10.0	75.5	80.7	107.4	107.8	1.8	1.5	23.6	67.4	9.0	15.2	68.3	16.5
Jordan	45.0	26.2	61.2	77.4	106.1	104.3	7.4	5.1	49.4	47.5	3.1	43.3	53.8	2.9
Kazakhstan	24.9	16.8	65.4	63.2	117.1	114.2	3.1	2.3	32.4	61.5	6.1	27.5	65.4	7.1
Korea, North	22.2	20.4	65.8	70.7	110.3	108.7	3.3	2.1	39.5	57.0	3.5	27.3	57.4	5.3
Korea, South	23.9	15.1	64.8	74.4	111.6	110.1	2.9	1.7	34.0	62.2	3.8	21.4	72.0	6.7
Kuwait	40.1	22.0	69.6	76.1	106.2	105.4	5.9	2.8	40.2	58.3	1.4	33.2	64.8	2.0
Kyrgyzstan	29.9	26.3	64.2	63.4	114.2	114.2	4.1	3.2	37.1	57.1	5.8	35.0	59.0	6.0
Laos	45.1	38.3	43.5	53.1	106.9	105.7	6.7	6.7	42.0	55.2	2.8	45.4	51.7	3.0
Lebanon	30.1	20.3	65.0	71.2	106.2	105.8	4.3	2.8	40.1	54.5	5.4	32.9	61.3	5.8
Malaysia	30.4	25.3	65.3	70.8	105.7	105.7	4.2	3.2	39.3	57.0	3.7	35.3	60.6	4.1
Mongolia	39.2	21.8	56.3	63.9	104.5	104.6	6.7	3.3	43.2	53.9	2.9	36.4	59.8	3.8
Myanmar (Burma)	37.5	20.6	51.2	54.9	106.4	105.0	5.3	3.3	39.6	56.4	4.0	34.0	61.4	4.6
Nepal	43.4	33.8	46.2	57.8	96.6	98.2	6.2	5.0	42.9	54.1	3.0	42.0	54.5	3.5
Oman	46.1	38.1	54.9	71.8	104.3	105.8	7.2	7.2	44.6	52.8	2.6	47.7	49.9	2.3
Pakistan	47.3	32.1	53.4	61.1	101.3	103.1	7.0	5.0	44.4	52.7	2.9	41.8	55.0	3.2
Philippines	35.9	27.9	59.9	67.5	105.5	104.4	5.0	3.6	41.9	55.3	2.8	36.7	59.7	3.6
Saudi Arabia	45.9	37.5	58.8	67.8	104.0	104.2	7.3	5.9	44.3	52.9	2.8	40.7	56.4	2.9
Singapore	17.2	12.8	70.8	80.0	106.6	105.3	1.9	1.8	27.0	68.2	4.7	22.6	70.3	7.1
Sri Lanka	28.5	16.8	66.8	71.8	105.4	105.6	3.8	2.1	35.3	60.4	4.3	26.2	67.2	6.6
Syria	46.0	31.1	60.1	68.5	106.2	105.9	7.4	4.0	48.5	48.3	3.2	40.8	56.1	3.1
Tajikistan	37.2	33.6	64.5	64.1	108.3	109.3	5.9	3.9	42.9	52.5	4.6	36.5	55.9	4.6
Thailand	31.6	16.9	61.2	68.5	106.6	109.0	4.3	1.7	40.0	56.5	3.5	25.2	69.1	5.8
Turkey	32.0	18.6	60.3	71.0	107.8	107.4	4.5	2.5	39.2	56.0	4.7	28.3	65.7	5.9
Turkmenistan	35.3	28.9	61.6	60.9	11.2	111.2	5.3	3.6	41.3	54.4	4.3	37.4	58.3	4.2
United Arab Emirates	30.5	18.0	66.8	74.1	106.5	102.7	5.7	3.5	28.6	70.1	1.3	28.1	69.4	2.5
Uzbekistan	34.4	26.2	65.1	63.7	111.0	110.9	5.1	3.5	40.9	54.0	5.1	37.5	57.9	4.6
Vietnam	38.3	21.6	55.8	69.3	108.2	107.7	5.6	3.0	42.5	52.7	4.8	34.3	60.5	5.2
Yemen	53.7	43.4	44.1	59.8	101.1	101.7	7.6	7.6	50.2	47.2	2.6	48.3	49.3	2.3
EUROPE	14.8	11.0	71.3	74.3	111.4	110.6	2.0	1.5	22.2	65.5	12.3	17.5	67.9	14.6
Albania	30.3	19.5	68.9	71.6	106.6	108.6	4.2	2.6	35.9	58.9	5.2	29.7	64.4	6.0
Austria	11.5	9.9	72.0	77.7	110.4	108.1	1.7	1.4	20.4	64.2	15.4	17.1	68.5	14.4
Belarus	16.1	9.3	71.1	68.0	114.7	119.3	2.1	1.4	22.9	66.4	10.7	18.8	67.3	13.8
Belgium	12.4	10.9	72.3	77.8	109.6	109.4	1.7	1.6	20.2	65.5	14.3	17.4	66.2	16.4
Bosnia-Herzegovina	19.6	12.9	69.9	71.5	107.0	107.0	2.2	1.4	27.3	66.7	6.1	18.9	71.2	9.8
Bulgaria	15.8	8.1	71.3	70.9	107.9	110.2	2.2	1.5	22.1	66.0	11.9	16.9	67.2	15.8
Croatia	15.6	12.8	70.6	73.7	110.4	111.6	2.0	1.6	21.2	67.2	11.7	17.3	68.0	14.6
Czech Republic	17.4	9.1	70.6	74.5	110.4	111.0	2.3	1.4	23.5	63.2	13.4	17.6	69.8	12.6
Denmark	12.3	12.2	74.2	76.5	108.4	106.8	1.7	1.8	20.8	64.7	14.4	18.6	66.7	14.7
Estonia	15.0	8.4	69.7	69.5	115.3	119.0	2.1	1.3	21.7	65.8	12.5	17.6	68.5	13.9
Finland	13.6	10.8	72.7	77.4	112.6	110.9	1.6	1.8	20.3	67.7	12.0	18.4	67.1	14.6
France	14.0	12.3	73.7	78.8	111.6	110.8	1.9	1.6	22.3	63.8	14.0	18.3	65.4	16.2
Germany	10.4	9.3	72.5	77.4	109.4	108.1	1.5	1.3	18.5	65.9	15.6	15.4	68.8	15.9
Greece	15.6	9.8	73.7	78.4	105.7	106.5	2.3	1.4	22.8	64.0	13.1	15.3	66.9	17.8
Hungary	16.2	9.3	69.4	71.4	109.8	111.9	2.1	14.0	21.9	64.6	13.4	17.0	68.5	14.5
Iceland	19.2	14.9	76.3	79.4	108.0	105.1	2.3	2.2	27.6	62.7	10.1	23.8	64.9	11.3

Table E *(Continued)*
World Countries: Basic Demographic Data, 1975–2000

COUNTRY	CRUDE BIRTHRATE (births per 1,000 population)		LIFE EXPECTANCY AT BIRTH (years)		LIFE EXPECTANCY OF FEMALES AS A PERCENTAGE OF MALES (years)		TOTAL FERTILITY RATE		PERCENTAGE OF POPULATION IN SPECIFIC AGE GROUPS					
									1980			2000		
	1975–80	1995-00	1975-80	1995-00	1975-80	1995-00	1975-80	1995-00	<15	15-65	65	<15	15-65	>65
Ireland	21.2	14.5	72.0	76.8	107.2	106.7	3.5	1.8	30.6	58.7	10.7	21.2	67.5	11.4
Italy	13.0	9.1	73.6	79.0	109.2	108.0	1.9	1.2	22.3	64.6	13.1	14.2	68.1	17.7
Latvia	14.0	7.8	69.2	68.4	115.6	119.3	2.0	1.4	20.4	66.5	13.0	18.2	67.5	14.4
Lithuania	15.4	9.8	70.8	69.1	114.2	118.7	2.1	1.5	23.6	65.1	11.3	19.5	67.0	13.5
Macedonia	22.6	13.7	69.6	73.8	105.2	105.6	2.7	1.9	28.6	64.5	6.9	22.3	68.3	9.4
Moldova	19.6	12.9	64.8	64.5	111.2	112.5	2.4	1.8	26.7	65.5	7.8	23.4	66.8	9.8
Netherlands	12.7	12.1	75.3	78.3	109.0	108.0	1.6	1.6	22.3	66.2	11.5	18.2	68.2	13.6
Norway	12.9	12.8	75.3	78.7	108.9	108.0	1.8	1.9	22.2	63.1	14.8	19.8	65.1	15.0
Poland	19.2	10.1	70.9	73.2	111.9	113.2	2.3	1.7	24.3	65.6	10.1	19.8	68.4	11.8
Portugal	18.2	11.5	70.2	75.8	110.6	109.7	2.4	1.5	25.9	63.6	10.5	16.7	67.5	15.7
Romania	19.1	10.8	69.6	69.9	107.0	112.1	2.6	1.4	26.7	63.1	10.3	18.5	68.4	13.2
Russia	15.8	9.0	67.4	67.2	118.1	119.6	1.9	1.4	21.6	68.1	10.2	18.0	69.3	12.7
Slovakia	20.6	10.0	70.4	73.7	110.9	111.5	2.5	1.5	26.1	63.5	10.4	20.0	68.8	11.1
Slovenia	16.6	9.3	71.0	74.9	111.6	109.8	2.2	1.3	23.4	65.3	11.4	15.7	70.2	14.2
Spain	17.4	9.2	74.3	78.8	108.4	109.3	2.6	1.2	26.6	62.7	10.7	15.0	68.4	16.5
Sweden	11.7	10.0	75.2	79.6	109.3	106.6	1.7	1.8	19.6	64.1	16.3	19.2	64.3	16.7
Switzerland	11.6	10.4	75.2	79.6	109.2	109.3	1.5	1.5	19.7	66.4	13.8	17.2	68.1	14.7
Ukraine	14.9	9.0	69.3	66.0	114.8	115.6	2.0	1.4	21.4	66.7	11.9	17.8	67.9	14.3
United Kingdom	12.4	11.7	72.8	77.7	109.0	106.6	1.7	1.7	20.9	64.0	15.1	18.9	65.5	15.8
Yugoslavia (Serbia-Montenegro)	18.4	13.5	70.3	73.9	106.5	107.1	2.4	1.8	24.1	66.1	9.8	19.8	66.9	13.3
OCEANIA	20.9	23.7	68.2	72.0	108.7	106.7	2.8	2.5	29.3	62.7	8.0	25.5	64.9	9.7
Australia	16.0	13.0	73.4	79.8	109.8	108.0	2.1	1.9	25.3	65.1	9.6	21.0	67.1	11.9
Fiji	33.2	23.5	67.1	67.9	105.3	105.6	4.0	2.8	39.1	58.0	2.8	31.3	64.3	4.5
New Zealand	17.2	14.3	72.4	77.8	109.2	108.1	2.2	2.0	26.7	63.3	10.0	23.0	65.6	11.3
Papua New Guinea	39.6	32.7	49.7	63.1	101.0	103.5	5.9	4.7	43.0	55.5	1.6	38.7	58.3	3.0
Solomon Islands	45.0	34.8	65.3	71.3	106.2	105.7	7.1	5.0	47.6	49.3	3.1	43.0	54.1	2.9
DEVELOPING COUNTRIES	32.8	24.1	56.7	62.3	103.8	104.8	4.7	3.1	39.3	56.6	4.1	32.8	62.2	5.0
DEVELOPED COUNTRIES	14.9	11.2	72.2	75.9	110.8	111.2	1.9	1.6	22.5	65.9	11.6	18.3	67.5	14.2

Sources: United Nations Population Division World Resources 2000–2001; U.S. Bureau of the Census International Database (2000).

Table F
World Countries: Mortality, Health, and Nutrition, 1980–1999

COUNTRY	MORTALITY						HEALTH			NUTRITION	
	Infant Mortality		Adult Mortality								
	Infant Mortality (per 1,000 live births)	Under-five Mortality (per 1,000)	Male (per 1,000)		Female (per 1,000)		Health Expenditure Per Capita ($US)	Physicians Per 1,000 People	Hospital Beds Per 1,000 People	Average daily per capita supply of calories (kilocalories)[c]	Average daily per capita supply of calories (kilocalories)[c]
	1999	1999	1980	1999	1980	1999	1990-1998[a]	1990-1998[a]	1990-1998[a]	1987	1997
Albania	24	–	140	175	82	84	36	1.4	3.2	2,556	2,961
Algeria	34	39	226	153	197	117	68	1.0	2.1	2,757	2,853
Angola	127	208	569	427	458	375	–	0.0[b]	1.3	2,063	2,183
Argentina	18	22	205	160	102	78	852	2.7	3.3	3,096	3,093
Armenia	14	18	158	159	85	77	27	3.0	0.7	–	2,371
Australia	5	5	178	108	85	55	1,692	2.5	8.5	3,159	3,224
Austria	4	5	197	121	92	59	2,162	3.0	8.9	3,419	3,536
Azerbaijan	16	21	262	205	127	98	9	3.8	9.7	–	2,236
Bangladesh	61	89	383	276	388	290	12	0.2	0.3	2,062	2,086
Belarus	11	14	255	335	95	115	83	4.3	12.2	–	3,226
Belgium	5	6	173	129	90	61	2,184	3.4	7.2	3,454	3,619
Benin	87	145	486	371	397	312	12	0.1	0.2	1,961	2,487
Bolivia	59	83	357	261	273	210	69	1.3	1.7	2,153	2,174
Bosnia-Herzegovina	13	18	181	166	108	90	–	0.5	1.8	–	2,266
Botswana	58	95	341	786	278	740	127	0.2	1.6	2,337	2,183
Brazil	32	40	221	256	161	139	309	1.3	3.1	2,745	2,974
Bulgaria	14	17	190	221	106	109	69	3.5	10.6	3,699	2,686
Burkina Faso	105	210	467	551	362	522	10	0.0[b]	1.4	2,181	2,121
Burundi	105	176	489	582	400	546	5	0.1	0.7	2,023	1,685
Cambodia	100	143	473	364	355	315	17	0.1	2.1	1,868	2,048
Cameroon	77	154	489	477	415	419	31	0.1	2.6	2,178	2,111
Canada	5	6	161	106	85	53	1,824	2.1	4.2	3,105	3,119
Central African Republic	96	151	540	608	424	555	9	0.1	0.9	1,914	2,016
Chad	101	189	556	438	449	383	7	0.0[b]	0.7	1,596	2,032
Chile	10	12	218	140	120	72	289	1.1	2.7	2,518	2,796
China	30	37	185	164	148	129	33	2.0	2.9	2,608[d]	2,897[d]
Colombia	23	28	237	210	162	115	227	1.1	1.5	2,341	2,597
Congo Republic	89	144	408	487	298	414	41	0.3	3.4	2,326	2,144
Costa Rica	12	14	159	116	100	68	267	0.9	1.7	2,717	2,649
Côte d'Ivoire	111	180	421	524	346	497	29	0.1	0.8	2,677	2,610
Croatia	8	9	233	194	106	76	428	2.0	5.9	–	2,445
Cuba	7	8	135	123	94	78	83	5.3	5.1	3,125	2,480
Czech Republic	5	5	225	173	102	81	392	3.0	8.7	–	3,244
Democratic Republic of the Congo (formerly Zaire)	85	161	–	515	–	482	–	0.1	1.4	2,132	1,755
Denmark	5	6	163	140	102	79	2,732	2.9	4.6	3,211	3,407
Dominican Republic	39	47	183	156	138	104	93	2.2	1.5	2,330	2,288
Ecuador	28	35	229	176	176	138	59	1.7	1.6	2,430	2,679
Egypt	47	61	257	193	204	168	48	1.6	2.1	3,120	3,287
El Salvador	30	36	410	207	178	125	143	1.0	1.6	2,312	2,562
Eritrea	60	105	–	484	–	431	–	0.0[b]	–	–	1,622
Estonia	10	12	291	288	110	94	219	3.1	7.4	–	2,849
Ethiopia	104	180	491	567	401	523	4	0.0[b]	0.2	1,677	1,858
Finland	4	5	206	136	74	59	1,722	3.0	7.8	2,941	3,100

Table F (Continued)
World Countries: Mortality, Health, and Nutrition, 1980–1999

COUNTRY	MORTALITY						HEALTH			NUTRITION	
	Infant Mortality		Adult Mortality								
	Infant Mortality (per 1,000 live births)	Under-five Mortality (per 1,000)	Male (per 1,000)		Female (per 1,000)		Health Expenditure Per Capita ($US) 1990-1998[a]	Physicians Per 1,000 People 1990-1998[a]	Hospital Beds Per 1,000 People 1990-1998[a]	Average daily per capita supply of calories (kilocalories)[c] 1987	Average daily per capita supply of calories (kilocalories)[c] 1997
	1999	1999	1980	1999	1980	1999					
France	5	5	190	124	85	50	2,377	3.0	8.5	3,543	3,518
Gabon	84	133	474	386	387	344	121	0.2	3.2	2,460	2,556
The Gambia	75	110	584	411	466	349	13	0.0[b]	0.6	2,498	2,350
Georgia	15	20	210	192	94	81	14	3.8	4.8	–	2,614
Germany	5	5	177	131	90	66	2,769	3.5	9.3	3,478	3,382
Ghana	57	109	400	316	334	272	19	–	1.5	1,979	2,611
Greece	6	7	134	114	86	61	957	4.0	5.0	3,481	3,649
Guatemala	40	52	336	288	266	186	78	0.9	1.0	2,384	2,339
Guinea	96	167	589	442	507	438	19	0.2	0.6	2,060	2,232
Guinea-Bissau	127	214	535	474	517	421	–	0.2	1.5	2,378	2,430
Haiti	70	118	348	438	275	344	21	0.2	0.7	1,848	1,869
Honduras	34	46	306	184	237	113	74	0.8	1.1	2,206	2,403
Hong Kong, China	3	5	150	106	87	54	1,134	1.3	–	–	–
Hungary	8	10	270	–	130	–	290	3.5	8.3	3,768	3,313
India	71	90	261	218	279	206	20	0.4	0.8	2,228	2,496
Indonesia	42	52	368	235	308	183	8	0.2	0.7	2,458	2,886
Iran	26	33	221	156	190	139	128	0.8	1.6	2,659	2,836
Iraq	101	128	207	196	191	169	–	0.6	1.5	3,418	2,619
Ireland	6	7	175	124	103	71	1,428	2.2	3.7	3,623	3,565
Israel	6	8	138	110	85	67	1,607	4.6	6.0	3,075	3,278
Italy	5	6	163	116	80	54	1,701	5.9	6.5	3,512	3,507
Jamaica	20	24	186	137	121	84	157	1.3	2.1	2,630	2,553
Japan	4	4	129	97	70	45	2,284	1.9	16.5	2,870	2,932
Jordan	26	31	–	156	–	118	123	1.7	1.8	2,780	3,014
Kazakhstan	22	28	312	380	140	166	86	3.5	8.5	–	3,085
Kenya	76	118	417	591	339	546	31	0.1	1.6	2,025	1,977
Korea, North	58	93	270	311	156	208	–	–	–	2,509	1,837
Korea, South	8	9	270	198	156	93	349	1.3	5.1	3,110	3,155
Kuwait	11	13	172	122	116	64	551	1.9	2.8	3,021	3,096
Kyrgyzstan	26	38	296	300	131	138	15	3.1	9.5	–	2,447
Laos	93	143	531	376	439	317	7	0.2	2.6	2,102	2,108
Latvia	14	18	281	297	106	98	167	3.4	10.3	–	2,864
Lebanon	26	32	241	175	181	132	361	2.3	2.7	3,040	3,277
Lesotho	92	141	371	518	279	486	–	0.1	–	2,216	2,244
Libya	22	28	276	183	218	125	–	1.3	4.3	3,308	3,289
Lithuania	9	12	243	261	92	86	183	3.9	9.6	–	3,261
Macedonia	16	17	–	160	–	102	113	2.3	5.2	–	2,664
Madagascar	90	149	353	327	278	287	5	0.3	0.9	2,292	2,022
Malawi	132	227	429	548	349	541	10	0.0[b]	1.3	2,027	2,043
Malaysia	8	10	230	183	149	111	81	0.5	2.0	2,616	2,977
Mali	120	223	454	470	362	406	11	0.1	0.2	1,967	2,030
Mauritania	88	142	505	346	416	294	19	0.1	0.7	2,509	2,622
Mauritius	19	23	277	207	181	113	120	0.9	3.1	–	–
Mexico	29	36	205	166	121	104	202	1.6	1.1	3,022	3,097
Moldova	17	22	289	310	173	173	33	3.6	12.1	–	2,567

-155-

COUNTRY	MORTALITY						HEALTH			NUTRITION	
	Infant Mortality		Adult Mortality								
	Infant Mortality (per 1,000 live births)	Under-five Mortality (per 1,000)	Male (per 1,000)		Female (per 1,000)		Health Expenditure Per Capita ($US) 1990-1998[a]	Physicians Per 1,000 People 1990-1998[a]	Hospital Beds Per 1,000 People 1990-1998[a]	Average daily per capita supply of calories (kilocalories)[c] 1987	Average daily per capita supply of calories (kilocalories)[c] 1997
	1999	1999	1980	1999	1980	1999					
Mongolia	58	73	320	199	273	168	24	2.6	11.5	2,034	1,917
Morocco	48	62	264	199	207	145	49	0.4	1.0	3,047	3,078
Mozambique	131	203	468	580	361	514	8	–	0.9	1,785	1,832
Myanmar (Burma)	77	120	284	278	313	228	102	0.3	0.6	2,697	2,862
Namibia	63	108	427	524	366	475	143	0.2	–	2,199	2,183
Nepal	75	109	376	264	395	275	11	0.0[b]	0.2	2,144	2,366
Netherlands	5	5	133	112	74	62	2,140	2.6	11.3	3,076	3,284
New Zealand	5	6	177	126	91	66	1,128	2.3	6.2	3,201	3,395
Nicaragua	34	43	277	200	189	137	54	0.8	1.5	2,330	2,186
Niger	116	252	562	468	453	374	5	0.0[b]	0.1	2,033	2,097
Nigeria	83	151	535	444	453	390	30	0.2	1.7	2,103	2,735
Norway	4	4	144	111	71	58	2,953	2.5	14.7	3,304	3,357
Oman	17	24	389	139	326	103	–	1.3	2.2	–	–
Pakistan	90	126	283	186	291	153	18	0.6	0.7	2,224	2,476
Panama	20	25	–	140	–	83	246	1.7	2.2	2,302	2,430
Papua New Guinea	58	77	514	369	478	330	27	0.1	4.0	2,137	2,224
Paraguay	24	27	198	185	144	130	86	1.1	1.3	2,564	2,566
Peru	39	48	287	199	229	140	141	0.9	1.5	2,276	2,302
Philippines	31	41	323	193	259	146	33	0.1	1.1	2,244	2,366
Poland	9	10	254	227	105	88	264	2.3	5.3	3,441	3,366
Portugal	6	6	199	152	95	70	803	3.1	4.0	3,400	3,667
Puerto Rico	10	–	159	152	78	58	–	1.8	3.3	–	–
Romania	20	24	216	262	119	119	63	1.8	7.6	2,944	3,253
Russia	16	20	341	382	120	138	133	4.6	12.1	–	2,904
Rwanda	123	203	503	604	409	566	10	0.0[b]	1.7	2,042	2,057
Saudi Arabia	19	25	283	160	241	129	611	1.7	2.3	2,488	2,783
Senegal	67	124	586	459	516	389	23	0.1	0.4	2,104	2,418
Sierra Leone	168	283	540	544	527	483	8	–	–	2,126	2,035
Singapore	3	4	199	130	115	72	841	1.4	3.6	–	–
Slovakia	8	10	226	206	105	87	285	3.0	7.5	–	2,984
Slovenia	5	6	250	165	105	72	746	2.1	5.7	–	3,101
South Africa	62	76	–	601	–	533	230	0.6	–	2,976	2,990
Spain	5	6	144	127	69	55	1,043	4.2	3.9	3,150	3,310
Sri Lanka	15	19	210	150	152	96	26	0.2	2.7	2,253	2,302
Sudan	67	109	537	384	462	338	126	0.1	1.1	2,208	2,395
Sweden	4	4	142	1,010	76	57	2,146	3.1	3.8	2,898	3,194
Switzerland	5	5	145	104	70	49	3,835	1.9	18.1	3,358	3,223
Syria	26	30	–	202	–	135	116	1.3	1.4	3,166	3,352
Tajikistan	20	34	190	232	129	140	13	2.1	8.8	–	2,001
Tanzania	95	152	451	542	370	500	8	0.0[b]	0.9	2,288	1,995
Thailand	28	33	280	240	210	147	112	0.4	2.0	2,133	2,360
Togo	77	143	457	478	375	435	8	0.1	1.5	1,946	2,469
Trinidad and Tobago	16	20	234	180	166	132	204	0.8	5.1	2,975	2,661
Tunisia	24	30	227	159	224	133	108	0.7	1.7	3,067	3,283
Turkey	36	45	–	177	–	145	177	1.2	2.5	3,496	3,525

COUNTRY	MORTALITY						HEALTH			NUTRITION	
	Infant Mortality		Adult Mortality								
	Infant Mortality (per 1,000 live births)	Under-five Mortality (per 1,000)	Male (per 1,000)		Female (per 1,000)		Health Expenditure Per Capita ($US) 1990-1998[a]	Physicians Per 1,000 People 1990-1998[a]	Hospital Beds Per 1,000 People 1990-1998[a]	Average daily per capita supply of calories (kilocalories)[c] 1987	Average daily per capita supply of calories (kilocalories)[c] 1997
	1999	1999	1980	1999	1980	1999					
Turkmenistan	33	45	263	281	154	158	31	0.2	11.5	–	2,306
Uganda	88	162	463	597	395	590	18	0.0[b]	0.9	2,113	2,085
Ukraine	14	17	282	346	112	134	42	4.5	11.8	–	2,795
United Arab Emirates	8	9	153	125	106	92	1,428	1.8	2.6	3,038	3,390
United Kingdom	6	6	160	119	96	66	1,597	1.7	4.2	3,215	3,276
United States	7	8	194	143	102	78	4,108	2.7	3.7	3,430	3,699
Uruguay	15	17	176	168	91	74	621	3.7	4.4	2,613	2,816
Uzbekistan	22	29	219	227	116	125	–	3.3	8.3	–	2,433
Venezuela	20	23	219	155	123	88	171	2.4	1.5	2,602	2,321
Vietnam	37	42	262	205	204	144	18	0.6	1.7	2,193	2,484
West Bank and Gaza	23	26	–	164	–	106	81	0.5	1.2	–	–
Yemen	79	97	382	307	304	283	18	0.2	0.6	2,126	2,051
Yugoslavia (Serbia-Montenegro)	12	16	164	176	106	106	–	2.0	5.3	–	3,031
Zambia	114	187	482	607	413	597	23	0.1	–	2,017	1,970
Zimbabwe	70	118	389	569	321	526	49	0.1	0.5	2,112	2,145

Sources: World Development Indicators (World Bank, Washington, DC, 2001); *World Resources 2000–2001* (World Resources Institute, Washington, DC)

a. Data are for the most recent year available
b. Less than 0.05
c. Has replaced calories available as a percentage of need as a benchmark for nutrition.
d. Data for China include Taiwan

Table G
World Countries: Education and Literacy, 1990–1999

COUNTRY	EDUCATIONAL INPUTS					OUTCOMES							
	Public Expenditure on Education 1994–1997[b] (% of GNI[c])	Primary Pupil-Teacher Ratio 1997 (pupils per teacher)	Primary School Enrollment 1997 (% of age group)[1]	Secondary School Enrollment 1997 (% of age group)[1]	Tertiary School Enrollment 1997 (% of age group)[1]	ADULT ILLITERACY (% above age 15)				YOUTH ILLITERACY (% aged 15–24)			
						Male		Female		Male		Female	
						1990	1999	1990	1999	1990	1999	1990	1999
Albania	3.1	18	107	38	12	14	9	32	23	3	1	7	3
Algeria	5.1	27	108	63	12	32	23	59	44	13	8	32	16
Angola	–	29	–	–	–	–	–	–	–	–	–	–	–
Argentina	3.5	17	111	73	36	4	3	4	3	2	2	2	1
Armenia	2.0	19	87	90	12	1	1	4	3	0[a]	0[a]	1	0[a]
Australia	5.4	18	101	153[d]	80	–	–	–	–	–	–	–	–
Austria	5.4	12	100	103	48	–	–	–	–	–	–	–	–
Azerbaijan	3.0	20	106	77	17								
Bangladesh	2.2	–	–	–	–	54	48	77	71	45	40	68	61
Belarus	5.9	19	98	93	44	0[a]	0[a]	1	1	0[a]	0[a]	0[a]	0[a]
Belgium	3.1	12	103	146[d]	56	–	–	–	–	–	–	–	–
Benin	3.2	50	78	18	3	59	45	84	76	37	23	74	63
Bolivia	4.9	–	–	–	–	13	8	30	21	4	2	11	7
Bosnia-Herzegovina	–	–	–	–	–	–	–	–	–	–	–	–	–
Botswana	8.6	28	108	65	6	34	26	30	21	21	16	13	8
Brazil	5.1	24	125	62	15	18	15	20	15	12	10	9	6
Bulgaria	3.2	17	99	77	41	2	1	4	2	1	0[a]	1	1
Burkina Faso	1.5	47	40	–	1	75	67	92	87	64	55	86	78
Burundi	4.0	42	51	7	–	50	44	73	61	42	36	55	40
Cambodia	2.9	44	113	24	1	49	41	86	79	34	25	73	59
Cameroon	–	49	85	27	–	28	19	46	31	10	6	15	7
Canada	6.9	16	102	105	88	–	–	–	–	–	–	–	–
Central African Republic	–	–	–	–	–	53	41	79	67	34	25	61	43
Chad	1.7	67	58	10	1	63	50	81	68	42	28	62	42
Chile	3.6	30	101	75	32	6	4	6	5	2	2	2	1
China	2.3	24	123	70	6	14	9	33	25	3	1	8	4
Colombia	4.1	25	113	67	17	11	9	12	9	6	4	4	3
Congo Republic	6.1	70	114	53	–	23	13	42	27	5	2	10	4
Costa Rica	5.4	29	104	48	30	6	5	6	5	3	2	2	1
Côte d'Ivoire	5.0	41	71	25	6	56	46	77	63	40	31	59	42
Croatia	5.3	19	87	82	28	1	1	5	3	0[a]	0[a]	0[a]	0[a]
Cuba	6.7	12	106	81	12	5	3	5	4	1	0[a]	1	0[a]
Czech Republic	5.1	18	104	99	24	–	–	–	–	–	–	–	–
Democratic Republic of the Congo (formerly Zaire)	–	45	72	26	2	38	28	66	51	19	12	42	27
Denmark	8.1	10	102	121	48	–	–	–	–	–	–	–	–
Dominican Republic	2.3	28	94	54	23	20	17	21	17	13	10	12	9
Ecuador	3.5	25	127	50	–	10	7	15	11	4	3	5	4
Egypt	4.8	23	101	78	20	40	34	66	57	29	24	49	38
El Salvador	2.5	33	97	37	18	24	19	31	24	15	11	17	13
Eritrea	1.8	44	53	20	1	42	33	72	61	27	20	54	39
Estonia	7.2	17	94	104	42	–	–	–	–	–	–	–	–
Ethiopia	4.0	43	43	12	1	64	57	80	68	52	46	64	48
Finland	7.5	18	99	118	74	–	–	–	–	–	–	–	–
France	6.0	19	105	111	51	–	–	–	–	–	–	–	–
Gabon	2.9	56	162	56	8	–	–	–	–	–	–	–	–
The Gambia	4.9	30	77	25	2	68	57	80	72	49	36	66	52

COUNTRY	EDUCATIONAL INPUTS					OUTCOMES							
	Public Expenditure on Education 1994–1997[b] (% of GNI[c])	Primary Pupil-Teacher Ratio 1997 (pupils per teacher)	Primary School Enrollment 1997 (% of age roup)[1]	Secondary School Enrollment 1997 (% of age group)[1]	Tertiary School Enrollment 1997 (% of age group)[1]	ADULT ILLITERACY (% above age 15)				YOUTH ILLITERACY (% aged 15–24)			
						Male		Female		Male		Female	
						1990	1999	1990	1999	1990	1999	1990	1999
Georgia	5.2	18	88	77	42	–	–	–	–	–	–	–	–
Germany	4.8	17	104	104	47	–	–	–	–	–	–	–	–
Ghana	4.2	33	79	–	–	30	21	53	39	12	7	25	13
Greece	3.1	14	93	95	47	2	2	8	4	1	0[a]	0[a]	0[a]
Guatemala	1.7	35	88	26	9	31	24	47	40	20	15	34	28
Guinea	1.9	49	54	14	1	–	–	–	–	–	–	–	–
Guinea-Bissau	–	–	62	–	–	54	42	89	82	30	19	79	68
Haiti	–	35	–	–	–	57	49	63	53	44	37	46	36
Honduras	3.6	35	111	–	10	31	26	32	26	23	19	20	16
Hong Kong, China	2.9	–	94	73	–	5	4	16	10.0	2	1	1	0[a]
Hungary	4.6	12	103	98	24	1	1	1	1	0[a]	0[a]	0[a]	0[a]
India	3.2	62	100	49	7	38	32	64	56	27	21	46	36
Indonesia	1.4	22	113	56	11	13	9	27	19	3	2	7	3
Iran	4.0	30	98	77	18	27	17	45	31	8	4	18	9
Iraq	–	20	85	42	–	43	35	67	55	29	23	48	34
Ireland	6.0	22	105	118	41	–	–	–	–	–	–	–	–
Israel	7.6	14	98	88	41	3	2	9	6	1	0a	2	0a
Italy	4.9	11	101	95	47	2	1	3	2	0[a]	0[a]	0[a]	0[a]
Jamaica	7.4	31	100	–	8	22	18	14	10	13	10	5	3
Japan	3.6	19	101	103	41	–	–	–	–	–	–	–	–
Jordan	6.8	21	71	57	18	10	6	28	17	2	1	4	0a
Kazakhstan	4.4	18	98	87	33	–	–	–	–	–	–	–	–
Kenya	6.5	31	85	24	–	19	12	39	25	7	4	13	6
Korea, North	–	–	–	–	–	–	–	–	–	–	–	–	–
Korea, South	3.7	31	94	102	68	2	1	7	4.0	0[a]	0[a]	0[a]	0[a]
Kuwait	5.0	14	77	65	19	20	16	27	21	12	9	13	7
Kyrgyzstan	5.3	20	104	79	12	–	–	–	–	–	–	–	–
Laos	2.1	30	112	29	3	47	37	80	68	28	18	62	44
Latvia	6.3	13	96	84	33	0[a]	0[a]	0[a]	0[a]	0[a]	0[a]	0[a]	0[a]
Lebanon	2.5	–	111	81	27	12	8	27	20	5	3	11	7
Lesotho	8.4	46	108	31	2	35	28	11	7	23	18	3	2
Libya	–	–	–	–	–	17	10	49	33	1	0a	17	7
Lithuania	5.4	16	98	86	31	1	0[a]	1	1	0[a]	0[a]	0[a]	0[a]
Macedonia	5.1	–	99	63	20	–	–	–	–	–	–	–	–
Madagascar	1.9	47	92	16	2	34	27	50	41	22	17	33	24
Malawi	5.4	59	134	17	1	31	26	64	55	24	20	49	40
Malaysia	4.9	19	101	64	12	13	9	25	17	5	3	6	3
Mali	2.2	80	49	13	1	67	53	81	67	46	29	63	42
Mauritania	5.1	50	79	16	4	53	48	74	69	44	39	65	60
Mauritius	4.6	24	106	65	6	15	12	25	19	9	7	9	6
Mexico	4.9	28	114	64	16	10	7	15	11	4	3	6	4
Moldova	10.6	23	97	81	27	1	1	4	2	0[a]	0[a]	0[a]	0[a]
Mongolia	5.7	31	88	56	17	35	27	59	48	21	16	39	27
Morocco	5.0	28	86	39	11	47	39	75	65	32	24	58	43
Mozambique	–	58	60	7	1	51	41	82	72	34	26	68	55
Myanmar (Burma)	1.2	46	121	30	5	13	11	26	20	10	9	14	10
Namibia	9.1	–	131	62	8	23	18	28	20	14	10	11	7
Nepal	3.2	39	113	42	5	53	42	86	77	34	25	73	59

Table G *(Continued)*
World Countries: Education and Literacy, 1990–1999

COUNTRY	EDUCATIONAL INPUTS					OUTCOMES							
	Public Expenditure on Education 1994–1997[b] (% of GNI[c])	Primary Pupil-Teacher Ratio 1997 (pupils per teacher)	Primary School Enrollment 1997 (% of age roup)[1]	Secondary School Enrollment 1997 (% of age group)[1]	Tertiary School Enrollment 1997 (% of age group)[1]	ADULT ILLITERACY (% above age 15)				YOUTH ILLITERACY (% aged 15–24)			
						Male		Female		Male		Female	
						1990	1999	1990	1999	1990	1999	1990	1999
Netherlands	5.1	14	108	132[d]	47	–	–	–	–	–	–	–	–
New Zealand	7.3	18	101	113	63	–	–	–	–	–	–	–	–
Nicaragua	3.9	36	102	55	12	36	33	34	30	32	29	28	24
Niger	2.3	41	29	7	–	82	77	95	92	75	68	91	87
Nigeria	0.7	34	98	33	–	41	29	62	46	19	11	34	18
Norway	7.4	7	100	119	62	–	–	–	–	–	–	–	–
Oman	4.5	26	76	67	8	33	21	62	40	5	1	25	5
Pakistan	2.7	40	–	–	–	50	41	79	70	36	24	67	52
Panama	5.1	–	106	69	32	10	8	12	9	4	3	5	4
Papua New Guinea	–	38	80	14	3	34	29	52	44	25	20	38	30
Paraguay	4.0	21	111	47	10	8	6	12	8	4	3	5	3
Peru	2.9	27	123	73	26	8	6	21	15	3	2	8	5
Philippines	3.4	35	117	78	29	7	5	8	5	3	2	3	1
Poland	7.5	15	96	98	25	0[a]	0[a]	1	0[a]	0[a]	0[a]	0[a]	0[a]
Portugal	5.8	12	128	111[d]	39	9	6	16	11	1	0[a]	0[a]	0[a]
Puerto Rico	–	–	–	–	–	9	7	9	6	5	3	3	2
Romania	3.6	20	104	78	23	1	1	5	3	1	1	1	0[a]
Russia	3.5	20	107	–	43	0[a]	0[a]	1	1	0[a]	0[a]	0[a]	0[a]
Rwanda	–	–	–	–	–	37	27	56	41	22	15	33	20
Saudi Arabia	7.5	13	76	61	16	22	17	49	34	9	5	21	10
Senegal	3.7	58	71	16	3	62	54	81	73	50	41	70	59
Sierra Leone	–	–	–	–	–	–	–	–	–	–	–	–	–
Singapore	3.0	25	94	74	39	6	4	17	12	1	0[a]	1	0[a]
Slovakia	5.0	20	102	94	22	–	–	–	–	–	–	–	–
Slovenia	5.7	14	98	92	36	0[a]	0[a]	1	0[a]	0[a]	0[a]	0[a]	0[a]
South Africa	7.9	45	133	95	19	18	14	20	16	11	9	12	9
Spain	5.0	15	107	120	51	2	2	5	3	0[a]	0[a]	0[a]	0[a]
Sri Lanka	3.4	28	109	75	5	7	6	15	11	4	3	6	4
Sudan	0.9	29	51	21	–	39	31	68	55	24	18	46	30
Sweden	8.3	12	107	140[d]	50	–	–	–	–	–	–	–	–
Switzerland	5.4	12	97	100	33	–	–	–	–	–	–	–	–
Syria	3.1	23	101	43	16	18	12	53	41	8	5	33	22
Tajikistan	2.2	24	95	78	20	1	1	3	1	0[a]	0[a]	0[a]	0[a]
Tanzania	–	37	67	6	1	23	16	49	34	10	7	22	12
Thailand	4.8	–	89	59	22	5	3	11	7	1	1	2	2
Togo	4.5	46	120	27	4	36	26	71	60	19	13	55	42
Trinidad and Tobago	3.6	25	99	74	8	6	5	11	8	3	2	4	3
Tunisia	7.7	24	118	64	14	28	20	54	41	7	3	25	12
Turkey	2.2	24	107	58	21	11	7	33	24	3	1	12	6
Turkmenistan	–	–	–	–	–	–	–	–	–	–	–	–	–
Uganda	2.6	35	74	12	2	31	23	57	45	20	15	39	29
Ukraine	7.3	21	–	–	42	0[a]	0[a]	1	1	0[a]	0[a]	0[a]	0[a]
United Arab Emirates	1.8	16	89	80	12	29	26	30	22	19	15	11	6
United Kingdom	5.3	19	116	129[d]	52	–	–	–	–	–	–	–	–
United States	5.4	16	102	97	81	–	–	–	–	–	–	–	–
Uruguay	3.3	20	109	85	30	4	3	3	2	1	1	1	0[a]
Uzbekistan	7.7	21	78	94	–	10	7	23	16	3	2	8	5

Table G *(Continued)*
World Countries: Education and Literacy, 1990–1999

COUNTRY	EDUCATIONAL INPUTS					OUTCOMES							
	Public Expenditure on Education 1994–1997[b] (% of GNI[c])	Primary Pupil-Teacher Ratio 1997 (pupils per teacher)	Primary School Enrollment 1997 (% of age roup)[1]	Secondary School Enrollment 1997 (% of age group)[1]	Tertiary School Enrollment 1997 (% of age group)[1]	ADULT ILLITERACY (% above age 15)				YOUTH ILLITERACY (% aged 15–24)			
						Male		Female		Male		Female	
						1990	1999	1990	1999	1990	1999	1990	1999
Venezuela	5.2	21	91	40	–	10	7	12	8	5	3	3	2
Vietnam	3.0	33	114	57	7	6	5	13	9	5	3	5	3
West Bank and Gaza	–	–	–	–	–	–	–	–	–	–	–	–	–
Yemen	7.0	30	70	34	4	45	33	87	76	26	18	75	56
Yugoslavia (Serbia-Montenegro)	–	–	69	62	22	–	–	–	–	–	–	–	–
Zambia	2.2	39	89	27	3	22	15	41	30	14	10	24	15
Zimbabwe	–	39	112	50	7	13	8	25	16	3	2	9	5

1. Large numbers of nontraditional students outside of age group may increase percentages enrolled to above 100 percent in certain countries.
a. Less than 0.5
b. Data are for the most recent year available
c. Gross National Income (GNI) has replaced GNP in the World Bank Atlas Method's estimate of national income
d. Includes training for the unemployed.

Source: World Development Indicators (World Bank, Washington, DC, 1998–1999).

Table H
World Countries: Agricultural Operations, 1996–2000

COUNTRY			AGRICULTURAL INPUTS				AGRICULTURAL OUTPUT AND PRODUCTIVITY			
			Agricultural Machinery							
	Arable Land (hectares per capita)	Irrigated Land (% of cropland)	Land Under Cereal Production (thousand hectares)	Fertilizer Consumption (hundreds of grams per hectare of arable land)	Tractors per Thousand Agricultural Workers	Tractors per Hundred Hectares of Agricultural Land	Crop Production Index (1989–91=100)	Food Production Index (1989–91=100)	Livestock Production Index (1989–91=100)	Agricultural Productivity (agricultural value added per worker in 1995$)
	1996-1998	1996-1998	1998-2000	1996-1998	1996-1998	1996-1998	1998-2000	1998-2000	1998-2000	1997-1999
Albania	0.17	48.5	214	212	10	141	–	–	–	1,934
Algeria	0.26	6.9	2,478	101	39	121	125.8	131.1	125.3	1,876
Angola	0.26	2.1	888	15	3	34	148.1	144	135.6	126
Argentina	0.70	5.7	10,261	330	190	112	159.5	137.9	105.6	9,983
Armenia	0.13	51.2	183	54	70	354	97.4	78	64.8	5,180
Australia	2.80	4.6	16,197	406	704	61	163.9	137.7	111.4	31,432
Austria	0.17	0.3	817	1,836	1,617	2,527	102.4	106	107.8	28,410
Azerbaijan	0.21	75.1	577	128	34	195	45.8	63.1	74.1	837
Bangladesh	0.06	44.8	11,227	1,460	0	7	110.4	114.5	136.2	292
Belarus	0.61	1.8	2,295	1,371	121	158	86.3	61.7	60.9	3,744
Belgium	0.08	4.2	333	3,834	1,186	1,326	138.6	114.7	114.3	48,529
Benin	0.29	0.6	836	212	0	1	175.6	152.4	123.6	558
Bolivia	0.24	6.2	776	53	4	31	151	137.3	127.1	1,054
Bosnia-Herzegovina	0.14	0.3	370	326	270	580	–	–	–	8,471
Botswana	0.22	0.3	87	110	20	175	70.6	97.6	101.3	681
Brazil	0.33	4.1	16,908	1,020	58	151	121.6	136.8	149.9	4,300
Bulgaria	0.51	17.9	1,938	417	68	58	67.8	72.2	64	6,007
Burkina Faso	0.32	0.7	2,999	115	0	6	146.1	136	136.2	162
Burundi	0.12	6.7	202	25	0	2	89.7	90.3	81.6	140
Cambodia	0.33	7.1	2,010	26	0	3	136	139.2	149.1	406
Cameroon	0.43	0.5	1045	63	0	1	124.6	126.2	119.6	1,072
Canada	1.52	1.6	17,444	591	1,678	156	128.9	128.8	132.8	34,922
Central African Republic	0.56	–	156	2	0	0	128.2	132.8	129.2	460
Chad	0.49	0.6	1,897	35	0	0	177.9	155.2	114	220
Chile	0.14	78.4	574	2,225	52	256	126.3	133.1	143.7	4,997
China	0.10	38.3	90,212	2,860	1	56	141.5	168.5	209.9	316
Colombia	0.05	20.7	1,041	2,826	6	105	100	118.5	125.8	3,454
Congo Republic	0.06	0.5	3	255	1	41	114.5	118	129	498
Costa Rica	0.06	24.9	91	7,972	22	311	133.6	132.3	120.9	4,973
Côte d'Ivoire	0.21	1.0	1,621	333	1	13	132.4	131.2	127.6	1,104
Croatia	0.30	0.2	608	1,606	13	21	87.2	72.8	54.4	7,123
Cuba	0.33	19.4	209	580	96	214	54.5	58.2	63.6	–
Czech Republic	0.30	0.7	1,614	1,048	167	275	90.3	83.6	76.6	5,091
Democratic Republic of Congo (formerly Zaire)	0.14	0.1	2,118	3	0	4	89	92.1	103.7	283
Denmark	0.44	20.2	1,509	1,827	1,133	597	95.6	107.9	118.5	52,809
Dominican Republic	0.13	17.1	149	937	4	23	89.2	100.7	121.1	2,710
Ecuador	0.13	28.8	861	655	7	57	122.4	134.1	150.1	1,789
Egypt	0.05	99.8	2,631	3,858	11	318	141.4	149.5	156	1,222
El Salvador	0.10	4.4	457	1,619	4	61	108.5	121.2	127.9	1,690
Eritrea	0.11	5.3	412	140	0	11	174.1	133.9	105.7	–
Estonia	0.77	0.4	357	247	519	451	66.8	45.2	39.7	3,646
Ethiopia	0.17	1.8	6,852	159	0	3	121.6	119.9	116.2	144
Finland	0.42	3.0	1,160	1,442	1,196	907	86.4	89.7	93.3	36,384
France	0.31	9.7	9,141	2,708	1,256	698	112.1	107.6	105	50,171
Gabon	0.28	3.0	18	8	7	46	118.2	113.9	118.3	1,889
The Gambia	0.16	1.0	122	59	0	2	114.1	115.9	117.3	222

-162-

Table H *(Continued)*
World Countries: Agricultural Operations, 1996–2000

COUNTRY			AGRICULTURAL INPUTS				AGRICULTURAL OUTPUT AND PRODUCTIVITY			
			Agricultural Machinery							
	Arable Land (hectares per capita)	Irrigated Land (% of cropland)	Land Under Cereal Production (thousand hectares)	Fertilizer Consumption (hundreds of grams per hectare of arable land)	Tractors per Thousand Agricultural Workers	Tractors per Hundred Hectares of Agricultural Land	Crop Production Index (1989–91=100)	Food Production Index (1989–91=100)	Livestock Production Index (1989–91=100)	Agricultural Productivity (agricultural value added per worker in 1995$)
	1996-1998	1996-1998	1998-2000	1996-1998	1996-1998	1996-1998	1998-2000	1998-2000	1998-2000	1997-1999
Georgia	0.14	43.8	372	442	31	212	61.1	78.3	83.5	1,952
Germany	0.14	4.0	6,904	2,423	960	950	114	94.4	86	28,924
Ghana	0.19	0.2	1,317	54	1	11	174.1	163.5	101.8	554
Greece	0.27	35.2	1,286	1,811	289	843	107	100.1	97.8	12,711
Guatemala	0.13	6.6	686	1,604	2	32	120.9	124.5	129	2,099
Guinea	0.13	6.4	743	35	0	6	142.4	143.1	139.6	284
Guinea-Bissau	0.26	4.9	134	13	0	1	119.4	119.9	120.5	306
Haiti	0.07	8.2	448	165	0	2	86.6	95.4	128.8	392
Honduras	0.28	3.7	500	720	7	30	116.5	112.4	130.1	1,008
Hong Kong, China	0.00	31	0	–	0	7	59.3	49.5	44.6	–
Hungary	0.47	4.2	2,598	929	162	191	79.3	74.7	69.2	4,860
India	0.17	33.6	101,190	976	6	91	122.1	124.6	133.5	395
Indonesia	0.09	15.5	15,298	1,434	1	39	117.4	119.1	125.4	742
Iran	0.28	39.8	7,954	691	38	136	151.3	151.4	146.1	3,679
Iraq	0.24	63.6	2,927	702	74	95	82.9	77.7	65	–
Ireland	0.37	–	283	5,135	993	1,239	110	110.3	110.6	–
Israel	0.06	45.5	57	3,423	323	699	105.6	110.7	116	–
Italy	0.14	24.5	4,128	2,169	950	1,774	105.6	104.8	104.2	23,906
Jamaica	0.07	9.1	2	1,353	11	177	122.9	120.8	119.6	1,229
Japan	0.04	54.6	2,054	3,278	681	4,830	88.3	92.4	94.2	30,620
Jordan	0.06	19.4	76	890	30	188	116.7	139.6	198.8	1,434
Kazakhstan	1.98	7.3	11,466	29	70	34	65.6	58.9	45.3	1,414
Kenya	0.14	1.5	1,851	357	1	36	107.1	104.5	104.1	226
Korea, North	0.07	73	1,330	826	19	441	–	–	–	–
Korea, South	0.04	60.5	1,173	5,358	50	779	106.4	112.3	150.3	12,252
Kuwait	0.00	81.0	1	2,944	11	129	163.8	185.7	180.5	–
Kyrgyzstan	0.29	75.4	645	293	36	142	129.1	114.5	78.7	3,430
Laos	0.17	18.9	708	91	0	11	139.5	144.3	161.6	558
Latvia	0.72	1.1	434	191	330	326	70.2	45.8	35.9	2,523
Lebanon	0.04	37.6	39	3,249	111	300	137.6	142.7	161.6	28,243
Lesotho	0.16	–	178	182	6	62	115.9	99.9	88.5	544
Libya	0.35	22.2	319	320	296	187	132.9	161.6	174.4	–
Lithuania	0.79	0.3	1,046	448	294	267	74.4	66.6	58.1	3,192
Macedonia	0.30	8.5	222	747	398	902	108.7	95.8	85.1	2,141
Madagascar	0.18	35.0	1,373	45	1	14	104.2	108.5	105.7	184
Malawi	0.18	1.4	1,547	294	0	8	149.1	153.2	112.4	138
Malaysia	0.08	4.8	702	6,940	23	238	111.2	134.1	149.1	6,578
Mali	0.46	3.0	2,422	92	1	6	145.1	127.2	123.3	279
Mauritania	0.20	9.8	235	60	1	7	152.3	107.3	100.9	469
Mauritius	0.09	17.6	0	3,480	6	37	93.3	103.1	135.3	5,330
Mexico	0.27	23.8	11,061	658	20	68	121.6	128.4	134.6	1,742
Moldova	0.41	14.1	865	668	82	257	54.7	45.2	36.3	1,277
Mongolia	0.57	6.4	261	33	21	53	35.4	87.8	92.3	1,193
Morocco	0.33	12.7	5,166	357	10	47	95.3	100.1	106.7	1,651
Mozambique	0.19	3.2	1,816	21	1	18	143.2	131	103.2	136
Myanmar (Burma)	0.22	15.5	6,302	182	0	9	152.8	148.6	143	–
Namibia	0.50	0.9	298	–	11	39	111.6	97.3	95.5	1,248
Nepal	0.13	38.2	3,283	383	0	16	120.8	121.4	123.5	189

Table H (Continued)
World Countries: Agricultural Operations, 1996–2000

COUNTRY			AGRICULTURAL INPUTS				AGRICULTURAL OUTPUT AND PRODUCTIVITY			
			Agricultural Machinery							
	Arable Land (hectares per capita)	Irrigated Land (% of cropland)	Land Under Cereal Production (thousand hectares)	Fertilizer Consumption (hundreds of grams per hectare of arable land)	Tractors per Thousand Agricultural Workers	Tractors per Hundred Hectares of Agricultural Land	Crop Production Index (1989–91=100)	Food Production Index (1989–91=100)	Livestock Production Index (1989–91=100)	Agricultural Productivity (agricultural value added per worker in 1995$)
	1996-1998	1996-1998	1998-2000	1996-1998	1996-1998	1996-1998	1998-2000	1998-2000	1998-2000	1997-1999
Netherlands	0.06	60.4	203	5,547	603	1,789	107.5	100.3	99.9	51,594
New Zealand	0.41	8.7	130	4,218	437	488	134.1	125.4	116.4	27,083
Nicaragua	0.53	3.2	378	198	7	11	134.2	131.2	116.7	1,919
Niger	0.51	1.3	7,532	7	0	0	151.3	140.1	120.4	205
Nigeria	0.24	0.8	18,440	59	2	10	155.4	152.2	125.6	641
Norway	0.21	–	334	2,203	1,306	1,584	87.8	96.3	100.1	32,848
Oman	0.01	98.4	3	3,779	1	94	113.8	114.9	104	–
Pakistan	0.17	81.2	12,489	1,178	12	150	125.4	143.3	150.4	626
Panama	0.18	4.9	188	753	20	100	97.2	107.2	121.9	2,580
Papua New Guinea	0.01	–	2	2,283	1	193	109.7	113.1	136.6	808
Paraguay	0.43	2.9	554	233	24	75	110.4	132.8	129.4	3,512
Peru	0.15	28.9	1091	498	3	25	162.9	161.7	150.5	1,569
Philippines	0.08	15.6	6,299	1,283	1	21	112.9	128.4	173.5	1,342
Poland	0.36	0.7	8,577	1,148	285	936	85.8	88.7	87.2	1,554
Portugal	0.19	24.0	590	1,288	219	802	85	94.1	117.1	7,621
Puerto Rico	0.01	51.3	0	–	–	–	62.9	81.7	87.5	–
Romania	0.41	30.6	5,496	392	88	176	90.3	93.4	87.9	3,228
Russia	0.86	3.8	46,809	112	101	72	66.8	60.8	51.6	2,282
Rwanda	0.10	0.4	219	4	0	1	88.1	91.5	108.8	234
Saudi Arabia	0.19	42.3	588	865	12	26	94.3	88.5	147.7	10,930
Senegal	0.25	3.1	1,218	108	0	2	102.9	114.2	138	307
Sierra Leone	0.10	5.4	279	62	0	2	81	85.3	109	379
Singapore	0.00	–	–	25,183	20	650	48.2	41.5	39.5	42,903
Slovakia	0.27	11.2	898	751	92	175	–	–	–	3,491
Slovenia	0.12	0.7	95	3,086	3,604	4,311	93.5	104.6	108.7	30,136
South Africa	0.36	8.5	4,742	534	60	68	105.3	103.3	96.5	4,070
Spain	0.36	19.0	6,652	1,474	576	583	107.6	110.7	122.5	21,687
Sri Lanka	0.05	32.1	890	2,517	2	81	113.9	115.7	133.5	734
Sudan	0.60	11.5	7,973	42	2	6	162	155.9	146.3	–
Sweden	0.32	–	1,221	1,068	989	590	93.8	101.1	104.1	34,285
Switzerland	0.06	5.6	188	4,529	635	2,675	99.4	97.1	94.1	–
Syria	0.32	21.3	3,075	737	66	188	158.4	150.5	133	–
Tajikistan	0.13	80.9	405	841	37	395	62.5	58.1	36.9	–
Tanzania	0.12	3.3	3,201	86	1	20	100.3	105.6	118.2	188
Thailand	0.28	23.1	11,425	925	10	123	112.9	113.1	127	939
Togo	0.52	0.3	765	76	0	0	148.1	140.7	128.8	543
Trinidad and Tobago	0.06	2.5	4	1,406	53	358	101.4	105	100.9	2,463
Tunisia	0.31	7.7	1,273	365	38	121	116.6	127	147.2	3,047
Turkey	0.42	14.8	13,655	751	00	330	114.7	113	100.1	1,858
Turkmenistan	0.35	96.2	612	982	81	307	80.3	132.1	134.4	856
Uganda	0.25	0.1	1,382	2	1	9	118.8	116.5	119.8	350
Ukraine	0.65	7.3	12,040	165	87	109	58.2	49.1	45.8	1,383
United Arab Emirates	0.02	88.9	1	7,775	4	69	275.3	250.6	170.8	–
United Kingdom	0.11	1.7	3,290	3,588	898	800	102.8	98.7	97.9	34,730
United States	0.65	12.0	59,953	1,135	1,515	271	121.9	122.9	120	–
Uruguay	0.39	13.5	578	1,102	173	262	149.7	137.8	122.2	8,679
Uzbekistan	0.19	88.3	1,455	1,623	59	380	88.4	116.5	115.8	1,621

Table H *(Continued)*
World Countries: Agricultural Operations, 1996–2000

COUNTRY			AGRICULTURAL INPUTS				AGRICULTURAL OUTPUT AND PRODUCTIVITY			
			Agricultural Machinery							
	Arable Land (hectares per capita)	Irrigated Land (% of cropland)	Land Under Cereal Production (thousand hectares)	Fertilizer Consumption (hundreds of grams per hectare of arable land)	Tractors per Thousand Agricultural Workers	Tractors per Hundred Hectares of Agricultural Land	Crop Production Index (1989-91=100)	Food Production Index (1989–91=100)	Livestock Production Index (1989–91=100)	Agricultural Productivity (agricultural value added per worker in 1995$)
	1996-1998	1996-1998	1998-2000	1996-1998	1996-1998	1996-1998	1998-2000	1998-2000	1998-2000	1997-1999
Venezuela	0.12	15.4	668	1,058	59	185	105.7	117.3	118.8	5,125
Vietnam	0.07	42.0	8,228	2,933	4	206	158.5	152.2	163.7	236
West Bank and Gaza	–	–	–	–	–	–	–	–	–	–
Yemen	0.09	31.0	694	111	2	40	129.1	130.9	138.6	355
Yugoslavia (Serbia-Montenegro)	–	–	–	–	–	–	–	–	–	–
Zambia	0.56	0.9	693	94	2	11	88.5	99.7	113.2	218
Zimbabwe	0.28	3.5	1,784	537	6	69	121.2	108.2	113.7	369

Source: World Development Indicators (World Bank, Washington, DC, 2001).

Table I
Land Use and Deforestation, 1980–2000

COUNTRY	LAND AREA	RURAL POPULATION DENSITY	LAND USE						FOREST AREA	AVERAGE ANNUAL DEFORESTATION
			Arable Land (% of land area)		Permanent Cropland (% of land area)		Other (% of land area)			
	(thousands of sq. km.)	(people per sq. km. of arable land)	Arable Land (% of land area)		Permanent Cropland (% of land area)		Other (% of land area)		(thousands of sq. km.)	Decline in forest area %
	1998	1998	1980	1998	1980	1998	1980	1998	2000	1990-2000
Albania	27	345	21.4	21.1	4.3	4.5	74.4	74.5	10	0.8
Algeria	2,382	159	2.9	3.2	0.3	0.2	96.8	96.6	21	-1.3
Angola	1,247	268	2.3	2.4	0.4	0.4	97.3	97.2	698	0.2
Argentina	2,737	15	9.1	9.1	0.8	0.8	90.1	90.1	346	0.8
Armenia	28	234	–	17.6	–	2.3	–	80.1	4	-1.3
Australia	7,682	5	5.7	7.0	0.0	0.0	94.2	93.0	1,581	0.0
Austria	83	205	18.6	16.9	1.2	1.0	80.2	82.1	39	-0.2
Azerbaijan	87	205	–	19.3	–	3.0	–	77.7	11	-1.3
Bangladesh	130	1,204	68.3	61.4	2.0	2.6	29.6	36.0	13	-1.3
Belarus	207	49	–	29.8	–	0.6	–	69.6	94	-3.2
Belgium[a]	33	35	23.2	24.8	0.4	0.6	76.4	74.6	7	0.2
Benin	111	207	12.2	15.4	4.0	1.4	83.8	83.3	27	2.3
Bolivia	1,084	156	1.7	1.8	0.2	0.2	98.1	98.0	531	0.3
Bosnia-Herzegovina	51	436	–	9.8	–	2.9	–	87.3	23	0.0
Botswana	567	231	0.7	0.6	0.0	0.0	99.3	99.4	124	0.9
Brazil	8,457	62	4.6	6.3	1.2	1.4	94.2	92.3	5,325	0.4
Bulgaria	111	60	34.6	38.8	3.2	2.0	62.2	59.2	37	-0.6
Burkina Faso	274	260	10.0	12.4	0.1	0.2	89.8	87.4	71	0.2
Burundi	26	779	35.8	30.0	10.1	12.9	54.0	57.2	1	9.0
Cambodia	177	263	11.3	21.0	0.4	0.6	88.3	78.4	93	0.6
Cameroon	465	127	12.7	12.8	2.2	2.6	85.1	84.6	239	0.9
Canada	9,221	15	4.9	4.9	0.0	0.0	95.0	95.0	2,446	0.0
Central African Republic	623	108	3.0	3.1	0.1	0.1	96.9	96.8	229	0.1
Chad	1,259	159	2.5	2.8	0.0	0.0	97.5	97.2	127	0.6
Chile	749	111	5.4	2.6	0.3	0.4	94.3	96.9	155	0.1
China[b]	9,327	689	10.4	13.3	0.4	1.2	89.3	85.5	1,635	-1.2
Colombia	1,039	529	3.6	2.0	1.4	2.0	95.0	96.0	496	0.4
Congo Republic	342	630	0.4	0.5	0.1	0.1	99.6	99.4	221	0.1
Costa Rica	51	824	5.5	4.4	4.4	5.5	90.1	90.1	20	0.8
Côte d'Ivoire	318	290	6.1	9.3	7.2	13.8	86.6	76.9	71	3.1
Croatia	56	133	–	26.1	–	2.3	–	71.6	18	-0.1
Cuba	110	77	23.9	33.1	6.4	7.6	69.7	59.3	23	-1.3
Czech Republic	77	84	–	40.1	–	3.0	–	56.9	26	0.0
Democratic Republic of the Congo (formerly Zaire)	2,267	506	3.0	3.0	0.3	0.5	96.6	96.5	1,352	0.4
Denmark	42	33	62.3	55.7	0.3	0.2	37.4	44.0	5	-0.2
Dominican Republic	48	280	22.1	22.1	7.2	9.9	70.6	68.0	14	0.0
Ecuador	277	302	5.6	5.7	3.3	5.2	91.1	89.2	106	1.2
Egypt	995	1,197	2.3	2.8	0.2	0.5	97.5	96.7	1	-3.4
El Salvador	21	582	27.0	27.0	8.0	12.1	65.0	60.9	1	4.6
Eritrea	101	638	–	4.9	–	0.0	–	95.0	16	0.3
Estonia	42	40	–	26.5	–	0.4	–	73.1	21	-0.6
Ethiopia	1,000	513	–	9.9	–	0.6	–	89.4	46	0.8
Finland	305	81	8.4	7.1	–	0.0	–	92.9	219	0.0
France	550	79	31.8	33.4	2.5	2.1	65.7	64.5	153	-0.4
Gabon	258	76	1.1	1.3	0.6	0.7	98.2	98.1	218	0.0

Table I (Continued)
Land Use and Deforestation, 1980–2000

COUNTRY	LAND AREA	RURAL POPULATION DENSITY	LAND USE						FOREST AREA	AVERAGE ANNUAL DEFORESTATION
	(thousands of sq. km.)	(people per sq. km. of arable land)	Arable Land (% of land area)		Permanent Cropland (% of land area)		Other (% of land area)		(thousands of sq. km.)	Decline in forest area %
	1998	1998	1980	1998	1980	1998	1980	1998	2000	1990-2000
The Gambia	10	430	15.9	19.5	–	0.5	–	80.0	5	-1.0
Georgia	70	279	–	11.3	–	4.1	–	84.6	30	0.0
Germany	349	89	34.4	34.0	1.4	0.7	64.1	65.3	107	0.0
Ghana	228	319	8.4	15.8	7.5	7.5	84.2	76.7	63	1.7
Greece	129	149	22.5	22.1	7.9	8.5	69.6	69.4	36	-0.9
Guatemala	108	482	11.7	12.5	4.4	5.0	83.9	82.4	29	1.7
Guinea	246	550	2.0	3.6	0.9	2.4	97.1	94.0	69	0.5
Guinea-Bissau	28	298	9.1	10.7	1.1	1.8	89.9	87.6	22	0.9
Haiti	28	895	19.8	20.3	12.5	12.7	67.7	67.0	1	5.7
Honduras	112	179	13.9	15.1	1.8	3.1	84.3	81.7	54	1.0
Hong Kong,China	1	0	7.0	5.1	1.0	1.0	92.0	93.9	–	–
Hungary	92	76	54.4	52.2	3.3	2.4	42.2	45.4	18	-0.4
India	2,973	438	54.8	54.3	1.8	2.7	43.4	43.0	641	-0.1
Indonesia	1,812	695	9.9	9.9	4.4	7.2	85.6	82.9	1,050	1.2
Iran	1,622	145	8.0	10.4	0.5	1.2	91.5	88.4	73	0.0
Iraq	437	104	12.0	11.9	0.4	0.8	87.6	87.3	8	0.0
Ireland	69	114	16.1	19.7	0.0	0.0	83.9	80.3	7	-3.0
Israel	21	153	15.8	17.0	4.3	4.2	80.0	78.8	1	-4.9
Italy	294	231	32.2	28.2	10.0	9.4	57.7	62.5	100	-0.3
Jamaica	11	664	16.6	16.1	5.5	9.2	77.8	74.7	3	1.5
Japan	377	599	11.4	12.0	1.6	1.0	87.0	87.0	241	0.0
Jordan	89	485	3.4	2.9	0.4	1.5	96.2	95.6	1	0.0
Kazakhstan	2,671	22	–	11.2	–	0.1	–	88.7	121	-2.2
Kenya	569	494	6.7	7.0	0.8	0.9	92.5	92.1	171	0.5
Korea,North	120	548	13.4	14.1	2.4	2.5	84.2	83.4	82	0.0
Korea,South	99	532	20.9	17.3	1.4	2.0	77.8	80.7	63	0.1
Kuwait	18	821	0.1	0.3	–	0.1	–	99.6	0	-5.2
Kyrgyzstan	192	235	–	7.0	–	0.4	–	92.6	10	-2.6
Laos	231	483	2.9	3.5	0.1	0.2	97.0	96.3	126	0.4
Latvia	62	41	–	29.7	–	0.5	–	69.8	29	-0.4
Lebanon	10	262	20.5	17.6	8.9	12.5	70.6	69.9	0	0.3
Lesotho	30	466	9.6	10.7	–	–	–	–	0	0.0
Libya	1,760	39	1.0	1.0	0.2	0.2	98.8	98.8	4	-1.4
Lithuania	65	40	–	45.4	–	0.9	–	53.6	20	-0.2
Macedonia	25	133	–	23.1	–	1.9	–	75.0	9	0.0
Madagascar	582	408	4.3	4.4	0.9	0.9	94.8	94.7	117	0.9
Malawi	94	437	13.3	19.9	0.9	1.3	85.8	78.7	26	2.4
Malaysia	329	537	3.0	5.5	11.6	17.6	85.4	76.9	193	1.2
Mali	1,220	160	1.6	3.8	0.0	0.0	98.3	96.2	132	0.7
Mauritania	1,025	233	0.2	0.5	0.0	0.0	99.8	99.5	3	2.7
Mauritius	2	684	49.3	49.3	3.4	3.0	47.3	47.8	0	0.6
Mexico	1,909	98	12.1	13.2	0.8	1.1	87.1	85.7	552	1.1
Moldova	33	129	–	54.5	–	11.7	–	33.8	3	-0.2
Mongolia	1,567	67	0.8	0.8	0.0	0.0	99.2	99.2	106	0.5
Morocco	446	140	16.6	20.2	1.1	2.1	82.3	77.6	30	0.0
Mozambique	784	339	3.6	4.0	0.3	0.3	96.1	95.7	306	0.2

Table I (Continued)
Land Use and Deforestation, 1980–2000

COUNTRY	LAND AREA	RURAL POPULATION DENSITY	LAND USE						FOREST AREA	AVERAGE ANNUAL DEFORESTATION
			Arable Land (% of land area)		Permanent Cropland (% of land area)		Other (% of land area)			Decline in forest area %
	(thousands of sq. km.)	(people per sq. km. of arable land)							(thousands of sq. km.)	
	1998	1998	1980	1998	1980	1998	1980	1998	2000	1990-2000
Myanmar (Burma)	658	340	14.6	14.5	0.7	0.9	84.8	84.6	344	1.4
Namibia	823	143	0.8	1.0	0.0	0.0	99.2	99.0	80	0.9
Nepal	143	700	16.0	20.3	0.2	0.5	83.8	79.2	39	1.8
Netherlands	34	186	23.3	26.7	0.9	1.0	75.8	72.3	4	-0.3
New Zealand	268	35	9.3	5.8	3.7	6.4	86.9	87.8	79	-0.5
Nicaragua	121	87	9.5	20.2	1.5	2.4	89.1	77.4	33	3.0
Niger	1,267	163	2.8	3.9	0.0	0.0	97.2	96.1	13	3.7
Nigeria	911	248	30.6	31.0	2.8	2.8	66.6	66.3	135	2.6
Norway	307	123	2.7	3.0	–	–	–	–	89	-0.4
Oman	212	2,785	0.1	0.1	0.1	0.2	99.8	99.7	0	0.0
Pakistan	771	394	25.9	27.8	0.4	0.8	73.7	71.4	25	1.1
Panama	74	244	5.8	6.7	1.6	2.1	92.5	91.2	29	1.6
Papua New Guinea	453	6,379	0.0	0.1	0.9	1.3	99.0	98.5	306	0.4
Paraguay	397	108	4.1	5.5	0.3	0.2	95.6	94.2	234	0.5
Peru	1,280	189	2.5	2.9	0.3	0.4	97.2	96.7	652	0.4
Philippines	298	573	14.5	18.4	14.8	15.1	70.8	66.5	58	1.4
Poland	304	97	48.0	46.0	1.1	1.2	50.9	52.8	93	-0.1
Portugal	92	206	26.5	20.5	7.8	7.7	65.7	71.8	37	-1.7
Puerto Rico	9	2,990	5.6	3.7	5.6	5.1	88.7	91.2	2	0.2
Romania	230	106	42.7	40.5	2.9	2.2	54.4	57.3	64	-0.2
Russia	16,889	27	–	7.5	–	0.1	–	92.4	8,514	0.0
Rwanda	25	929	30.8	33.2	10.3	10.1	58.9	56.6	3	3.9
Saudi Arabia	2,150	82	0.9	1.7	0.0	0.1	99.1	98.2	15	0.0
Senegal	193	219	12.2	11.6	0.0	0.2	87.8	88.2	62	0.7
Sierra Leone	72	649	6.3	6.8	0.7	0.8	93.0	92.5	11	2.9
Singapore	1	0	3.3	1.6	9.8	0.0	86.9	98.4	0	0.0
Slovakia	48	157	–	30.6	–	2.8	–	66.6	20	-0.3
Slovenia	20	427	–	11.5	–	2.7	–	85.8	11	-0.2
South Africa	1,221	140	10.2	12.1	0.7	0.8	89.1	87.1	89	0.1
Spain	499	63	31.1	28.6	9.9	9.6	59.0	61.8	144	-0.6
Sri Lanka	65	1,664	13.2	13.4	15.9	15.8	70.8	70.8	19	1.6
Sudan	2,376	112	5.2	7.0	0.0	0.1	94.8	92.9	616	1.4
Sweden	412	53	7.2	6.8	–	–	–	–	271	0.0
Switzerland	40	553	9.9	10.5	0.5	0.6	89.6	88.9	12	-0.4
Syria	184	151	28.5	25.6	2.5	4.2	69.1	70.2	5	0.0
Tajikistan	141	583	–	5.4	–	0.9	–	93.7	4	-0.5
Tanzania	884	595	2.5	4.2	1.0	1.0	96.5	94.7	388	0.2
Thailand	511	281	32.3	32.9	3.5	7.0	64.2	60.1	148	0.7
Togo	54	137	36.8	40.4	6.6	1.8	56.6	57.7	5	3.4
Trinidad and Tobago	5	460	13.6	14.6	9.0	9.2	77.4	76.2	3	0.8
Tunisia	155	116	20.5	18.7	9.7	12.9	69.7	68.5	5	-0.2
Turkey	770	70	32.9	31.8	4.1	3.3	63.0	65.0	102	-0.2
Turkmenistan	470	160	–	3.5	–	0.1	–	96.4	38	0.0
Uganda	200	357	20.4	25.3	8.0	8.8	71.6	65.9	42	2.0
Ukraine	579	49	–	56.7	–	1.7	–	41.6	96	-0.3
United Arab Emirates	84	1,017	0.2	0.5	0.1	0.5	99.7	99.0	3	-2.8

Table I *(Continued)*
Land Use and Deforestation, 1980–2000

COUNTRY	LAND AREA	RURAL POPULATION DENSITY	LAND USE						FOREST AREA	AVERAGE ANNUAL DEFORESTATION
	(thousands of sq. km.)	(people per sq. km. of arable land)	Arable Land (% of land area)		Permanent Cropland (% of land area)		Other (% of land area)		(thousands of sq. km.)	Decline in forest area %
	1998	1998	1980	1998	1980	1998	1980	1998	2000	1990-2000
United Kingdom	242	100	28.7	25.9	0.3	0.2	71.1	73.9	26	-0.8
United States	9,159	36	20.6	19.3	0.2	0.2	79.2	80.5	2,260	-0.2
Uruguay	175	24	8.0	7.2	0.3	0.3	91.7	92.5	13	-5.0
Uzbekistan	414	335	–	10.8	–	0.9	–	88.3	20	-0.2
Venezuela	882	117	3.2	3.0	0.9	1.0	95.9	96.0	495	0.4
Vietnam	325	1,080	18.2	17.5	1.9	4.8	79.8	77.7	98	-0.5
West Bank and Gaza	–	–	–	–	–	–	–	–	–	–
Yemen	528	838	2.6	2.8	0.2	0.2	97.2	96.9	4	1.8
Yugoslavia (Serbia-Montenegro)	–	–	–	–	–	–	–	–	29	0.0
Zambia	743	111	6.9	7.1	0.0	0.0	93.1	92.9	312	2.4
Zimbabwe	387	240	6.4	8.3	0.3	0.3	93.4	91.3	190	1.5

a. Includes Luxembourg.
b. Includes Taiwan.

Source: World Development Indicators (World Bank, Washington, DC, 2001).

Table J
World Countries: Energy Production and Consumption, 1980–1998

COUNTRY	COMMERCIAL ENERGY PRODUCTION		COMMERCIAL ENERGY USE			COMMERCIAL ENERGY USE PER CAPITA			NET ENERGY IMPORTS[a]	
	Thousand Metric Tons of Oil Equivalent		Thousand Metric Tons of Oil Equivalent		Average Annual % Growth	Kg. of Oil Equivalent		Average Annual % Growth	% of Commercial Energy Use	
	1980	1998	1980	1998	1980-98	1980	1998	1980-98	1980	1998
Albania	3,428	864	3,049	947	−7.0	1,142	284	−8.0	−12	9
Algeria	66,741	132,332	12,089	26,506	3.7	648	898	1.0	−452	−399
Angola	11,301	43,035	4,437	7,147	2.9	632	595	−0.2	−155	−502
Argentina	38,813	80,657	41,868	62,349	2.3	1490	1,726	0.9	7	−29
Armenia	1,263	547	1,070	1,939	−	346	511	−	−	72
Australia	86,096	212,012	70,372	105,009	2.4	4,790	5,600	1.0	−22	−102
Austria	7,655	8,999	23,450	28,815	1.5	3,105	3,567	1.1	67	69
Azerbaijan	14,821	16,178	15,001	12,372	−	2,433	1,564	−	−	−31
Bangladesh	9,234	16,725	10,930	19,965	3.6	126	159	1.4	16	16
Belarus	2,566	3,395	2,385	26,470	−	247	2,614	−	−	87
Belgium	7,986	12,810	46,100	58,349	1.8	4,682	5,719	1.5	83	78
Benin	1,212	1,947	1,363	2,240	2.6	394	377	−0.5	11	13
Bolivia	4,241	5,837	2,287	4,621	3.0	427	581	0.7	−85	−26
Bosnia-Herzegovina	−	684	−	1,950	−	−	517	−	−	65
Botswana	−	−	−	−	−	−	−	−	−	−
Brazil	62,083	126,065	111,262	174,964	2.6	914	1,055	0.9	44	28
Bulgaria	7,737	10,116	28,673	19,963	−2.5	3,235	2,418	−2.1	73	49
Burkina Faso	−	−	−	−	−	−	−	−	−	−
Burundi	−	−	−	−	−	−	−	−	−	−
Cambodia	−	−	−	−	−	−	−	−	−	−
Cameroon	6,707	12,965	3,676	6,183	2.6	425	432	−0.3	−82	−110
Canada	207,417	365,674	193,000	234,325	1.6	7,848	7,747	0.4	−7	−56
Central African Republic	−	−	−	−	−	−	−	−	−	−
Chad	−	−	−	−	−	−	−	−	−	−
Chile	5,801	7,905	9,662	23,630	5.8	867	1,594	4.1	40	67
China	608,625	1,020,270	593,118	1,031,410	3.8	604	830	2.4	−3	1
Colombia	18,040	74,422	19,349	30,713	2.8	680	753	0.8	7	−142
Congo Republic	3,970	14,160	845	1,206	2.0	506	433	−0.8	−370	−1074
Costa Rica	767	1102	1,527	2,781	4.0	669	789	1.5	50	60
Côte d'Ivoire	−	−	−	−	−	−	−	−	−	−
Croatia	−	3,956	−	8,136	−	−	1,808	−	−	51
Cuba	4,227	4,448	14,910	11,858	−2.0	1,536	1,066	−2.8	72	62
Czech Republic	41,000	30,555	47,252	41,034	−1.1	4,618	3,986	−1.2	13	26
Democratic Republic of the Congo (formerly Zaire)	8,697	13,546	8,706	13,711	3.0	322	284	−0.3	0	1
Denmark	896	20,177	19,734	20,804	0.9	3,852	3,925	0.7	95	3
Dominican Republic	1,332	1,433	3,464	5,583	2.4	608	676	0.3	62	74
Ecuador	11,755	22,514	5,191	8,973	2.6	652	737	0.3	−126	−151
Egypt	34,168	57,464	15,970	41,798	4.7	391	679	2.4	−114	−37
El Salvador	1,913	1,987	2,537	3,860	2.0	553	640	0.5	25	49
Eritrea	−	−	−	−	−	−	−	−	−	−
Estonia	6951	2,920	6275	4,835	−	4240	3,335	−	−	40
Ethiopia	10,588	16,379	11,157	17,429	2.4	296	284	−0.4	5	6
Finland	6,912	13,591	25,413	33,459	1.7	5,317	6,493	1.3	73	59
France	46,829	125,528	190,111	255,674	2.0	3,528	4,378	1.6	75	51
Gabon	9,441	18,892	1,493	1,668	−0.4	2,160	1,413	−3.4	−532	−1,033
The Gambia	−	−	−	−	−	−	−	−	−	−
Georgia	1,504	729	4,474	2,526	−	882	464	−	−	71
Germany	185,628	131,412	360,441	344,506	−0.1	4,603	4,199	−0.5	48	62

COUNTRY	COMMERCIAL ENERGY PRODUCTION		COMMERCIAL ENERGY USE			COMMERCIAL ENERGY USE PER CAPITA			NET ENERGY IMPORTS[a]	
	Thousand Metric Tons of Oil Equivalent		Thousand Metric Tons of Oil Equivalent		Average Annual % Growth	Kg. of Oil Equivalent		Average Annual % Growth	% of Commercial Energy Use	
	1980	1998	1980	1998	1980-98	1980	1998	1980-98	1980	1998
Ghana	3,305	5,705	4,027	7,270	3.7	375	396	0.6	18	22
Greece	3,696	9,892	15,960	26,976	3.1	1,655	2,565	2.6	77	63
Guatemala	2,503	4,739	3,754	6,258	3.1	550	579	0.5	33	24
Guinea	–	–	–	–	–	–	–	–	–	–
Guinea-Bissau	–	–	–	–	–	–	–	–	–	–
Haiti	1,877	1,626	2,099	2,072	–0.1	392	271	–2.0	11	22
Honduras	1,315	1,897	1,892	3,333	3.1	530	542	0.0	31	43
Hong Kong, China	39	48	5,439	16,593	5.9	1,079	2,497	4.5	99	100
Hungary	14,957	11,849	28,961	25,255	–1.0	2,705	2,497	–0.7	48	53
India	222,418	413,055	242,592	475,788	3.9	353	486	1.9	8	13
Indonesia	128,403	211,522	59,561	123,074	4.7	402	604	2.9	–116	–72
Iran	84,001	232,481	38,918	102,148	6.1	995	1,649	3.5	–116	–128
Iraq	136,643	110,824	12,030	29,972	4.8	925	1,342	1.7	–1,036	–270
Ireland	1,894	2,465	8,485	13,251	2.5	2,495	3,570	2.2	78	81
Israel	153	619	8,563	18,873	5.2	2,208	3,165	2.6	98	97
Italy	19,644	29,049	138,629	167,933	1.4	2,456	2,916	1.3	86	83
Jamaica	224	655	2,378	4,058	3.8	1,115	1,575	2.8	91	84
Japan	43,247	109,965	346,492	510,106	2.7	2,967	4,035	2.3	88	78
Jordan	1	295	1,714	4,887	5.1	786	1,063	0.6	100	94
Kazakhstan	76,799	64,086	76,799	39,037	–	5,163	2,590	–	–	–64
Kenya	7,891	11,609	9,791	14,527	2.2	589	505	–0.9	19	20
Korea, North	–	–	–	–	–	–	–	–	–	–
Korea, South	9644	27,738	41,238	163,375	9.5	1082	3,519	8.3	77	83
Kuwait	91,636	114,225	12,248	14,598	–0.2	8,908	7,823	–0.7	–648	–682
Kyrgyzstan	2,190	1,227	1,717	2,921	–	473	609	–	–	58
Laos	–	–	–	–	–	–	–	–	–	–
Latvia	261	1774	566	4,275	–	222	1,746	–	–	59
Lebanon	178	200	2,480	5,288	4.6	826	1,256	2.6	93	96
Lesotho	–	–	–	–	–	–	–	–	–	–
Libya	96,550	76,524	7,193	12,420	3.8	2,364	2,343	0.7	–1,242	–516
Lithuania	–	4,510	–	9,347	–	–	2,524	–	–	52
Macedonia	–	–	–	–	–	–	–	–	–	–
Madagascar	–	–	–	–	–	–	–	–	–	–
Malawi	–	–	–	–	–	–	–	–	–	–
Malaysia	18,202	74,912	12,215	43,623	7.9	888	1,967	5.1	–49	–72
Mali	–	–	–	–	–	–	–	–	–	–
Mauritania	–	–	–	–	–	–	–	–	–	–
Mauritius	–	–	–	–	–	–	–	–	–	–
Mexico	149,359	228,187	98,898	147,834	2.1	1,464	1,552	0.2	–51	–54
Moldova	35	63	–	4,053	–	–	943	–	–	98
Mongolia	–	–	–	–	–	–	–	–	–	–
Morocco	877	753	4,778	9,344	4.2	247	336	2.1	82	92
Mozambique	7,413	6,945	8,074	6,863	–1.0	668	405	–2.6	8	–1
Myanmar	9,513	12,405	9,430	13,631	2.0	279	307	0.5	–1	9
Namibia	–	–	–	–	–	–	–	–	–	–
Nepal	4,630	6,886	4,805	7,831	2.7	331	343	0.2	4	12
Netherlands	71,830	62,495	65,000	74,408	1.5	4,594	4,740	0.9	–11	16
New Zealand	5,488	13,837	9,251	17,159	3.8	2,972	4,525	2.7	41	19
Nicaragua	910	1,458	1,566	2,651	2.8	536	553	0.0	42	45
Niger	–	–	–	–	–	–	–	–	–	–

COUNTRY	COMMERCIAL ENERGY PRODUCTION		COMMERCIAL ENERGY USE			COMMERCIAL ENERGY USE PER CAPITA			NET ENERGY IMPORTS[a]	
	Thousand Metric Tons of Oil Equivalent		Thousand Metric Tons of Oil Equivalent		Average Annual % Growth	Kg. of Oil Equivalent		Average Annual % Growth	% of Commercial Energy Use	
	1980	1998	1980	1998	1980-98	1980	1998	1980-98	1980	1998
Nigeria	148,479	184,847	52,846	86,489	2.6	743	716	−0.3	−181	−114
Norway	55,716	206,667	18,792	25,423	1.7	4,593	5,736	1.3	−196	−713
Oman	15,090	52,202	996	7,285	11.5	905	3,165	7.0	−1,415	−617
Pakistan	20,997	42,351	25,472	57,854	4.8	308	440	2.2	18	27
Panama	529	642	1,865	2,383	1.7	957	862	−0.2	72	73
Papua New Guinea	–	–	–	–	–	–	–	–	–	–
Paraguay	1,605	6,868	2,089	4,277	4.5	671	819	1.5	23	−61
Peru	14,655	11,964	11,700	14,400	1.1	675	581	−0.9	−25	17
Philippines	10,670	17,818	21,212	38,313	3.7	439	526	1.4	50	53
Poland	121,848	86,703	123,465	96,440	−1.3	3,470	2,494	−1.8	1	10
Portugal	1,481	2,315	10,291	21,849	4.4	1,054	2,192	4.4	86	89
Puerto Rico	–	–	–	–	–	–	–	–	–	–
Romania	52,587	28,241	65,110	39,611	−3.0	2,933	1,760	−3.0	19	29
Russia	748,647	928,987	763,707	581,774	–	5,494	3,963	–	–	−60
Rwanda	–	–	–	–	–	–	–	–	–	–
Saudi Arabia	533,071	505,121	35,357	103,230	5.2	3,773	5,244	0.9	−1,408	−389
Senegal	1,046	1,653	1,921	2,822	2.2	347	312	−0.5	46	41
Sierra Leone	–	–	–	–	–	–	–	–	–	–
Singapore	–	24	6,062	24,299	9.8	2,656	7,681	7.8	–	100
Slovakia	3,416	4,833	20,810	16,906	−1.4	4,175	3,136	−1.8	84	71
Slovenia	1,623	2,891	4,313	6,649	–	2,269	3,354	–	–	57
South Africa	73,169	144,405	65,417	110,986	2.2	2,372	2,681	−0.1	−12	−30
Spain	15,644	31,920	68,583	112,782	3.1	1,834	2,865	2.8	77	72
Sri Lanka	3,209	4,319	4,536	7,300	2.3	308	389	0.9	29	41
Sudan	7,089	13,527	8,406	14,899	2.8	450	526	0.5	16	9
Sweden	16,133	34,155	40,984	52,472	1.3	4,932	5,928	0.8	61	35
Switzerland	7,030	11,163	20,861	26,605	1.5	3,301	3,742	0.8	66	58
Syria	9,502	35,411	5,348	17,346	5.6	614	1,133	2.3	−78	−104
Tajikistan	1,986	1,268	1,650	3,255	–	416	532	–	–	61
Tanzania	9,502	13,931	10,280	14,660	2.0	553	456	−1.1	8	5
Thailand	11,182	39,347	22,808	68,971	7.9	488	1,153	6.4	51	43
Togo	–	–	–	–	–	–	–	–	–	–
Trinidad and Tobago	13,141	14,651	3,873	8,950	3.9	3,580	6,964	3.0	−239	−64
Tunisia	6,966	7,113	3,907	7,582	3.7	612	812	1.5	−78	6
Turkey	17,190	28,649	31,314	72,512	4.9	704	1,144	2.9	45	60
Turkmenistan	8,034	17,411	7,948	11,122	–	2,778	2,357	–	–	−57
Uganda	–	–	–	–	–	–	–	–	–	–
Ukraine	109,708	80,415	97,893	142,939	–	1,956	2,842	–	–	44
United Arab Emirates	89,716	144,935	6,112	27,336	8.7	5,860	10,035	3.1	−1,368	−430
United Kingdom	197,864	274,230	201,299	232,879	1.1	3,574	3,930	0.8	2	−18
United States	1,553,260	1,695,430	1,811,650	2,181,800	1.4	7,973	7,937	0.4	14	22
Uruguay	763	1,262	2,641	2,992	1.3	906	910	0.6	71	58
Uzbekistan	4,615	50,334	4,821	46,278	–	302	1,930	–	–	−9
Venezuela	139,392	230,563	34,962	56,543	2.5	2,317	2,433	0.0	−299	−308
Vietnam	18,364	42,668	19,573	33,695	3.0	364	440	0.9	6	−27
West Bank and Gaza	–	–	–	–	–	–	–	–	–	–
Yemen	60	19,565	1,424	3,333	4.5	167	201	0.5	96	−487
Yugoslavia (Serbia-Montenegro)	–	–	–	–	–	–	–	–	–	–
Zambia	4,198	5,657	4,551	6,088	1.3	793	630	−1.6	8	7
Zimbabwe	5,793	8,235	6,570	10,065	2.8	937	861	−0.2	12	18

Table J (Continued)
World Countries: Energy Production and Consumption, 1980–1998

COUNTRY	COMMERCIAL ENERGY PRODUCTION		COMMERCIAL ENERGY USE			COMMERCIAL ENERGY USE PER CAPITA			NET ENERGY IMPORTS[a]	
	Thousand Metric Tons of Oil Equivalent		Thousand Metric Tons of Oil Equivalent		Average Annual % Growth	Kg. of Oil Equivalent		Average Annual % Growth	% of Commercial Energy Use	
	1980	1998	1980	1998	1980-98	1980	1998	1980-98	1980	1998
World[1]	6,882,644	9,611,004	6,902,381	9,345,307	2.9	1,627	1,659	0.9	–	–
Low income	797,751	1,290,575	648,676	1,178,897	5.1	442	550	2.5	–23	–10
Middle income	3,302,896	4,605,397	2,481,018	3,409,502	4.4	1,246	1,311	2.3	–33	–35
Lower middle income	1,944,378	2,867,598	1,779,108	2,282,178	5.4	1,126	1,116	3.2	–9	–26
Upper middle income	1,358,518	1,737,798	701,910	1,127,324	2.7	1,713	2,025	1.0	–94	–54
Low & middle income	4,100,647	5,895,972	3,129,694	4,588,399	4.6	905	967	2.3	–31	–29
East Asia & Pacific	814,603	1,446,679	779,155	1,516,091	4.5	571	857	3.0	–5	5
Europe & Central Asia	1,241,543	1,380,292	1,332,941	1,215,898	7.8	3,349	2,637	–	7	–14
Latin America & Carib.	475,245	830,882	379,775	585,082	2.4	1,070	1,183	0.5	–24	–42
Middle East & N. Africa	988,969	1,237,344	145,929	378,338	5.1	838	1,344	2.3	–577	–228
South Asia	260,487	483,335	288,334	568,738	4.0	325	445	1.9	10	–15
Sub-Saharan Africa	319,801	517,440	203,560	324,252	2.3	727	700	–0.5	–57	–60
High income	2,781,997	3,715,032	3,772,688	4,756,908	1.7	4,796	5,366	1.0	27	22
Europe EMU	365,532	420,629	940,146	1,114,343	1.2	3,408	3,834	0.9	61	62

a. A negative value indicates that a country is a net exporter.
1. World Bank estimations; missing data are imputed whenever possible.

Source: World Development Indicators (World Bank, Washington, DC, 2001).

Table K
World Countries: Energy Efficiency and Emissions, 1980–1998

COUNTRY	GDP PER UNIT OF ENERGY USE		TRADITIONAL FUEL USE		CARBON DIOXIDE EMISSIONS					
	PPP $ per Kg. of Oil Equivalent		% of Total Energy Use		Total Million Metric Tons		Per Capita Metric Tons		Kg. per PPP $ of GDP	
	1980	1998	1980	1997	1980	1997	1980	1997	1980	1997
Albania	–	10.3	13.1	7.3	5.3	1.7	2.0	0.5	–	0.2
Algeria	5.0	5.4	1.9	1.5	68.2	98.7	3.7	3.4	1.1	0.7
Angola	–	3.8	64.9	69.7	5.4	5.3	0.8	0.5	–	0.2
Argentina	4.7	7.3	5.9	4.0	111.0	140.6	4.0	3.9	0.6	0.3
Armenia	–	4.3	–	0.0	–	2.9	–	0.8	–	0.4
Australia	2.1	4.1	3.8	4.4	205.5	319.6	14.0	17.2	1.4	0.8
Austria	3.5	6.7	1.2	4.7	55.1	62.6	7.3	7.8	0.7	0.3
Azerbaijan	–	1.5	–	0.0	–	32.0	–	4.1	–	1.9
Bangladesh	4.5	8.9	81.3	46.0	7.8	24.6	0.1	0.2	0.2	0.1
Belarus	–	2.5	–	0.8	–	62.3	–	6.1	–	1.0
Belgium	2.4	4.3	0.2	1.6	131.0	106.5	13.3	10.5	1.2	0.4
Benin	1.3	2.4	85.4	89.2	0.7	1.0	0.2	0.2	0.4	0.2
Bolivia	3.4	4.0	19.3	14.0	4.7	11.3	0.9	1.4	0.6	0.6
Bosnia-Herzegovina	–	–	–	10.1	–	4.5	–	1.2	–	–
Botswana	–	–	35.7	–	1.0	3.4	1.1	2.2	0.6	0.3
Brazil	4.4	6.5	35.5	28.7	197.0	307.2	1.6	1.9	0.4	0.3
Bulgaria	0.9	2.0	0.5	1.3	77.9	50.3	8.8	6.1	3.0	1.2
Burkina Faso	–	–	91.3	87.1	0.4	1.0	0.1	0.1	0.1	0.1
Burundi	–	–	97.0	94.2	0.1	0.2	0.0	0.0	0.1	0.1
Cambodia	–	–	100.0	89.3	0.3	0.5	0.0	0.0	–	0.0
Cameroon	2.8	3.5	51.7	69.2	4.1	2.7	0.5	0.2	0.4	0.1
Canada	1.5	3.2	0.4	4.7	426.1	496.6	17.3	16.6	1.5	0.7
Central African Republic	–	–	88.9	87.5	0.1	0.2	0.0	0.1	0.1	0.1
Chad	–	–	95.9	97.6	0.2	0.1	0.0	0.0	0.1	0.0
Chile	3.1	5.4	12.3	11.3	28.3	60.1	2.5	4.1	0.9	0.5
China	0.8	4.0	8.4	5.7	1516.6	3593.5	1.5	2.9	3.3	0.9
Colombia	4.1	7.9	15.9	17.7	42.0	71.9	1.5	1.8	0.5	0.3
Congo Republic	0.8	1.8	77.8	53.0	0.4	0.3	0.2	0.1	0.6	0.1
Costa Rica	5.7	9.5	26.3	54.2	2.7	5.4	1.2	1.6	0.3	0.2
Côte d'Ivoire	–	–	52.8	91.5	5.3	13.3	0.6	0.9	0.5	0.6
Croatia	–	3.9	–	3.2	–	20.1	–	4.4	–	0.6
Cuba	–	–	27.9	30.2	32.4	26.0	3.3	2.3	–	–
Czech Republic	–	3.2	0.6	1.6	–	125.2	–	12.2	–	0.9
Democratic Republic of the Congo (formerly Zaire)	3.5	2.8	73.9	91.7	3.7	2.3	0.1	0.1	0.1	0.1
Denmark	–	6.4	0.4	5.9	63.9	57.7	12.5	10.9	–	0.4
Dominican Republic	3.7	7.5	27.5	14.3	6.9	14.0	1.2	1.7	0.5	0.4
Ecuador	3.0	4.3	26.7	17.5	14.1	21.7	1.8	1.8	0.9	0.6
Egypt	3.5	4.7	4.7	3.2	46.7	118.3	1.1	2.0	0.8	0.6
El Salvador	4.3	6.5	52.9	34.5	2.4	5.9	0.5	1.0	0.2	0.2
Eritrea	–	–	–	96.0	–	–	–	–	–	–
Estonia	–	2.5	–	13.8	–	19.1	–	13.1	–	1.6
Ethiopia	1.4	2.1	89.6	95.9	1.9	3.8	0.0	0.1	0.1	0.1
Finland	1.8	3.4	4.3	6.5	55.8	56.6	11.7	11.0	1.2	0.5
France	2.9	5.0	1.3	5.7	497.2	349.8	9.2	6.0	0.9	0.3
Gabon	1.9	4.5	30.8	32.9	5.0	3.4	7.2	3.0	1.7	0.5

World Countries: Energy Efficiency and Emissions, 1980–1998

COUNTRY	GDP PER UNIT OF ENERGY USE		TRADITIONAL FUEL USE		CARBON DIOXIDE EMISSIONS					
	PPP $ per Kg. of Oil Equivalent		% of Total Energy Use		Total Million Metric Tons		Per Capita Metric Tons		Kg. per PPP $ of GDP	
	1980	1998	1980	1997	1980	1997	1980	1997	1980	1997
The Gambia	–	–	72.7	78.6	0.2	0.2	0.2	0.2	0.3	0.1
Georgia	–	7.1	–	1.0	–	4.5	–	0.8	–	0.3
Germany	–	5.5	0.3	1.3	–	851.5	–	10.4	–	0.5
Ghana	2.9	4.6	43.7	78.1	2.6	4.8	0.2	0.3	0.2	0.1
Greece	4.2	5.7	3.0	4.5	58.1	87.2	6.0	8.3	0.9	0.6
Guatemala	4.1	6.1	54.6	62.0	4.8	8.3	0.7	0.8	0.3	0.2
Guinea	–	–	71.4	74.2	0.9	1.1	0.2	0.2	–	0.1
Guinea-Bissau	–	–	80.0	57.1	0.1	0.2	0.2	0.2	0.5	0.2
Haiti	3.7	5.3	80.7	74.7	0.9	1.4	0.2	0.2	0.1	0.1
Honduras	2.9	4.5	55.3	54.8	2.3	4.6	0.6	0.8	0.4	0.3
Hong Kong, China	6.4	8.5	0.9	0.7	17.1	23.8	3.4	3.7	0.5	0.2
Hungary	2.0	4.3	2.0	1.6	84.8	59.6	7.9	5.9	1.4	0.6
India	1.9	4.3	31.5	20.7	356.1	1065.4	0.5	1.1	0.8	0.5
Indonesia	2.2	4.6	51.5	29.3	97.5	251.5	0.7	1.3	0.8	0.4
Iran	2.9	3.3	0.4	0.7	120.0	296.9	3.1	4.9	1.1	0.9
Iraq	–	–	0.3	0.1	46.7	92.3	3.6	4.2	–	–
Ireland	2.3	6.4	0.0	0.2	26.1	37.3	7.7	10.2	1.3	0.5
Israel	3.6	5.7	0.0	0.0	22.1	60.4	5.7	10.4	0.7	0.6
Italy	3.9	7.4	0.8	1.0	392.7	424.7	7.0	7.4	0.7	0.3
Jamaica	1.9	2.2	5.0	6.0	8.5	11.0	4.0	4.3	1.9	1.2
Japan	3.3	6.0	0.1	1.6	964.2	1204.2	8.3	9.6	0.8	0.4
Jordan	3.3	3.6	0.0	0.0	5.2	15.7	2.4	3.5	0.9	0.9
Kazakhstan	–	1.8	–	0.2		123.0	–	8.0	–	1.7
Kenya	1.1	2.0	76.8	80.3	6.8	7.2	0.4	0.3	0.7	0.2
Korea, North	–	–	3.1	1.4	128.9	260.5	7.3	11.4	–	–
Korea, South	2.8	4.0	4.0	2.4	132.9	457.4	3.5	9.9	1.2	0.6
Kuwait	1.3	2.1	0.0	0.0	25.4	51.0	18.5	28.2	1.6	1.8
Kyrgyzstan	–	4.0	–	0.0	–	6.8	–	1.4	–	0.6
Laos	–	–	72.3	88.7	0.2	0.4	0.1	0.1	–	0.1
Latvia	19.6	3.4	–	26.2	–	8.3	–	3.3	–	0.6
Lebanon	–	3.7	2.4	2.5	6.9	17.7	2.3	4.3	–	1.0
Lesotho	–	–	–	–	–	–	–	–	–	–
Libya	–	–	2.3	0.9	28.5	43.5	9.4	8.4	–	–
Lithuania	–	2.7	–	6.3	–	15.1	–	4.1	–	0.6
Macedonia	–	–	–	6.1	–	10.9	–	5.5	–	1.2
Madagascar	–	–	78.4	84.3	1.6	1.2	0.2	0.1	0.3	0.1
Malawi	–	–	90.6	88.6	0.8	0.8	0.1	0.1	0.3	0.1
Malaysia	2.7	3.9	15.7	5.5	29.1	137.2	2.1	6.3	0.9	0.7
Mali	–	–	86.7	88.9	0.4	0.5	0.1	0.0	0.1	0.1
Mauritania	–	–	0.0	0.0	0.6	3.0	0.4	1.2	0.4	0.8
Mauritius	–	–	59.1	36.1	0.6	1.7	0.6	1.5	0.3	0.2
Mexico	3.1	5.2	5.0	4.5	259.6	379.7	3.8	4.0	0.8	0.5
Moldova	–	2.2	–	0.5	–	10.4	–	2.4	–	1.1
Mongolia	–	–	14.4	4.3	6.9	7.8	4.1	3.3	3.6	2.1
Morocco	6.8	10.2	5.2	4.0	17.7	35.9	0.9	1.3	0.5	0.4
Mozambique	0.6	2.0	43.7	91.4	3.3	1.2	0.3	0.1	0.7	0.1
Myanmar (Burma)	–	–	69.3	60.5	5.0	8.8	0.1	0.2	–	–

Table K (Continued)
World Countries: Energy Efficiency and Emissions, 1980–1998

COUNTRY	GDP PER UNIT OF ENERGY USE		TRADITIONAL FUEL USE		CARBON DIOXIDE EMISSIONS					
	PPP $ per Kg. of Oil Equivalent		% of Total Energy Use		Total Million Metric Tons		Per Capita Metric Tons		Kg. per PPP $ of GDP	
	1980	1998	1980	1997	1980	1997	1980	1997	1980	1997
Namibia	–	–	–	–	–	–	–	–	–	–
Nepal	1.5	3.5	94.2	89.6	0.6	2.2	0.0	0.1	0.1	0.1
Netherlands	2.2	4.9	0.0	1.1	154.5	163.6	10.9	10.5	1.1	0.5
New Zealand	–	4.0	0.2	0.8	17.9	31.6	5.8	8.4	–	0.5
Nicaragua	3.6	4.0	49.2	42.2	2.1	3.2	0.7	0.7	0.4	0.3
Niger	–	–	79.5	80.6	0.6	1.1	0.1	0.1	0.1	0.2
Nigeria	0.8	1.2	66.8	67.8	69.1	83.7	1.0	0.7	1.6	0.9
Norway	2.4	4.8	0.4	1.1	91.5	68.5	22.4	15.6	2.0	0.6
Oman	–	–	0.0	–	5.9	18.4	5.3	8.2	–	–
Pakistan	2.1	4.0	24.4	29.5	33.3	98.2	0.4	0.8	0.6	0.4
Panama	3.2	6.5	26.6	14.4	3.7	8.0	1.9	2.9	0.6	0.5
Papua New Guinea	–	–	65.4	62.5	1.8	2.5	0.6	0.5	0.5	0.2
Paraguay	4.2	5.4	62.0	49.6	1.6	4.1	0.5	0.8	0.2	0.2
Peru	4.6	7.8	15.2	24.6	24.7	30.1	1.4	1.2	0.5	0.3
Philippines	5.6	7.0	37.0	26.9	38.8	81.7	0.8	1.1	0.3	0.3
Poland	–	3.2	0.4	0.8	465.4	357.0	13.1	9.2	–	1.2
Portugal	5.6	7.0	1.2	0.9	29.9	53.8	3.1	5.4	0.5	0.4
Puerto Rico	–	–	0.0	–	14.7	17.1	4.6	4.5	–	–
Romania	1.6	3.5	1.3	5.7	199.6	111.3	9.0	4.9	1.9	0.8
Russia	–	1.7	–	0.8	–	1444.5	–	9.8	–	1.4
Rwanda	–	–	89.8	88.3	0.3	0.5	0.1	0.1	0.1	0.1
Saudi Arabia	3.0	2.1	0.0	0.0	132.2	273.7	14.1	14.3	1.2	1.3
Senegal	2.3	4.4	50.8	56.2	3.0	3.5	0.5	0.4	0.7	0.3
Sierra Leone	–	–	90.0	86.1	0.6	0.5	0.2	0.1	0.3	0.2
Singapore	2.3	3.1	0.4	0.0	31.1	81.9	12.9	21.9	2.2	1.1
Slovakia	–	3.2	–	0.5	–	38.1	–	7.1	–	0.7
Slovenia	–	4.4	–	1.5	–	15.5	–	7.8	–	0.5
South Africa	2.7	3.3	4.9	43.4	214.9	321.5	7.8	7.9	1.2	0.9
Spain	3.8	5.9	0.4	1.3	214.0	257.7	5.7	6.6	0.8	0.4
Sri Lanka	3.5	8.0	53.5	46.5	3.7	8.1	0.3	0.4	0.2	0.1
Sudan	–	–	86.9	75.1	3.4	3.8	0.2	0.1	–	–
Sweden	2.1	3.6	7.7	17.9	72.6	48.6	8.7	5.5	0.8	0.3
Switzerland	4.4	7.0	0.9	6.0	43.0	42.6	6.8	6.0	0.5	0.2
Syria	2.9	3.3	0.0	0.0	20.3	49.9	2.3	3.3	1.3	1.0
Tajikistan	–	–	–	–	–	5.6	–	0.9	–	–
Tanzania	–	1.1	92.0	91.4	2.0	2.9	0.1	0.1	–	0.2
Thailand	3.0	5.1	40.3	24.6	42.7	226.8	0.9	3.8	0.6	0.6
Togo	–	–	35.7	71.9	0.8	1.0	0.3	0.2	0.2	0.2
Trinidad and Tobago	1.3	1.1	1.4	0.8	16.8	22.3	15.5	17.4	3.4	2.4
Tunisia	4.0	6.9	16.1	12.5	10.3	18.8	1.6	2.0	0.7	0.4
Turkey	3.6	5.8	20.5	3.1	82.8	216.0	1.9	3.5	0.7	0.5
Turkmenistan	–	1.2	–	–	–	31.0	–	6.7	–	2.5
Uganda	–	–	93.6	89.7	0.6	1.2	0.1	0.1	0.1	0.1
Ukraine	–	1.2	–	0.5	–	370.5	–	7.3	–	2.1
United Arab Emirates	4.4	1.8	0.0	–	37.1	82.5	35.6	32.0	1.4	1.6
United Kingdom	–	5.4	0.0	3.3	591.2	527.1	10.5	8.9	–	0.4
United States	1.6	3.8	1.3	3.8	4609.4	5467.1	20.3	20.1	1.6	0.7

Table K *(Continued)*
World Countries: Energy Efficiency and Emissions, 1980–1998

COUNTRY	GDP PER UNIT OF ENERGY USE		TRADITIONAL FUEL USE		CARBON DIOXIDE EMISSIONS					
	PPP $ per Kg. of Oil Equivalent		% of Total Energy Use		Total Million Metric Tons		Per Capita Metric Tons		Kg. per PPP $ of GDP	
	1980	1998	1980	1997	1980	1997	1980	1997	1980	1997
Uruguay	5.0	9.9	11.1	21.0	6.2	5.7	2.1	1.8	0.5	0.2
Uzbekistan	–	1.1	–	0.0	–	104.8	–	4.4	–	2.1
Venezuela	1.7	2.4	0.9	0.7	92.0	191.2	6.1	8.4	1.5	1.4
Vietnam	–	4.0	49.1	37.8	17.0	45.5	0.3	0.6	–	0.4
West Bank and Gaza	–	–	–	–	–	–	–	–	–	–
Yemen	–	3.7	0.0	1.4	–	16.7	–	1.0	–	1.3
Yugoslavia (Serbia-Montenegro)	–	–	–	1.5	–	50.2	–	4.7	–	–
Zambia	0.7	1.2	37.4	72.7	3.6	2.6	0.6	0.3	0.9	0.4
Zimbabwe	1.5	3.3	27.6	25.2	9.9	18.8	1.4	1.6	1.0	0.6
WORLD[1]	**2.1**	**4.2**	**7.4**	**8.2**	**14014.6**	**23868.2**	**3.5**	**4.1**	**1.2**	**0.6**
Low income	–	3.4	46.4	29.8	794.9	2527.5	0.5	1.1	0.7	0.6
Middle income	2.2	3.9	10.4	7.3	4304.9	10006.0	2.4	3.8	1.3	0.8
Lower middle income	1.6	3.6	10.7	5.7	2457.8	6957.9	1.7	3.4	1.7	0.9
Upper middle income	3.3	4.3	8.6	10.6	1847.0	3048.1	4.6	5.5	1.0	0.6
Low & middle income	–	3.7	18.5	12.9	5099.8	12533.6	1.5	2.5	1.2	0.7
East Asia & Pacific	–	–	15.1	9.7	2019.6	5075.6	1.4	2.8	2.0	0.8
Europe & Central Asia	–	2.3	3.2	1.3	915.8	3285.6	–	6.9	2.1	1.2
Latin America & Carib.	3.7	5.7	18.4	16.0	884.7	1356.4	2.5	2.8	0.6	0.4
Middle East & N. Africa	3.4	3.5	1.6	1.1	517.8	1113.6	3.1	4.0	1.1	0.9
South Asia	2.0	4.5	34.2	23.8	403.4	1200.5	0.4	0.9	0.7	0.5
Sub-Saharan Africa	–	–	47.2	63.5	358.4	501.8	1.0	0.8	0.9	0.6
High Income	2.2	4.6	1.0	3.4	8914.8	11334.6	12.6	12.8	1.2	0.5
Europe EMU	3.1	5.6	0.7	2.5	1569.7	2378.6	7.9	8.2	0.9	0.4

Source: World Development Indicators (World Bank, Washington, DC, 2001).

1. World Bank estimations. Missing data are inputed wherever possible.

Table L
World Countries: Power and Transportation

	Electric power						Goods transported by road Millions of ton-km hauled		Goods transported by rail Ton-km per $ million of GDP (PPP)		Air passengers carried Thousands
	Consumption per capita Kilowatt-hours		Transmission and distribution losses % of output		Paved roads % of total						
Country	1980	1996	1980	1996	1990	1997	1990	1997	1990	1997	1996
AFRICA											
Algeria	265	524	11	18	67	69	14,000	X	25,161	X	3,494
Angola	67	61	25	28	25	25	X	X	X	X	585
Benin	36	48	220	87	20	20	X	X	X	X	75
Botswana	X	X	X	X	32	24	X	X	X	X	104
Burkina Faso	X	X	X	X	17	16	X	X	X	X	138
Burundi	X	X	X	X	18	7	X	X	X	X	9
Cameroon	167	171	7	20	11	13	X	X	33,209	34,023	362
Central African Republic	X	X	X	X	X	X	144	60	X	X	75
Chad	X	X	X	X	1	1	X	X	X	X	93
Congo (Zaire)	147	130	8	3	X	X	X	X	32,198	X	178
Congo Republic	94	207	1	0	10	10	X	X	144,851	X	253
Côte d'Ivoire	192	174	7	16	9	10	X	X	15,791	13,486	179
Egypt	380	924	13	0	72	78	31,400	31,500	23,310	X	4,282
Equatorial Guinea	X	X	X	X	X	X	X	X	X	X	X
Eritrea	X	X	X	X	19	22	X	X	X	X	X
Ethiopia	16	18	8	1	15	15	X	X	2,467	X	743
Gabon	X	X	X	X	X	X	X	X	X	X	X
Gambia	X	X	X	X	X	X	X	X	X	X	X
Ghana	X	X	X	X	X	X	X	X	X	X	X
Guinea	X	X	X	X	15	17	X	X	X	X	36
Guinea-Bissau	X	X	X	X	X	X	X	X	X	X	X
Kenya	92	126	16	16	13	14	X	X	75,496	X	779
Lesotho	X	X	X	X	18	18	X	X	X	X	17
Liberia	X	X	X	X	X	X	X	X	X	X	X
Libya	X	X	X	X	X	X	X	X	X	X	X
Madagascar	X	X	X	X	15	12	X	X	X	X	542
Malawi	X	X	X	X	22	19	X	X	14,881	10,003	153
Mali	X	X	X	X	11	12	X	X	53,882	X	75
Mauritania	X	X	X	X	11	11	X	X	X	X	235
Mauritius	X	X	X	X	X	X	X	X	X	X	X
Morocco	223	408	10	4	49	52	2,638	2,086	72,108	55,523	2,301
Mozambique	370	76	0	0	17	19	X	110	X	X	163
Namibia	X	X	X	X	11	8	X	X	308,833	139,137	237
Niger	X	X	X	X	29	8	X	X	X	X	75
Nigeria	68	85	36	32	30	19	X	X	3,009	X	221
Rwanda	X	X	X	X	9	9	X	X	X	X	9
Senegal	97	103	11	16	27	29	X	X	51,209	X	155
Sierra Leone	X	X	X	X	11	8	X	X	X	X	15
Somalia	X	X	X	X	X	X	X	X	X	X	X
South Africa	3,213	3,719	8	8	30	42	X	X	430,594	337,153	7,183
Sudan	X	X	X	X	X	X	X	X	X	X	X
Swaziland	X	X	X	X	X	X	X	X	X	X	X
Tanzania	50	59	14	12	37	4	X	X	77,466	91,623	224
Togo	X	X	X	X	21	32	X	X	X	X	75
Tunisia	379	674	12	11	76	79	X	X	58,795	53,343	1,371
Uganda	X	X	X	X	X	X	X	X	12,582	11,567	100
Zambia	1,016	560	7	11	17	X	X	X	73,728	56,426	235
Zimbabwe	990	765	14	7	14	47	X	X	247,759	196,429	654

Table L (Continued)
World Countries: Power and Transportation

	Electric power										
	Consumption per capita Kilowatt-hours		Transmission and distribution losses % of output		Paved roads % of total		Goods transported by road Millions of ton-km hauled		Goods transported by rail Ton-km per $ million of GDP (PPP)		Air passengers carried Thousands
Country	1980	1996	1980	1996	1990	1997	1990	1997	1990	1997	1996
NORTH AMERICA											
Canada	12,329	15,129	9	7	35	35	54,700	71,473	433,765	X	22,856
United States	8,914	11,796	9	7	58	61	1,073,100	1,439,532	360,699	361,911	571,072
CENTRAL AMERICA											
Belize	X	X	X	X	X	X	X	X	X	X	X
Costa Rica	860	1,349	0	12	15	17	2,243	3,070	X	X	918
Cuba	X	X	X	X	X	X	X	X	X	X	X
Dominican Republic	433	608	21	25	45	49	X	X	X	X	30
El Salvador	293	516	13	13	14	20	X	X	X	X	1,800
Guatemala	212	364	6	13	25	28	X	X	X	X	300
Haiti	41	34	26	54	22	24	X	X	X	X	X
Honduras	225	350	14	27	21	20	X	X	X	X	X
Jamaica	482	2,108	17	11	64	71	X	X	X	X	1,388
Mexico	846	1,381	11	15	35	37	108,884	165,000	64,884	53,917	14,678
Nicaragua	303	256	14	28	11	10	X	X	X	X	51
Panama	828	1,140	13	18	32	34	X	X	X	X	689
Trinidad and Tobago	X	X	X	X	X	X	X	X	X	X	X
SOUTH AMERICA											
Argentina	1,170	1,541	13	18	29	29	X	X	36,412	x	7,913
Bolivia	226	371	10	12	4	6	X	X	37,118	X	1,784
Brazil	974	1,660	12	17	10	9	X	X	56,068	X	22,012
Chile	876	1,864	12	9	14	14	X	X	15,882	5,998	3,622
Colombia	561	922	16	22	12	12	6,227	X	2,400	X	8,342
Ecuador	361	616	14	21	13	19	2,638	3,558	X	X	1,925
Guyana	X	X	X	X	X	X	X	X	X	X	X
Paraguay	233	914	6	7	9	10	X	X	X	X	261
Peru	502	598	13	15	10	10	X	X	7,486	X	2,328
Suriname	X	X	X	X	X	X	X	X	X	X	X
Uruguay	977	1,605	15	20	74	90	X	X	10,455	16,125	504
Venezuela	2,037	2,498	12	20	36	39	X	X	X	X	4,487
ASIA											
Afghanistan	X	X	X	X	X	X	X	X	X	X	X
Armenia	2,729	905	10	38	99	100	1,533	479	X	X	X
Azerbaijan	2,440	1,822	14	22	X	X	3,287	497	X	X	1,233
Bangladesh	16	97	35	30	7	12	X	X	8,032	X	1,252
Bhutan	X	X	X	X	X	X	X	X	X	X	X
Cambodia	X	X	X	X	8	8	X	1,200	X	X	X
China	253	687	8	7	X	X	X	X	671,824	364,633	51,770
Hong Kong, China	2,167	5,013	11	14	100	100	X	X	X	X	X
Georgia	1,910	1,020	16	23	94	94	7,370	98	X	X	152
India	130	347	18	18	47	46	X	X	248,469	176,217	13,395
Indonesia	44	296	19	12	46	46	X	X	8,619	X	17,139
Iran	491	1,142	10	20	X	50	X	X	40,223	X	7,610
Iraq	X	X	X	X	X	X	X	X	X	X	X
Israel	2,826	5,081	5	4	100	100	X	X	16,663	11,947	3,695
Japan	4,395	7,083	4	4	69	74	274,444	305,510	11,603	8,664	95,914
Jordan	387	1,187	19	10	100	100	X	X	78,625	47,242	1,299
Kazakhstan	0	2,865	X	15	55	83	44,775	6,481	5,042,201	X	568
Korea, North	X	X	X	X	X	X	X	X	X	X	X
Korea, South	841	4,453	6	5	72	74	31,841	74,504	40,875	24,826	33,033

Table L *(Continued)*
World Countries: Power and Transportation

	Electric power						Goods transported by road Millions of ton-km hauled		Goods transported by rail Ton-km per $ million of GDP (PPP)		Air passengers carried Thousands
	Consumption per capita Kilowatt-hours		Transmission and distribution losses % of output		Paved roads % of total						
Country	1980	1996	1980	1996	1990	1997	1990	1997	1990	1997	1996
Kuwait	4,749	12,808	10	0	73	81	X	X	X	X	2,133
Kyrgystan	1,556	1,479	6	33	90	91	5,627	350	X	X	488
Laos	X	X	X	X	24	14	120	X	X	X	125
Lebanon	789	1,651	10	13	95	95	X	X	X	X	775
Malaysia	630	2,078	9	11	70	75	X	X	16,313	9,416	15,118
Mongolia	X	X	X	X	10	3	1,871	X	1,324,119	X	X
Myanmar (Burma)	31	58	22	36	11	12	X	X	X	X	335
Nepal	13	39	29	28	38	42	X	X	X	X	755
Oman	X	X	X	X	X	X	X	X	X	X	X
Pakistan	125	333	29	23	54	58	352	84,174	43,586	26,582	5,375
Philippines	353	405	2	17	0	0	X	X	X	X	7,263
Saudi Arabia	1,356	3,980	9	8	41	43	X	X	4,634	4,206	11,706
Singapore	2,412	7,196	5	4	97	97	X	X	X	X	11,841
Sri Lanka	96	203	15	17	32	40	19	30	5,926	X	1,171
Syria	345	755	18	0	72	23	X	X	48,075	29,655	599
Tajikistan	2,217	2,292	7	12	72	83	X	X	X	X	594
Thailand	279	1,289	10	9	55	98	X	X	14,869	X	14,078
Turkey	439	1,161	12	17	X	25	X	139,789	30,838	17,747	8,464
Turkmenistan	1,720	1,020	12	11	74	81	X	X	X	X	523
United Arab Emirates	X	X	X	X	X	X	X	X	X	X	X
Uzbekistan	2,085	1,657	9	9	79	87	X	X	X	X	1,566
Vietnam	50	177	18	19	24	25	X	X	13,526	16,352	2,108
Yemen	59	99	6	26	9	8	X	X	X	X	X
EUROPE											
Albania	1,083	904	4	52	X	30	1,195	80	85,396	5,523	13
Austria	4,371	5,952	6	6	100	100	13,300	16,600	89,362	78,423	4,719
Belarus	2,455	2,476	9	16	96	98	22,128	9,065	1,297,626	624,045	843
Belgium	4,402	6,878	5	5	81	80	32,100	42,800	46,189	31,976	5,174
Bosnia-Herzegovina	X	X	X	X	X	X	X	X	X	X	X
Bulgaria	3,349	3,577	10	13	92	92	13,823	483	360,291	210,161	718
Croatia	0	2,291	X	16	80	82	2,458	470	190,170	86,593	727
Czech Republic	3,595	4,875	7	8	100	100	X	43,088	X	207,099	1,394
Denmark	4,245	6,113	7	5	100	100	9,400	9,400	19,119	14,518	5,892
Estonia	3,433	3,293	5	19	52	51	4,510	2,773	516,391	536,100	149
Finland	7,779	12,979	6	4	61	64	26,300	24,100	99,052	68,994	5,598
France	3,881	6,091	7	6	X	100	137,000	158,200	49,908	39,109	41,253
Germany	5,005	5,596	4	5	99	99	245,700	281,300	X	39,350	40,118
Greece	2,064	3,395	7	7	92	92	12,600	12,800	6,395	1,913	6,396
Hungary	2,335	2,814	10	13	50	43	1,836	770	247,428	104,327	1,563
Ireland	2,528	4,363	10	9	94	94	5,100	5,500	14,322	9,132	7,677
Italy	2,831	4,196	9	7	100	100	177,900	197,600	20,795	18,420	25,839
Latvia	2,664	1,783	26	47	13	38	5,853	800	1,209,517	1,114,210	276
Lithuania	2,715	1,785	12	11	82	89	7,019	8,622	915,522	545,100	214
Macedonia	0	2,443	X	X	59	64	1,708	1,210	X	X	287
Moldova	1,495	1,314	8	23	87	87	6,305	780	X	X	190
Netherlands	4,057	5,555	4	4	88	90	22,900	27,600	12,779	9,751	17,114
New Zealand	6,269	8,420	13	11	57	58	X	X	51,927	X	9,597
Norway	18,289	23,487	9	8	69	74	7,940	11,838	X	X	12,727
Poland	2,470	2,420	10	13	62	66	49,800	95,500	475,103	284,381	1,806
Portugal	1,469	3,044	12	10	X	X	10,900	11,200	13,976	13,598	4,806

Table L *(Continued)*
World Countries: Power and Transportation

	Electric power										
	Consumption per capita Kilowatt-hours		Transmission and distribution losses % of output		Paved roads % of total		Goods transported by road Millions of ton-km hauled		Goods transported by rail Ton-km per $ million of GDP (PPP)		Air passengers carried Thousands
Country	1980	1996	1980	1996	1990	1997	1990	1997	1990	1997	1996
Romania	2,434	1,757	6	12	*51*	51	13,800	22,400	507,379	*231,838*	913
Russian Federation	4,706	4,165	8	9	74	X	300	138	2,725,816	X	22,117
Slovak Republic	3,817	4,450	8	6	99	99	4,180	3,779	X	*297,426*	63
Slovenia	4,089	4,766	8	6	72	83	3,440	1,775	*142,879*	112,529	393
Spain	2,401	3,749	9	9	74	99	151,000	186,700	22,427	*15,984*	27,759
Sweden	10,216	14,239	9	7	71	77	26,500	*31,200*	127,826	103,299	9,879
Switzerland	5,579	6,919	7	7	X	X	10,400	13,000	X	X	10,468
Ukraine	3,598	2,640	8	10	94	95	79,668	20,532	2,109,937	*1,411,737*	1,151
United Kingdom	4,160	5,198	8	9	100	100	136,300	*153,900*	17,191	X	64,209
Yugoslavia (Serbia-Montenegro)	X	X	X	X	X	X	X	X	X	X	X
OCEANIA											
Australia	5,393	8,086	10	7	35	*39*	X	X	82,122	X	30,075
Fiji	X	X	X	X	X	X	X	X	X	X	X
New Zealand	6,269	8,420	13	11	57	*58*	X	X	51,927	X	9,597
Papua New Guinea	X	X	X	X	3	*4*	X	X	X	X	970
Solomon Islands	X	X	X	X	X	X	X	X	X	X	X
World	1,576w	**2,027w**	**8w**	**8w**	**39m**	**44m**					**1,389,943s**
Low income	188	433	12	12	17	*19*					103,110
Excl. China & India	155	218	14	19	17	*18*					37,945
Middle income	1,585	1,902	9	12	52	*51*					238,360
Lower middle income	1,835	1,771	8	11	54	*51*					102,609
Upper middle income	1,188	2,106	10	13	52	*47*					135,751
Low and middle income	633	886	9	12	29	*30*					341,470
East Asia & Pacific	260	724	8	9	24	*12*					143,204
Europe & Central Asia	2,925	2,795	8	11	77	*83*					46,014
Latin America & Carib.	854	1,347	12	16	22	*26*					76,275
Middle East & N. Africa	483	1,162	10	9	67	*50*					37,484
South Asia	116	313	19	19	38	*41*					22,445
Sub-Saharan Africa	444	439	9	10	17	*16*					16,049
High income	5,783	8,121	8	6	86	*92*					1,048,473

Note: Figures in italics are for years other than those specified.

Source: Entering the 21st Century: World Development Report 1999/2000 (Oxford University Press, World Bank, 2000).

Table M
World Countries: Communications, Information, and Science and Technology

	Per 1,000 People						Internet hosts Per 10,000 people	Scientists and engineers in R & D Per million people	High-technology exports % of mfg. exports	No. of patent applications filed[a]	
	Daily newspapers	Radio	Television sets	Telephone main lines	Mobile telephones	Personal computers					
Country	1996	1996	1997	1997	1997	1997	January 1999	1985–95	1997	Residents	Nonresidents
AFRICA											
Algeria	38	239	67	48	1	4.2	0.01	X	22	48	150
Angola	12	54	91	5	1	0.7	0.00	X	X	X	X
Benin	2	108	91	6	1	0.9	0.02	177	X	X	X
Botswana	27	155	27	56	0	13.4	4.18	X	X	5	56
Burkina Faso	1	32	6	3	0	0.7	0.16	X	X	X	X
Burundi	3	68	10	3	0	X	0.00	32	X	1	4
Cameroon	7	162	81	5	0	1.5	0.00	X	3	X	X
Central African Republic	2	84	5	3	0	X	0.00	55	0	X	X
Chad	0	249	2	1	0	X	0.00	X	X	X	X
Congo (Zaire)	3	98	43	1	0	X	0.00	X	X	2	27
Congo Republic	8	124	8	8	0	X	0.00	X	16	X	X
Côte d'Ivoire	16	157	61	9	2	3.3	0.16	X	X	X	X
Egypt	38	316	127	56	0	7.3	0.31	458	7	504	706
Equatorial Guinea	X	X	X	X	X	X	X	X	X	X	X
Eritrea	X	101	11	6	0	X	0.00	X	X	X	X
Ethiopia	2	194	5	3	0	X	0.01	X	0	3	X
Gabon	X	X	X	X	X	X	X	X	X	X	X
Gambia	X	X	X	X	X	X	X	X	X	X	X
Ghana	14	238	109	6	1	1.6	0.10	X	X	X	33
Guinea	X	47	41	3	0	0.3	0.00	X	X	X	X
Guinea-Bissau	X	X	X	X	X	X	X	X	X	X	X
Kenya	9	108	19	8	0	2.3	0.23	X	11	15	39,034
Lesotho	7	48	24	10	1	X	0.09	X	X	2	37,043
Liberia	X	X	X	X	X	X	X	X	X	X	X
Libya	X	X	X	X	X	X	X	X	X	X	X
Madagascar	4	192	45	3	0	1.3	0.04	11	2	7	20,800
Malawi	3	256	2	4	0	X	0.00	X	3	3	39,031
Mali	1	49	10	2	0	0.6	0.00	X	X	X	X
Mauritania	1	150	89	5	0	5.3	0.06	X	X	X	X
Mauritius	X	X	X	X	X	X	X	X	X	X	X
Morocco	26	241	160	50	3	2.5	0.20	X	27	90	237
Mozambique	3	39	4	4	0	1.6	0.08	X	8	X	X
Namibia	19	143	32	58	8	18.6	15.79	X	X	X	X
Niger	0	69	26	2	0	0.2	0.02	X	X	X	X
Nigeria	24	197	61	4	0	5.1	0.03	15	X	X	X
Rwanda	0	102	X	3	0	X	0.00	24	X	X	X
Senegal	5	141	41	13	1	11.4	0.21	X	55	X	X
Sierra Leone	5	251	20	4	0	X	0.03	X	X	X	X
Somalia	X	X	X	X	X	X	X	X	X	X	X
South Africa	30	316	125	107	37	41.6	34.67	938	X	X	X
Sudan	X	X	X	X	X	X	X	X	X	X	X
Swaziland	X	X	X	X	X	X	X	X	X	X	X
Tanzania	4	278	21	3	1	1.6	0.04	X	X	X	X
Togo	4	217	19	6	1	5.8	0.24	X	X	X	X
Tunisia	31	218	182	70	1	8.6	0.07	388	11	46	128
Uganda	2	123	26	2	0	1.4	0.05	X	X	X	38,497
Zambia	14	121	80	9	0	X	0.31	X	X	6	93
Zimbabwe	18	96	29	17	1	9.0	0.87	X	6	30	181

World Countries: Communications, Information, and Science and Technology

	Daily newspapers	Radio	Television sets	Telephone main lines	Mobile telephones	Personal computers	Internet hosts Per 10,000 people	Scientists and engineers in R & D Per million people	High-technology exports % of mfg. exports	No. of patent applications filed[a]	
	Per 1,000 People										
Country	1996	1996	1997	1997	1997	1997	January 1999	1985–95	1997	Residents	Nonresidents
NORTH AMERICA											
Canada	159	1,078	708	609	139	270.6	364.25	2,656	25	3,316	45,938
United States	212	2,115	847	644	206	406.7	1,131.52	3,732	*44*	111,883	111,536
CENTRAL AMERICA											
Belize	X	X	X	X	X	X	X	X	X	X	X
Costa Rica	91	271	403	169	19	X	9.20	X	*14*	X	X
Cuba	X	X	X	X	X	X	X	X	X	X	X
Dominican Republic	52	177	*84*	88	16	X	5.79	X	23	X	X
El Salvador	48	461	*250*	56	7	X	1.33	19	16	3	64
Guatemala	31	73	126	41	6	*3.0*	0.83	99	13	2	102
Haiti	3	55	*5*	8	0	X	0.00	X	X	3	6
Honduras	55	409	*90*	37	2	X	0.16	X	4	10	126
Jamaica	64	482	*323*	140	22	4.6	1.24	8	67	X	X
Mexico	97	324	251	96	18	37.3	11.64	213	33	389	30,305
Nicaragua	32	283	190	29	2	X	1.47	214	38	X	X
Panama	62	299	*187*	134	6	X	2.66	X	14	31	142
Trinidad and Tobago	X	X	X	X	X	X	X	X	X	X	X
SOUTH AMERICA											
Argentina	123	677	289	191	56	39.2	18.28	671	15	X	X
Bolivia	55	672	*115*	69	15	X	0.78	250	9	17	106
Brazil	40	435	316	107	28	26.3	12.88	168	18	2,655	29,451
Chile	*99*	354	233	180	28	54.1	20.18	X	19	189	1,771
Colombia	49	565	217	148	35	33.4	3.93	X	20	87	1,172
Ecuador	70	342	294	75	13	*13.0*	1.26	169	12	7	354
Guyana	X	X	X	X	X	X	X	X	X	X	X
Paraguay	50	182	*101*	43	17	X	2.18	X	4	X	X
Peru	43	271	143	68	18	12.3	1.91	625	10	52	565
Suriname	X	X	X	X	X	X	X	X	X	X	X
Uruguay	116	610	*242*	232	46	21.9	46.61	688	8	25	182
Venezuela	206	471	172	116	46	36.6	3.37	208	10	182	1,822
ASIA											
Afghanistan	X	X	X	X	X	X	X	X	X	X	X
Armenia	*23*	5	218	150	2	X	1.01	X	X	162	20,268
Azerbaijan	*28*	20	211	87	5	X	0.21	X	X	165	16,470
Bangladesh	9	50	*7*	3	0	X	X	X	0	70	156
Bhutan	X	X	X	X	X	X	X	X	X	X	X
Cambodia	X	127	124	2	3	0.9	0.06	X	X	X	X
China	X	195	270	56	10	6.0	0.14	350	21	11,698	41,016
Hong Kong, China	800	695	412	565	343	230.8	122.71	98	29	41	2,059
Georgia	X	553	*473*	114	6	X	1.27	X	X	289	21,124
India	X	105	69	19	1	2.1	0.13	149	*11*	1,660	6,632
Indonesia	23	155	134	25	5	8.0	0.75	X	20	40	3,957
Iran	24	237	148	107	4	*32.7*	0.04	521	X	X	X
Iraq	X	X	X	X	X	X	X	X	X	X	X
Israel	291	530	321	450	283	186.1	161.96	X	33	1,363	12,172
Japan	580	957	708	479	304	202.4	133.53	6,309	38	340,861	60,390
Jordan	45	287	43	70	*2*	8.7	0.80	106	26	X	X
Kazakhstan	30	384	*234*	108	1	X	0.94	X	X	1,024	20,064
Korea, North	X	X	X	X	X	X	X	X	X	X	X
Korea, South	394	1,037	341	444	150	150.7	40.00	2,636	39	68,446	45,548

-183-

Table M *(Continued)*
World Countries: Communications, Information, and Science and Technology

	Per 1,000 People						Internet hosts Per 10,000 people	Scientists and engineers in R & D Per million people	High-technology exports % of mfg. exports	No. of patent applications filed[a]	
	Daily newspapers	Radio	Television sets	Telephone main lines	Mobile telephones	Personal computers					
Country	1996	1996	1997	1997	1997	1997	January 1999	1985–95	1997	Residents	Nonresidents
Kuwait	376	688	*491*	227	116	82.9	32.80	X	4	X	X
Kyrgyzstan	13	115	44	76	*0*	X	4.04	703	24	126	20,179
Laos	4	139	4	5	1	*1.1*	0.00	X	X	X	X
Lebanon	141	892	354	179	135	31.8	5.56	X	X	X	X
Malaysia	163	432	166	195	113	46.1	21.36	87	*67*	X	X
Mongolia	27	139	63	37	1	5.4	0.08	943	*2*	114	20,882
Myanmar (Burma)	10	89	*7*	5	0	X	0.00	X	X	X	X
Nepal	11	*37*	4	8	0	X	0.07	X	0	X	X
Oman	X	X	X	X	X	X	X	X	X	X	X
Pakistan	*21*	92	65	19	1	4.5	0.23	54	4	16	782
Philippines	82	159	109	29	18	13.6	1.21	157	56	163	2,634
Saudi Arabia	59	319	*260*	117	17	43.6	0.15	X	29	27	810
Singapore	324	739	354	543	273	399.5	210.02	2,728	71	215	38,403
Sri Lanka	29	210	91	17	6	4.1	0.29	173	X	50	21,138
Syria	20	274	*68*	88	0	1.7	0.00	X	1	X	X
Tajikistan	20	X	281	38	0	X	0.12	709	X	32	19,570
Thailand	65	204	234	80	33	19.8	3.35	119	43	203	4,355
Turkey	111	178	286	250	26	20.7	*4.30*	261	9	367	19,668
Turkmenistan	X	96	*175*	78	0	X	0.55	X	X	66	18,948
United Arab Emirates	X	X	X	X	X	X	X	X	X	X	X
Uzbekistan	3	452	*273*	63	0	X	0.10	1,760	X	914	21,088
Vietnam	4	106	*180*	21	2	4.6	0.00	308	X	37	22,206
Yemen	15	64	*273*	13	1	1.2	0.01	X	0	X	X
EUROPE											
Albania	34	235	*161*	23	1	X	0.30	X	1	1	18,761
Austria	294	740	496	492	144	210.7	176.79	1,631	*24*	2,506	75,985
Belarus	*174*	290	314	227	1	X	0.70	2,339	X	701	20,347
Belgium	160	792	510	468	95	235.3	162.39	1,814	23	1,356	59,099
Bosnia-Herzegovina	X	X	X	X	X	X	X	X	X	X	X
Bulgaria	253	531	*366*	323	8	29.7	9.05	X	X	318	22,235
Croatia	114	333	*267*	335	27	22.0	12.84	1,978	19	259	356
Czech Republic	256	806	447	318	51	82.5	71.79	1,159	13	623	24,856
Denmark	311	1,146	568	633	273	360.2	526.77	2,647	27	2,452	72,151
Estonia	173	680	479	321	99	15.1	152.98	2,018	24	12	21,144
Finland	455	1,385	534	556	417	310.7	1,058.13	2,812	26	3,262	61,556
France	218	943	606	575	99	174.4	82.91	2,584	*31*	17,090	81,418
Germany	311	946	570	550	99	255.5	160.23	2,843	26	56,757	98,338
Greece	*153*	477	466	516	89	44.8	48.81	774	12	434	52,371
Hungary	189	697	436	304	69	49.0	82.74	1,033	39	832	24,147
Ireland	153	703	*455*	411	146	241.3	148.70	1,871	62	925	52,407
Italy	104	874	483	447	204	113.0	58.80	1,325	15	8,860	71,992
Latvia	246	699	*592*	302	31	7.9	42.59	1,189	15	197	21,498
Lithuania	92	292	*377*	283	41	6.5	27.48	X	21	101	21,249
Macedonia	19	184	*252*	204	6	X	2.56	X	X	53	18,934
Moldova	59	720	302	145	1	3.8	1.17	1,539	*9*	290	20,245
Netherlands	305	963	541	564	110	280.3	358.51	2,656	44	4,884	61,958
Norway	593	920	579	621	381	360.8	717.53	3,678	*24*	1,550	25,638
Poland	113	518	413	194	22	36.2	28.07	1,299	12	2,414	24,902
Portugal	75	306	523	402	152	74.4	50.01	1,185	11	105	71,544
Romania	X	317	*226*	167	9	8.9	7.42	1,382	7	1,831	22,139

Table M (Continued)
World Countries: Communications, Information, and Science and Technology

	Per 1,000 People					Internet hosts	Scientists and engineers in R & D	High-technology exports	No. of patent applications filed[a]	
	Daily newspapers	Radio	Television sets	Telephone main lines	Mobile telephones	Personal computers	Per 10,000 people	Per million people	% of mfg. exports		
Country	1996	1996	1997	1997	1997	1997	January 1999	1985–95	1997	Residents	Nonresidents
Russian Federation	105	344	390	183	3	32.0	10.04	3,520	19	18,138	28,149
Slovak Republic	185	580	401	259	37	241.6	33.27	1,821	15	201	22,865
Slovenia	206	416	353	364	47	188.9	89.83	2,544	16	301	21,686
Spain	99	328	506	403	110	122.1	67.21	1,210	*17*	2,689	81,294
Sweden	446	907	531	679	358	350.3	487.13	3,714	34	7,077	76,364
Switzerland	330	969	536	661	147	394.9	315.52	X	28	2,699	75,576
Ukraine	54	872	493	186	1	*5.6*	3.13	3,173	X	3,640	22,862
United Kingdom	332	1,445	641	540	151	242.4	240.99	2,417	41	25,269	104,084
Yugoslavia (Serbia-Montenegro)	X	X	X	X	X	X	X	X	X	X	X
OCEANIA											
Australia	297	1,385	638	505	264	362.2	420.57	3,166	39	9,196	34,125
Fiji	X	X	X	X	X	X	X	X	X	X	X
New Zealand	223	1,027	*501*	486	149	263.9	360.44	1,778	11	1,421	26,947
Papua New Guinea	15	91	24	*11*	1	X	0.25	X	X	X	X
Solomon Islands	X	X	X	X	X	X	X	X	X	X	X
WORLD	Xw	380w	280w	144w	40w	58.4w	75.22w				
Low income	X	147	162	32	5	4.4	0.17				
Excl. China & India	13	133	*59*	16	1	X	0.23				
Middle income	75	383	272	136	24	32.4	10.15				
Lower middle income	63	327	247	108	11	*12.2*	4.91				
Upper middle income	95	469	302	179	43	45.5	19.01				
Low and middle income	X	218	194	65	11	12.3	3.08				
East Asia & Pacific	X	206	237	60	15	11.3	1.66				
Europe & Central Asia	99	412	380	189	13	*17.7*	13.00				
Latin America & Carib.	71	414	263	110	26	31.6	9.64				
Middle East & N. Africa	33	265	*140*	71	6	9.8	0.25				
South Asia	X	99	69	18	1	2.1	0.14				
Sub-Saharan Africa	12	172	*44*	16	4	7.2	2.39				
High income	286	1,300	664	552	188	269.4	470.12				

Note: Figures in italics are for years other than those specified.

a. Other patent applications filed in 1996 include those filed under the auspices of the African Intellectual Property Organization (75 by residents, 20,863 by nonresidents), the African Regional Industrial Property Organization (10 by residents, 20,347 by nonresidents), the European Patent Office (38,546 by residents, 48,068 by nonresidents), and the Eurasian Patent Organization (39 by residents, 18,055 by nonresidents). The original information was provided by the World Intellectual Property Organization (WIPO). The International Bureau of WIPO assumes no liability or responsibility with regard to the transformation of these data.

Source: Entering the 21st Century: World Development Report 1999/2000 (Oxford University Press, World Bank, 2000).

Table N
World Countries: Water Resources

COUNTRY	ANNUAL RENEWABLE WATER RESOURCES[a]		ANNUAL AVERAGE GROUNDWATER RESOURCES[b]		SECTORAL WITHDRAWALS (%)[c]					
	Supply Per Capita (cubic meters) 2000	Withdrawal Per Capita (cubic meters) 2000	Recharge Per Capita (cubic meters) 2000	Withdrawal Per Capita (cubic meters) 2000	Domestic		Industry		Agriculture	
					Surface	Ground	Surface	Ground	Surface	Ground
WORLD	**7,045**	**664**	**–**	**–**	**9**	**65**	**19**	**15**	**67**	**20**
AFRICA	**5,159**	**307**	**–**	**–**	**8**	**–**	**4**	**–**	**63**	**–**
Algeria	442	180	54	117.1	25	46	15	5	60	49
Angola	14,288	57	5,591	–	14	–	10	–	76	–
Benin	1,689	28	295	–	23	–	10	–	67	–
Botswana	1,788	83	1,048	–	30	–	19	–	46[e]	–
Burkina Faso	1,466	39	796	–	19	–	0	–	81	–
Burundi	538	20	314	–	36	–	0	–	64	–
Cameroon	17,766	38	6,629	–	46	–	19	–	35	–
Central African Republic	39,001	26	15,490	–	21	–	6	–	74	–
Chad	1,961	34	1,503	15.7	16	29	2	–	82	71
Congo Republic	75,387	20	67,268	–	62	–	27	–	11	–
Côte d'Ivoire	5,187	66	2,550	–	22	–	11	–	67	–
Democratic Republic of the Congo (formerly Zaire)	18,101	8	8,150	–	61	–	16	–	23	–
Egypt	26	920	19	85.1	6	58	8	0	86	42[k]
Equatorial Guinea	66,275	30	22,092	–	81	–	13	–	6	–
Eritrea	727	–	–	–	–	–	–	–	–	–
Ethiopia	1,758	47	703	–	11	–	3	–	86	–
Gabon	133,754	70	50,566	0.6	72	100	22	0	6	0
The Gambia	2,298	33	383	–	7	–	2	–	91[f]	–
Ghana	1,499	35	1,301	–	35	–	13	–	52	–
Guinea	30,416	141	5,114	–	10	–	3	–	87	–
Guinea-Bissau	13,189	17	11,541	–	60	–	4	–	36	–
Kenya	672	87	100	–	20	–	4	–	76	–
Lesotho	2,430	31	232	–	22	–	22	–	56	–
Liberia	63,412	54	19,023	–	27	–	13	–	60	–
Libya	143	783	116	734.9	9	9	4	4	87	87l
Madagascar	21,139	1,694	3,450	482.9	1	0	0	–	99[g]	–
Malawi	1,605	98	128	–	10	–	3	–	86	–
Mali	5,341	164	1,780	11.6	2	–	1	–	97	–
Mauritania	150	923	112	498.3	6	–	2	–	92	–
Mauritius	–	–	–	–	–	–	–	–	–	–
Morocco	1,058	446	317	97.9	5	16	3	–	92	–
Mozambique	5,081	40	864	–	9	–	2	–	89	–
Namibia	3,592	185	1,217	–	28	–	3	–	68	–
Niger	326	69	233	17.9	16	58	2	4	82	39
Nigeria	1,982	45	780	–	31	–	15	–	54	–
Rwanda	815	134	466	–	5	–	2	–	94	–
Senegal	2,784	202	802	39.2	5	24	3	–	92	72

Table N
World Countries: Water Resources

COUNTRY	ANNUAL RENEWABLE WATER RESOURCES[a]		ANNUAL AVERAGE GROUNDWATER RESOURCES[b]		SECTORAL WITHDRAWALS (%)[c]					
	Supply Per Capita (cubic meters) 2000	Withdrawal Per Capita (cubic meters) 2000	Recharge Per Capita (cubic meters) 2000	Withdrawal Per Capita (cubic meters) 2000	Domestic		Industry		Agriculture	
					Surface	Ground	Surface	Ground	Surface	Ground
Sierra Leone	32,960	98	10,300	–	7	–	4	–	89	–
Somalia	594	115	327	45.8	3	–	0	–	97	–
South Africa	1,110	391	119	64.9	17	11	11	6	72	84
Sudan	1,187	669	237	13	5	–	1	–	94	–
Swaziland	–	–	–	–	–	–	–	–	–	–
Tanzania	2,387	40	895	–	9	–	2	–	89	–
Togo	2,484	28	1,231	–	62	–	13	–	25	–
Tunisia	367	295	433	181.8	14	10	3	4	83	86
Uganda	1,791	20	1,332	–	32	–	8	–	60	–
Zambia	8,747	214	5,137	–	16	–	7	–	77	–
Zimbabwe	1,208	136	428	–	14	–	7	–	79	–
NORTH AMERICA	**21,583**	**1,907**	**–**	**–**	**10**	**–**	**39**	**–**	**46**	**–**
Canada	87,971	1,623	11,879	37.3	11	34	68	11	7	34[m]
United States	8,838	1,844	5,439	432.3	11	20	44	5	40	62[n]
CENTRAL AMERICA	**6,290**	**715**	**–**	**–**	**–**	**–**	**–**	**–**	**–**	**–**
Belize	66,470	469	–	–	12	–	88	–	0	–
Costa Rica	27,936	1540	5,219	–	13	–	7	–	80	–
Cuba	3,393	475	714	408.3	49	–	0	–	51	–
Dominican Republic	2,472	1085	353	–	11	–	0	–	89	–
El Salvador	2,820	137	–	–	34	–	20	–	46	–
Guatemala	11,805	126	2,723	–	9	–	17	–	74	–
Haiti	1,473	139	304	–	5	–	1	–	94	–
Honduras	14,818	293	6,013	–	4	–	5	–	91	–
Jamaica	3,640	371	–	–	15	–	7	–	77	–
Mexico	4,136	812	1,406	275.4	17	13	5	23	78	64[o]
Nicaragua	37,484	267	11,627	–	14	–	2	–	84	–
Panama	51,616	685	14,708	–	28	–	2	–	70	–
Trinidad and Tobago	–	–	–	–	–	–	–	–	–	–
SOUTH AMERICA	**34,791**	**518**	**–**	**–**	**20**	**–**	**11**	**–**	**60**	**–**
Argentina	9,721	822	3,456	180.4	16	11	9	19	75	70
Bolivia	37,941	197	15,609	–	10	–	3	–	87	–
Brazil	31,849	359	11,016	57	21	38	18	25	61	38
Chile	61,007	1,629	9,204	–	5	–	11	–	84	–
Colombia	50,400	228	12,051	–	59	–	4	–	37	–
Ecuador	34,952	1423	10,596	–	12	–	6	–	82	–
Guyana	279,799	1,811	119,582	–	1	–	0	–	98	–
Paraguay	17,102	112	7,459	–	15	–	7	–	78	–
Peru	68,039	849	11,807	139.4	7	25	7	15	86	60

Table N
World Countries: Water Resources

COUNTRY	ANNUAL RENEWABLE WATER RESOURCES[a]		ANNUAL AVERAGE GROUNDWATER RESOURCES[b]		SECTORAL WITHDRAWALS (%)[c]					
	Supply Per Capita (cubic meters) 2000	Withdrawal Per Capita (cubic meters) 2000	Recharge Per Capita (cubic meters) 2000	Withdrawal Per Capita (cubic meters) 2000	Domestic		Industry		Agriculture	
					Surface	Ground	Surface	Ground	Surface	Ground
Suriname	479,467	1,171	191,787	–	6	–	5	–	89	–
Uruguay	17,680	1352	6,892	–	6	–	3	–	91[h]	–
Venezuela	35,002	382	9,393	–	44	–	10	–	46	–
ASIA	**3,668**	**627**	**–**	**–**	**7**	**–**	**9**	**–**	**81**	**–**
Afghanistan	2,421	1,846	1,276	–	1	–	0	–	99	–
Armenia	2,577	817	1,193	–	30	–	4	–	66	–
Azerbaijan	1,049	2,186	842	–	5	–	25	–	70	–
Bangladesh	813	134	163	97.6	12	13	2	1	86	88[p]
Bhutan	44,728	13	–	–	36	–	10	–	54	–
Cambodia	10,795	66	1,576	–	5	–	1	–	94	–
China	2,201	439	649	47.1	5	–	18	–	77	54
Georgia	11,702	635	3,469	549.5	21	–	20	–	59	–
India	1,244	588	413	223.3	5	9	3	2	92	89[q]
Indonesia	13,380	407	2,145	–	6	–	1	–	93	–
Iran	1,898	1,165	620	738.8	6	–	2	–	92	–
Iraq	1,523	2,368	562	13.1	3	50	5	40	92	–
Israel	312	292	80	204.5	29	18	7	2	64[i]	80[r]
Japan	3,393	735	213	108.2	19	29	17	41	64	30[s]
Jordan	102	187	87	100.7	22	30	3	4	75	66[t]
Kazakhstan	4,649	2,019	2,211	143.9	2	21	17	71	81	8[u]
Korea, North	2,787	726	874	–	11	–	16	–	73	–
Korea, South	1,384	531	284	55.1	26	–	11	–	63	17[nn]
Kuwait	–	307	–	142.7	37	0	2	0	60	100[v]
Kyrgyzstan	9,884	2,219	2,894	132	3	50	3	25	94	25
Laos	35,049	260	6,994	–	8	–	10	–	82	–
Lebanon	1,463	444	1,463	153.2	28	13	4	9	68	78
Malaysia	26,074	633	2,877	19	11	62	13	33	76	5
Mongolia	13,073	182	2,291	149.1	20	–	27	–	53	–
Myanmar (Burma)	19,306	102	3,420	–	7	–	3	–	90	–
Nepal	8,282	1397	–	–	1	–	0	–	99	–
Oman	388	658	376	280.7	5	–	2	–	94	–
Pakistan	541	1,269	351	489.5	2	–	2	–	97	90[w]
Philippines	6,305	811	2,369	82.8	8	50	4	50	88	–
Saudi Arabia	111	1,002	44	899.3	9	10	1	–	90	90
Singapore	–	–	–	–	–	–	–	–	–	–
Sri Lanka	2,656	573	414	–	2	–	2	–	96	–
Syria	434	1,069	409	133.5	4	13	2	4	94	83[x]
Tajikistan	10,714	2,095	970	398.7	4	–	4	–	92	–
Thailand	3,420	596	682	15	5	60	4	26	91	14
Turkey	2,943	560	300	124	16	31	11	9	73	60[x]
Turkmenistan	305	5,947	753	100.3	1	53	1	9	98	38

Table N
World Countries: Water Resources

COUNTRY	ANNUAL RENEWABLE WATER RESOURCES[a]		ANNUAL AVERAGE GROUNDWATER RESOURCES[b]		SECTORAL WITHDRAWALS (%)[c]					
	Supply Per Capita (cubic meters) 2000	Withdrawal Per Capita (cubic meters) 2000	Recharge Per Capita (cubic meters) 2000	Withdrawal Per Capita (cubic meters) 2000	Domestic		Industry		Agriculture	
					Surface	Ground	Surface	Ground	Surface	Ground
United Arab Emirates	61	954	49	724.1	24	–	10	19[d]	67	81[y]
Uzbekistan	672	2,626	809	334.3	4	–	2	11	94	57[z]
Vietnam	4,591	814	601	11.9	4	–	10	–	86	–
Yemen	226	253	84	139.2	7	–	1	–	92	–
EUROPE	**3,981**	**704**	**–**	**–**	**14**	**–**	**45**	**–**	**39**	**–**
Albania	8,646	441	2,248	193.6	29	48	0	–	71	52
Austria	6,699	281	2,716	172.5	31	52	60	43	9	5[aa]
Belarus	3,634	266	1,758	115.7	22	52	43	13	35	28[bb]
Belgium	1,181	917	89	79	–	55	–	22	–	4
Bosnia-Herzegovina	8,938	–	–	–	–	–	–	–	–	–
Bulgaria	2,188	1,574	1,629	566.1	3	–	77	–	22	–
Croatia	8,429	170	2459	–	50	–	50	–	0	–
Czech Republic	1,464	244	–	48	39	–	57	–	1	–
Denmark	1,134	170	5,668	169.8	53	40	9	22	16	38[cc]
Estonia	9,105	106	2,865	–	56	–	39	–	5	–
Finland	20,673	477	367	47.8	17	65	82	11	–	24[dd]
France	3,047	704	1,693	103.8	15	56	73	27	12	17
Germany	1,301	583	556	89 4	14	48	86	47	0	4[ee]
Greece	5,073	688	968	195.7	16	37	3	5	81	58
Hungary	598	612	678	96.5	14	35	70	48	5	18[ff]
Iceland	605,049	611	85,419	558.9	50	–	6	–	–	–
Ireland	13,136	326	928	62.3	40	35	21	38	15	29[gg]
Italy	2,804	840	750	243.2	14	39	33	4	53	58
Latvia	7,104	111	934	–	55	–	32	–	13	–
Lithuania	4,239	68	327	55.1	81	–	16	–	3	–
Macedonia	2,965	–	–	–	–	–	–	–	–	–
Moldova	228	677	91	–	9	–	65	–	26	–
Netherlands	697	522	285	70.2	16	32	68	45	–	23[hh]
Norway	85,560	488	21,502	97.5	27	27	68	73	3	–
Poland	1,419	313	929	51.5	20	70	64	30	2	–
Portugal	3,747	739	516	311	8	39	40	23	53	39[ii]
Romania	2,195	851	372	158	–	61	–	38	–	1[jj]
Russia	29,351	520	5,363	85.5	19	–	62	–	20	–
Slovakia	2,413	263	–	113	39	–	50	–	8	–
Slovenia	9,317	250	–	88.9	50	–	50	–	0	–
Spain	2,821	897	729	137.2	13	18	18	2	68	80
Sweden	19,977	310	2,245	72.8	35	92	30	8	4	–
Switzerland	5,416	363	366	126.3	42	72	58	40	0	–
Ukraine	1,052	501	396	77.5	18	30	52	18	30	52[kk]
United Kingdom	2,465	160	167	42.4	65	51	8	47	2	2[ll]

Table N
World Countries: Water Resources

COUNTRY	ANNUAL RENEWABLE WATER RESOURCES[a]		ANNUAL AVERAGE GROUNDWATER RESOURCES[b]		SECTORAL WITHDRAWALS (%)[c]					
	Supply Per Capita (cubic meters) 2000	Withdrawal Per Capita (cubic meters) 2000	Recharge Per Capita (cubic meters) 2000	Withdrawal Per Capita (cubic meters) 2000	Domestic		Industry		Agriculture	
					Surface	Ground	Surface	Ground	Surface	Ground
Yugoslavia (Serbia-Montenegro)	4,135	–	–	–	–	–	–	–	–	–
OCEANIA	**78,886**	**1178**	–	–	–	–	–	–	–	–
Australia	18,638	839	3812	143.2	12	–	6	20[d]	70[j]	67[mm]
Fiji	–	–	–	–	–	–	–	–	–	–
New Zealand	84,673	545	51,270	–	9	–	13	–	55	–
Papua New Guinea	166,644	28	–	–	29	–	22	–	49	–
Solomon Islands	–	–	–	–	–	–	–	–	–	–

a. Annual renewable water resources usually include river flows from other countries.
b. Withdrawal data from most recent year available; varies by country from 1987 to 1995.
c. Total withdrawals may exceed 100% because of groundwater withdrawals or river inflows.
d. Domestic and industrial withdrawals have been combined.
e. An additional 5% is reserved for overflow.
f. Sectoral figures are for 1982.
g. Sectoral figures are for 1984.
h. Sectoral figures are for 1965.
i. Sectoral withdrawals reflect percentages of a total water withdrawal (1.959 cubic km) for 1997, including recycled water.
J. Sectoral figures are for 1985.
k. Sectoral figures are for 1992, Margat Blue Plan.
l. Sectoral percentages are calculated using groundwater withdrawal of 3.81 km^3, which is an estimate provided with sectoral data for 1995 in the Margat Blue Plan.
m. Sectoral data for Canada are calculated using a groundwater withdrawal value of 1.6 km^3 from 1985 as reported by Margat 1990.
n. Sectoral data for the U.S. are calculated using a groundwater withdrawal value of 101.3 km^3 from 1985 as reported by Margat 1990.
o. Sectoral data for Mexico are calculated using a groundwater withdrawal value of 23.5 km^3 from 1985 as reported by Margat 1990.
p. Sectoral data for Bangladesh are calculated using a groundwater withdrawal value of 3.4 km^3 from 1979 as reported by Margat 1990.
q. Sectoral data are from around 1990 as provided by Shiklomonov; total withdrawal data are also from 1990, but are from FAO.
r. Sectoral figures are for 1994, Margat Blue Plan.
s. Sectoral data are from around 1987 as provided by Shiklomonov 1997 based on groundwater withdrawal of 12.88 km^3.
t. Ground water withdrawal and sectoral data are estimated from a bar graph from 1993 from FAO Water Report.
u. Both withdrawal and sectoral data are estimated from a bar graph from FAO Report.
v. Ground water withdrawal and sectoral data are estimated from a bar graph from 1994 from FAO Water Report.
w. Sectoral data for Pakistan are from Shiklomonov who reports approximately 90% for agriculture share; total withdrawal also is approximately 60 km^3 per year for 1990.
x. Sectoral figures are for 1990, Margat Blue Plan.
y. Sectoral percentages for UAE are a combination of data from text and a bar graph for 1995 from FAO Water Report.
z. Sectoral data for Uzbekistan are from 1994 FAO irrigation in the former Soviet Union in figures estimated from a bar graph.
aa. Sectoral data for Austria are calculated using a groundwater withdrawal value of 1.17 km^3 from 1985 as reported by Margat 1990.
bb. Sectoral data for Belarus are calculated using a groundwater withdrawal value of 1.06 km^3 from 1985 as reported by Margat 1990.
cc. Sectoral data calculated using a groundwater withdrawal value of 1.32 km^3 from 1977 as reported by Margat 1990.
dd. Sectoral data calculated using a groundwater withdrawal value of .37 km^3 from 1980 as reported by Margat 1990.
ff. Sectoral data calculated using a groundwater withdrawal value of 1.6 km^3 from 1972 as reported by Margat 1990.
gg. Sectoral data calculated using a groundwater withdrawal value of .17 km^3 from 1980 as reported by Margat 1990.
hh. Sectoral data calculated using a groundwater withdrawal value of 1.28 km^3 from 1981 as reported by Margat 1990.
ii. Sectoral data calculated using a groundwater withdrawal value of 2.0 km^3 from 1980 as reported by Margat 1990.
jj. Sectoral data calculated using a groundwater withdrawal value of 1.18 km^3 from 1975 as reported by Margat 1990.
kk. Sectoral data calculated using a groundwater withdrawal value of 4.22 km^3 from 1985 as reported by Margat 1990.
ll. Sectoral data calculated using a groundwater withdrawal value of 2.38 km^3 from 1975 as reported by Margat 1990.
mm. Sectoral data calculated using a groundwater withdrawal value of 2.46 km^3 from 1983 as reported by Margat 1990.
nn. Sectoral data calculated using a groundwater withdrawal value of 1.2 km^3 from 1985 as reported by Margat 1990.

Source: *World Resources 1998–99* (Washington, DC, World Resources Institute)

Table O
World Countries: Globally Threatened Plant and Animal Species

COUNTRY	Mammals Threatened Species	Mammals Number of Species per 10,000 km²	Birds Threatened Species	Birds Number of Species per 10,000 km²	Reptiles Threatened Species	Reptiles Number of Species per 10,000 km²	Amphibians Threatened Species	Amphibians Number of Species per 10,000 km²	Freshwater Fish Threatened Species	Plants Rare and Threatened Species	Plants Number of Species per 10,000 km²
AFRICA											
Algeria	15	15	8	32	1	X	0	X	1	145	509
Angola	17	56	13	156	5	X	0	X	0	25	1,017
Benin	9	85	1	138	2	X	0	X	0	3	899
Botswana	5	43	7	101	0	41	0	X	0	4	X
Burkina Faso	6	49	1	112	1	X	0	10	0	0	369
Burundi	5	76	6	322	0	X	0	X	0	1	1,783
Cameroon	32	83	14	193	3	X	1	X	26	74	2,237
Central African Republic	11	53	2	137	1	X	0	X	0	0	921
Chad	14	27	3	75	1	X	0	X	0	12	322
Congo (Zaire)	38	69	26	153	3	X	0	X	1	7	1,817
Congo Republic	10	62	3	140	2	X	0	X	0	3	1,356
Côte d'Ivoire	16	73	12	170	4	X	1	X	0	66	1,118
Egypt	15	21	11	33	6	18	0	1	0	84	452
Equatorial Guinea	12	131	4	194	2	X	1	X	0	9	2,135
Eritrea	6	49	3	140	3	X	0	X	0	X	X
Ethiopia	35	54	20	133	1	X	0	X	0	153	1,378
Gabon	12	64	4	157	3	X	0	X	0	78	2,197
Gambia	4	104	1	269	1	X	0	X	0	0	928
Ghana	13	78	10	186	4	X	0	X	0	32	1,264
Guinea	11	66	12	142	3	X	1	X	0	35	1,043
Guinea-Bissau	4	71	1	159	3	X	0	X	0	0	655
Kenya	43	94	24	221	5	49	0	23	20	158	1,571
Lesotho	2	23	5	40	0	X	0	X	1	7	1,093
Liberia	11	87	13	168	3	28	1	17	0	1	1,037
Libya	11	14	2	17	3	X	0	X	0	57	327
Madagascar	46	27	28	53	17	66	2	38	13	189	2,347
Malawi	7	86	9	230	0	55	0	31	0	61	1,592
Mali	13	28	6	81	1	3	0	X	0	14	355
Mauritania	14	13	3	59	3	X	0	X	0	3	239
Mauritius	4	7	10	46	6	19	0	0	0	222	1,183
Morocco	18	30	11	60	2	X	0	X	1	195	1,028
Mozambique	13	42	14	117	5	X	0	15	2	92	1,294
Namibia	11	36	8	109	3	X	1	7	3	23	729
Niger	11	27	2	60	1	X	0	X	0	0	237
Nigeria	26	62	9	153	4	X	0	X	0	9	1,036
Rwanda	9	110	6	373	0	X	0	X	0	0	1,662
Senegal	13	58	6	144	7	X	0	X	0	32	771
Sierra Leone	9	77	12	243	3	X	0	X	0	12	1,091
Somalia	18	43	8	107	2	49	0	7	3	57	761
South Africa	33	51	16	122	19	61	9	19	27	953	4,711
Sudan	21	43	9	110	3	X	0	X	0	8	506
Swaziland	5	37	6	303	0	85	0	33	0	41	2,197
Tanzania	33	70	30	183	4	63	0	28	19	406	229
Togo	8	110	1	220	3	X	0	X	0	0	1,128
Tunisia	11	31	6	69	2	X	0	X	0	24	855
Uganda	18	118	10	290	1	52	0	17	28	6	1,762
Zambia	11	55	10	145	0	X	0	20	0	9	1,105
Zimbabwe	9	81	9	159	0	46	0	36	0	94	1,253

COUNTRY	Mammals		Birds		Reptiles		Amphibians		Freshwater Fish	Plants	
	Threatened Species	Number of Species per 10,000 km²	Threatened Species	Number of Species per 10,000 km²	Threatened Species	Number of Species per 10,000 km²	Threatened Species	Number of Species per 10,000 km²	Threatened Species	Rare and Threatened Species	Number of Species per 10,000 km²
NORTH AMERICA											
Canada	7	20	5	44	3	4	1	4	13	649	299
United States	35	45	50	68	28	29	24	24	123	1,845	1,679
CENTRAL AMERICA											
Belize	5	95	1	271	5	81	0	24	0	41	2,090
Costa Rica	14	120	13	350	7	125	1	95	0	456	6,421
Cuba	9	14	13	62	7	46	0	19	4	811	2,714
Dominican Republic	4	12	11	81	10	62	1	21	0	73	2,965
El Salvador	2	106	0	196	6	57	0	18	0	35	1,956
Guatemala	8	114	4	208	9	105	0	45	0	315	3,638
Haiti	4	2	11	54	6	73	1	33	0	28	3,345
Honduras	7	78	4	190	7	68	0	25	0	55	2,252
Jamaica	4	23	7	110	8	35	4	20	0	371	2,662
Mexico	64	79	36	135	18	120	3	50	86	1,048	4,382
Nicaragua	4	86	3	207	7	69	0	25	0	78	3,003
Panama	17	112	10	376	7	116	0	84	1	561	4,618
Trinidad and Tobago	1	125	3	324	5	87	0	32	0	16	2,470
SOUTH AMERICA											
Argentina	27	50	41	140	5	34	5	23	1	170	1,407
Bolivia	23	67	27	X	3	44	0	24	0	49	3,500
Brazil	71	43	103	161	15	51	5	54	12	463	5,935
Chile	16	22	18	71	1	17	3	10	4	292	1,229
Colombia	35	75	64	355	15	122	0	123	5	376	10,479
Ecuador	28	100	53	460	12	124	0	133	1	375	6,052
Guyana	10	70	3	246	8	X	0	X	0	47	2,180
Paraguay	10	90	26	164	3	35	0	25	0	12	2,208
Peru	46	69	64	310	9	60	1	63	0	377	3,448
Suriname	10	72	2	240	6	60	0	38	0	48	1,870
Uruguay	5	31	11	92	0	X	0	X	0	11	845
Venezuela	24	69	22	266	14	58	0	45	5	107	4,510
ASIA											
Afghanistan	11	31	13	59	1	26	1	2	0	6	882
Armenia	4	X	5	X	3	32	0	4	0	0	X
Azerbaijan	11	X	8	X	3	26	0	4	5	1	X
Bangladesh	18	45	30	122	13	49	0	8	0	24	2,074
Bhutan	20	59	14	269	1	11	0	14	0	20	3,268
Cambodia	23	47	18	118	9	32	0	11	5	7	X
China	75	41	90	114	15	35	1	27	28	343	3,112
Georgia	10	X	5	X	7	24	0	6	3	1	X
India	75	47	73	136	16	57	3	29	4	1,256	2,216
Indonesia	128	77	104	260	19	90	0	48	60	281	4,864
Iran	20	26	14	60	8	30	2	2	7	1	X
Iraq	7	23	12	49	2	23	0	2	2	2	X
Israel	13	72	8	141	5	X	0	X	0	38	X
Japan	29	40	33	X	8	20	10	16	7	704	1,418
Jordan	7	34	4	68	1	X	0	X	0	10	1,069
Kazakhstan	15	X	15	X	1	6	1	2	5	0	X
Korea, North	7	X	19	51	0	8	0	6	0	7	1,274
Korea, South	6	23	19	53	0	12	0	7	0	69	1,360
Kuwait	1	17	3	17	2	24	0	2	0	0	193

Table O *(Continued)*
World Countries: Globally Threatened Plant and Animal Species

COUNTRY	Mammals		Birds		Reptiles		Amphibians		Freshwater Fish	Plants	
	Threatened Species	Number of Species per 10,000 km²	Threatened Species	Number of Species per 10,000 km²	Threatened Species	Number of Species per 10,000 km²	Threatened Species	Number of Species per 10,000 km²	Threatened Species	Rare and Threatened Species	Number of Species per 10,000 km²
Kyrgystan	6	X	5	X	1	9	0	1	0	1	X
Laos	30	61	27	171	7	23	0	13	4	5	X
Lebanon	5	53	5	152	2	X	0	X	0	4	X
Malaysia	42	90	34	158	14	85	0	50	14	510	4,732
Mongolia	12	25	14	X	0	4	0	2	0	1	429
Myanmar (Burma)	31	62	44	216	20	51	0	19	1	29	1,742
Nepal	28	70	27	255	5	33	0	15	0	21	2,716
Oman	9	20	5	39	4	23	0	X	3	4	371
Pakistan	13	36	25	88	6	41	0	4	1	12	1,163
Philippines	49	50	86	129	7	62	2	21	26	371	2,604
Saudi Arabia	9	13	11	26	2	14	0	X	0	6	294
Singapore	6	113	9	295	1	X	0	X	1	14	5,007
Sri Lanka	14	47	11	134	8	77	0	21	8	436	1,613
Syria	4	24	7	78	3	X	0	X	0	10	X
Tajikistan	5	X	9	X	1	16	0	1	1	0	X
Thailand	34	72	45	168	16	81	0	29	14	382	2,999
Turkey	15	28	14	72	12	24	2	4	18	1827	2,012
Turkmenistan	11	X	12	X	2	22	0	1	5	1	X
United Arab Emirates	3	12	4	33	2	18	0	X	1	0	X
Uzbekistan	7	X	11	X	0	15	0	1	3	5	X
Vietnam	38	67	47	168	12	57	1	25	3	350	X
Yemen	5	18	13	39	2	21	0	X	0	X	X
EUROPE											
Albania	2	48	7	162	1	22	0	9	7	50	2,093
Austria	7	41	5	106	1	7	0	10	7	22	1,462
Belarus	4	X	4	81	0	3	0	4	0	0	X
Belgium	6	40	3	125	0	6	0	12	1	3	969
Bosnia-Herzegovina	10	X	2	X	X	X	1	X	6	0	X
Bulgaria	13	37	12	108	1	15	0	8	8	94	1,584
Croatia	10	X	4	126	X	X	1	X	20	0	X
Czech Republic	7	X	6	101	X	X	X	X	6	X	X
Denmark	3	27	2	121	0	3	0	9	0	6	741
Estonia	4	40	2	130	0	3	0	7	1	2	992
Finland	4	18	4	78	0	2	0	2	1	11	325
France	13	25	7	72	3	9	2	9	3	117	1,198
Germany	8	23	5	73	0	4	0	6	7	X	X
Greece	13	41	10	107	6	22	1	6	16	539	2,091
Hungary	8	34	10	98	1	7	0	8	11	24	1,029
Iceland	1	5	0	41	0	0	0	0	0	1	157
Ireland	2	13	1	75	0	1	0	2	1	9	469
Italy	10	29	7	76	4	13	4	11	9	273	1,776
Latvia	4	45	6	117	0	4	0	7	1	0	623
Lithuania	5	37	4	109	0	4	0	7	1	0	646
Macedonia	10	X	3	X	1	X	X	X	4	X	X
Moldova	2	46	7	119	1	6	0	9	9	1	X
Netherlands	6	35	3	120	0	4	0	10	1	1	758
Norway	4	17	3	77	0	2	0	2	1	20	524
Poland	10	27	6	72	0	3	0	6	2	27	738
Portugal	13	30	7	99	0	14	1	8	9	240	1,200

World Countries: Globally Threatened Plant and Animal Species

COUNTRY	Mammals		Birds		Reptiles		Amphibians		Freshwater Fish	Plants	
	Threatened Species	Number of Species per 10,000 km^2	Threatened Species	Number of Species per 10,000 km^2	Threatened Species	Number of Species per 10,000 km^2	Threatened Species	Number of Species per 10,000 km^2	Threatened Species	Rare and Threatened Species	Number of Species per 10,000 km^2
Romania	16	29	11	87	2	9	0	7	11	122	1,116
Russian Federation	31	23	38	54	5	5	0	2	13	127	X
Slovak Republic	8	X	4	124	0	X	0	X	7	X	X
Slovenia	10	55	3	164	0	17	1	X	5	11	X
Spain	19	22	10	76	6	15	3	7	10	896	X
Sweden	5	17	4	71	0	2	0	4	1	19	1,400
Switzerland	6	47	4	121	0	9	0	11	4	9	1,033
Ukraine	15	X	10	85	2	6	0	5	12	16	756
United Kingdom	4	17	2	80	0	3	0	2	1	28	539
Yugoslavia (Serbia-Montenegro)	12	X	8	X	1	X	X	X	13	X	X
OCEANIA											
Australia	58	28	45	72	37	83	25	23	37	1,597	1,672
Fiji	4	3	9	61	6	20	1	2	0	72	1,071
New Zealand	3	3	44	51	11	13	1	1	8	236	727
Papua New Guinea	57	60	31	182	10	79	0	56	13	95	2,821
Solomon Islands	20	37	18	115	4	43	0	12	0	43	1,959

Source: World Conservation Monitoring Centre and World Conservation Union; World Resources Institute, *World Resources 1998–99, 1998.*

Part IX

Geographic Index

Name/Description	Latitude & Longitude	Page
Abidjan,Cote d'Ivoire (city,nat. cap.)	5N 4W	108
Abu Dhabi, U.A.E. (city, nat. cap.)	24N 54E	114
Accra, Ghana (city, nat. cap.)	64N 0	108
Aconcagua, Mt. 22,881	38S 78W	95
Acre (st., Brazil)	9S 70W	94
Addis Ababa, Ethiopia (city, nat. cap.)	9N 39E	108
Adelaide, S. Australia (city, st. cap.,Aust.)	35S 139E	123
Aden, Gulf of	12N 46E	115
Aden, Yemen (city)	13N 45E	114
Admiralty Islands	1S 146E	124
Adriatic Sea	44N 14E	101
Aegean Sea	39N 25E	101
Afghanistan (country)	35N 65E	114
Aguascalientes (st., Mex.)	22N 110W	95
Aguascalientes, Aguas. (city, st. cap., Mex.)	22N 102W	95
Agulhas, Cape	35S 20E	109
Ahaggar Range	23N 6E	109
Ahmadabad, India (city)	23N 73E	114
Akmola, Kazakhstan (city)	51N 72E	114
Al Fashir, Sudan (city)	14N 25E	108
Al Fayyum, Egypt (city)	29N 31E	108
Al Hijaz Range	30N 40E	115
Al Khufra Oasis	24N 23E	109
Alabama (st., US)	33N 87W	86
Alagoas (st., Brazil)	9S 37W	94
Alaska (st., US)	63N 153W	86 inset
Alaska, Gulf of	58N 150W	86 inset
Alaska Peninsula	57N 155W	87 inset
Alaska Range	60N 150W	87 inset
Albania (country)	41N 20E	100
Albany, Australia (city)	35S 118E	123
Albany, New York (city, st. cap., US)	43N 74W	86
Albert Edward, Mt. 13,090	8S 147E	124
Albert, Lake	2N 30E	109
Alberta (prov., Can.)	55N 117W	86
Albuquerque, NM (city)	35N 107W	86

The geographic index contains approximately 1,500 names of cities, states, countries, rivers, lakes, mountain ranges, oceans, capes, bays, and other geographic features. The name of each geographical feature in the index is accompanied by a geographical coordinate (latitude and longitude) in degrees and by the page number of the primary map on which the geographical feature appears. Where the geographical coordinates are for specific places or points, such as a city or a mountain peak, the latitude and longitude figures give the location of the map symbol denoting that point. Thus, Los Angeles, California, is at 34N and 118W and the location of Mt. Everest is 28N and 87E.

The coordinates for political features (countries or states) or physical features (oceans, deserts) that are areas rather than points are given according to the location of the name of the feature on the map, except in those cases where the name of the feature is separated from the feature (such as a country's name appearing over an adjacent ocean area because of space requirements). In such cases, the feature's coordinates will indicate the location of the center of the feature. The coordinates for the Sahara Desert will lead the reader to the place name "Sahara Desert" on the map; the coordinates for North Carolina will show the center location of the state since the name appears over the adjacent Atlantic Ocean. Finally, the coordinates for geographical features that are lines rather than points or areas will also appear near the center of the text identifying the geographical feature.

Alphabetizing follows general conventions; the names of physical features such as lakes, rivers, mountains are given as: proper name, followed by the generic name. Thus "Mount Everest" is listed as "Everest, Mt." Where an article such as "the," "le," or "al" appears in a geographic name, the name is alphabetized according to the article. Hence, "La Paz" is found under "L" and not under "P."

Geographic Index

Name/Description	Latitude & Longitude	Page
Aldabra Islands	9S 44E	109
Aleppo, Syria (city)	36N 37E	114
Aleutian Islands	55N 175W	87
Alexandria, Egypt (city)	31N 30E	108
Algeria (country)	28N 15E	108
Algiers, Algeria (city, nat. cap.)	37N 3E	108
Alice Springs, Aust. (city)	24S 134E	123
Alma Ata, Kazakhstan (city, nat. cap.)	43N 77E	114
Alps Mountains	46N 6E	101
Altai Mountains	49N 87E	115
Altun Shan	45N 90E	115
Amapa (st., Brazil)	2N 52W	94
Amazon Riv.	2S 53W	95
Amazonas (st., Brazil)	2S 64W	94
Amman, Jordan (city, nat. cap.)	32N 36E	114
Amsterdam, Netherlands (city)	52N 5E	100
Amu Darya (riv., Asia)	40N 62E	115
Amur (riv., Asia)	52N 156E	115
Anchorage, AK (city)	61N 150W	86 inset
Andaman Islands	12N 92E	114
Andes Mountains	25S 70W	95
Angara (riv., Asia)	60N 100E	115
Angola (country)	11S 18E	108
Ankara, Turkey (city, nat. cap.)	40N 33E	114
Annapolis, Maryland (city, st. cap., US)	39N 76W	86
Antananarivo, Madagascar (city, nat. cap.)	19S 48E	108
Antofagasta, Chile (city)	24S 70W	94
Antwerp, Belgium (city)	51N 4E	100
Appalachian Mountains	37N 80W	87
Appenines Mountains	32N 14E	101
Arabian Desert	25N 33E	101
Arabian Peninsula	23N 40E	124
Arabian Sea	18N 61E	115
Aracaju, Sergipe (city, st. cap., Braz.)	11S 37W	94
Arafura Sea	9S 133E	124
Araguaia, Rio (riv., Brazil)	13S 50W	95
Aral Sea	45N 60E	115
Arctic Ocean	75N 160W	94
Arequipa, Peru (city)	16S 71W	94
Argentina (country)	39S 67W	94
Aripuana, Rio (riv., S.Am.)	11S 60W	95
Arizona (st., US)	34N 112W	86
Arkansas (riv., N.Am.)	38N 100W	87
Arkansas (st., US)	37N 94W	86
Arkhangelsk, Russia (city)	75N 160W	100
Armenia (country)	40N 45E	100
Arnhem, Cape	11S 139E	124
Arnhem Land	12S 133E	124
As Sudd	9N 26E	109

Geographic Index

Name/Description	Latitude & Longitude	Page
Ascension (island)	9S 13W	109
Ashburton (rlv., Australasia)	23S 115W	115
Ashkhabad, Turkmenistan (city, nat. cap.)	38N 58E	114
Asia Minor	39N 33E	108
Asmera, Eritrea (city, nat. cap.)	15N 39E	108
Astrakhan, Russia (city)	46N 48E	100
Asuncion, Paraguay (city, nat. cap.)	25S 57W	94
Aswan, Egypt (city)	24N 33E	108
Asyuf, Egypt (city)	27N 31E	108
Atacama Desert	23S 70W	95
Athabasca (lake, N.Am.)	60N 109W	87
Athabaska (riv., N.Am.)	58N 114W	87
Athens, Greece (city, nat. cap.)	38N 24E	100
Atlanta, Georgia (city, st. cap., US)	34N 84W	86
Atlantic Ocean	30N 40W	87
Atlas Mountains	31N 6W	109
Auckland, New Zealand (city)	37S 175E	123
Augusta, Maine (city, st. cap., US)	44N 70W	86
Austin, Texas (city, st. cap., US)	30N 98W	86
Australia (country)	20S 135W	123
Austria (country)	47N 14E	100
Ayers Rock 2844	25S 131E	124
Azerbaijan (country)	38N 48E	100
Azov, Sea of	48N 36E	101
Bab el Mandeb (strait)	13N 42E	109
Baffin Bay	74N 65W	87
Baffin Island	70N 72W	87
Baghdad, Iraq (city, nat. cap.)	33N 44E	101
Bahamas (island)	25N 75W	87
Bahia (st., Brazil)	13S 42W	94
Bahia Blanca, Argentina (city)	39S 62W	94
Baikal, Lake	52N 105E	115
Baja California (st., Mex.)	30N 110W	86
Baja California Sur (st., Mex.)	25N 110W	86
Baku, Azerbaijan (city, nat. cap.)	40N 50E	100
Balearic Islands	29N 3E	100
Balkash, Lake	47N 75E	115
Ballarat, Aust. (city)	38S 144E	123
Baltic Sea	56N 18E	101
Baltimore, MD (city)	39N 77W	115
Bamako, Mali (city, nat. cap.)	13N 8W	108
Bandiera Peak 9,843	20S 42W	95
Bangalore, India (city)	13N 75E	114
Bangeta, Mt. 13,520	6S 147E	124
Banghazi, Libya (city)	32N 20E	108
Bangkok, Thailand (city, nat. cap.)	14N 100E	114
Bangladesh (country)	23N 92E	114
Bangui, Cent. African Rep. (city, nat. cap.)	4N 19E	108
Banjul, Gambia (city, nat. cap.)	13N 17W	108

Geographic Index

Geographic Index

Name/Description	Latitude & Longitude	Page
Boise, Idaho (city, st. cap., US)	44N 116W	86
Bolivia (country)	17S 65W	94
Boma, Congo Republic (city)	5S 13E	108
Bombay, (Mumbai) India (city)	19N 73E	114
Bonn, Germany (city, nat. cap.)	51N 7E	100
Boothia Peninsula	71N 94W	87
Borneo (island)	0 11E	114
Bosnia-Herzegovina (country)	45N 18E	100
Bosporus, Strait of	41N 29E	101
Boston, Massachusetts (city, st. cap., US)	42N 71W	86
Botany Bay	35S 153E	123
Bothnia, Gulf of	62N 20E	101
Botswana (country)	23S 25E	108
Brahmaputra (riv., Asia)	30N 100E	115
Branco, Rio (riv., S.Am.)	3N 62W	95
Brasilia, Brazil (city, nat. cap.)	16S 48W	94
Bratislava, Slovakia (city, nat. cap.)	48N 17E	100
Brazil (country)	10S 52W	94
Brazilian Highlands	18S 45W	95
Brazzaville, Congo (city, nat. cap.)	4S 15E	108
Brisbane, Queensland (city, st. cap., Aust.)	27S 153E	123
Bristol Bay	58N 159W	87 inset
British Columbia (prov., Can.)	54N 130W	86
Brooks Range	67N 155W	87
Bruce, Mt. 4052	22S 117W	124
Brussels, Belgium (city, nat. cap.)	51N 4E	100
Bucharest, Romania (city, nat. cap.)	44N 26E	100
Budapest, Hungary (city, nat. cap.)	47N 19E	100
Buenos Aires, Argentina (city, nat. cap.)	34S 58W	94
Buenos Aires (st., Argentina)	36S 60W	94
Buffalo, NY (city)	43N 79W	86
Bujumbura, Burundi (city, nat. cap.)	3S 29E	108
Bulgaria (country)	44N 26E	100
Bur Sudan, Sudan (city)	19N 37E	108
Burdekin (riv., Australasia)	19S 146W	124
Burkina Faso (country)	11N 2W	108
Buru (island)	4S 127E	124
Burundi (country)	4S 30E	108
Cairns, Aust. (city)	17S 145E	123
Cairo, Egypt (city, nat. cap.)	30N 31E	108
Calcutta, (Kolkota) India (city)	23N 88E	114
Calgary, Canada (city)	51N 114W	86
Calicut, India (city)	11N 76E	114
California (st., US)	35N 120W	86
California, Gulf of	29N 110W	86
Callao, Peru (city)	13S 77W	94
Cambodia (country)	10N 106E	114
Cameroon (country)	5N 13E	108
Campeche (st., Mex.)	19N 90W	86

Geographic Index

Name/Description	Latitude & Longitude	Page
Campeche Bay	20N 92W	87
Campeche, Campeche (city, st. cap., Mex.)	19N 90W	86
Campo Grande, M.G.S. (city, st. cap., Braz.)	20S 55W	94
Canada (country)	52N 100W	86
Canadian (riv., N.Am.)	30N 100W	87
Canary Islands	29N 18W	109
Canberra, Australia (city, nat. cap.)	35S 149E	123
Cape Breton Island	46N 60W	87
Cape Town, South Africa (city)	34S 18E	108
Caracas, Venezuela (city, nat. cap.)	10N 67W	94
Caribbean Sea	18N 75W	94
Carnarvon, Australia (city)	25S 113E	123
Carpathian Mountains	48N 24E	101
Carpentaria, Gulf of	14S 140E	124
Carson City, Nevada (city, st. cap., US)	39N 120W	86
Cartagena, Colombia (city)	10N 76W	94
Cascade Range	45N 120W	87
Casiquiare, Rio (riv., S.Am.)	4N 67W	95
Caspian Depression	49N 48E	101
Caspian Sea	42N 48E	101
Catamarca (st., Argentina)	25S 70W	94
Catamarca, Catamarca (city, st. cap., Argen.)	28S 66W	94
Cauca, Rio (riv., S.Am.)	8N 75W	95
Caucasus Mountains	42N 40E	101
Cayenne, French Guiana (city, nat. cap.)	5N 52W	94
Ceara (st., Brazil)	4S 40W	94
Celebes (island)	0 120E	115
Celebes Sea	2N 120E	115
Central African Republic (country)	5N 20E	108
Ceram (island)	3S 129E	123
Chaco (st., Argentina)	25S 60W	94
Chad (country)	15N 20E	108
Chad, Lake	12N 12E	108
Changchun, China (city)	44N 125E	114
Chari (riv., Africa)	11N 16E	109
Charleston, SC (city)	33N 80W	86
Charleston, West Virginia (city, st. cap., US)	38N 82W	86
Charlotte, NC (city)	35N 81W	86
Charlotte Waters, Aust. (city)	26S 135E	123
Charlottetown, P.E.I. (city, prov. cap., Can.)	46N 63W	86
Chelyabinsk, Russia (city)	55N 61E	100
Chengdu, China (city)	30N 104E	114
Chesapeake Bay	36N 74W	87
Chetumal, Quintana Roo (city, st. cap., Mex.)	19N 88W	86
Cheyenne, Wyoming (city, st. cap., US)	41N 105W	86
Chiapas (st., Mex.)	17N 92W	86
Chicago, IL (city)	42N 87W	86
Chiclayo, Peru (city)	7S 80W	94
Chidley, Cape	60N 65W	87

Geographic Index

Name/Description	Latitude & Longitude	Page
Chihuahua (st., Mex.)	30N 110W	86
Chihuahua, Chihuahua (city, st. cap., Mex.)	29N 106W	86
Chile (country)	32S 75W	94
Chiloe (island)	43S 74W	95
Chilpancingo, Guerrero (city, st. cap., Mex.)	19N 99W	86
Chimborazo, Mt. 20,702	2S 79W	95
China (country)	38N 105E	114
Chisinau, Moldova (city, nat. cap.)	47N 29E	100
Chongqing, China (city)	30N 107E	114
Christchurch, New Zealand (city)	43S 173E	123
Chubut (st., Argentina)	44S 70W	94
Chubut, Rio (riv., S.Am.)	44S 71W	95
Cincinnati, OH (city)	39N 84W	86
Cleveland (city)	41N 82W	86
Coahuila (st., Mex.)	30N 105W	86
Coast Mountains (Can.)	55N 130W	87
Coast Ranges (US)	40N 120W	87
Coco Island	8N 88W	87
Cod, Cape	42N 70W	87
Colima (st., Mex.)	18N 104W	86
Colima, Colima (city, st. cap., Mex.)	19N 104W	86
Colombia (country)	4N 73W	94
Colombo, Sri Lanka (city, nat. cap.)	7N 80E	114
Colorado (riv., N.Am.)	36N 110W	87
Colorado (st., US)	38N 104W	86
Colorado, Rio (riv., S.Am.)	38S 70W	95
Colorado (Texas) (riv., N.Am.)	30N 100W	87
Columbia (riv., N.Am.)	45N 120W	86
Columbia, South Carolina (city, st. cap., US)	34N 81W	86
Columbus, Ohio (city, st. cap., US)	40N 83W	86
Comodoro Rivadavia, Argentina (city)	68S 70W	94
Comoros (country)	12S 44E	108
Conakry, Guinea (city, nat. cap.)	9N 14W	100
Concord, New Hampshire (city, st. cap., US)	43N 71W	86
Congo (country)	3S 15E	108
Congo (riv., Africa)	3N 22E	109
Congo Basin	4N 22E	109
Congo, Democratic Republic of (country)	5S 15E	108
Connecticut (st., US)	43N 76W	86
Connecticut (riv., N.Am.)	43N 76W	87
Cook, Mt. 12,316	44S 170E	124
Cook Strait	42S 175E	124
Copenhagen, Denmark (city, nat. cap.)	56N 12E	100
Copiapo, Chile (city)	27S 70W	94
Copiapo , Mt. 19,947	26S 70W	95
Coquimbo, Chile (city)	30S 70W	94
Coral Sea	15S 155E	124
Cordilleran Highlands	45N 118W	87
Cordoba (st., Argentina)	32S 67W	94

Geographic Index

Name/Description	Latitude & Longitude	Page
Cordoba, Cordoba (city, st. cap., Argen.)	32S 64W	94
Corrientes (st., Argentina)	27S 60W	94
Corrientes, Corrientes (city, st. cap., Argen.)	27S 59W	94
Corsica (island)	42N 9E	100
Cosmoledo Islands	9S 48E	109
Costa Rica (country)	15N 84W	86
Cote d'Ivoire (country)	7N 86W	108
Cotopaxi, Mt. 19,347	1S 78W	95
Crete (island)	36N 25W	101
Croatia (country)	46N 20W	100
Cuango (riv., Africa)	10S 16E	109
Cuba (country)	22N 78W	86
Cuiaba, Mato Grosso (city, st. cap., Braz.)	16S 56W	94
Cuidad Victoria, Tamaulipas (city, st. cap., Mex.)	24N 99W	86
Culiacan, Sinaloa (city, st. cap., Mex.)	25N 107W	86
Curitiba, Parana (city, st. cap., Braz.)	26S 49W	94
Cusco, Peru (city)	14S 72W	94
Cyprus (island)	36N 34E	101
Czech Republic (country)	50N 16E	100
d'Ambre, Cape	12S 50E	108
Dakar, Senegal (city, nat. cap.)	15N 17W	108
Dakhla, Western Sahara (city)	24N 16W	108
Dallas, TX (city)	33N 97W	86
Dalrymple , Mt. 4190	22S 148E	124
Daly (riv., Australasia)	14S 132E	124
Damascus, Syria (city, nat. cap.)	34N 36E	100
Danube (riv., Europe)	44N 24E	101
Dar es Salaam, Tanzania (city, nat. cap.)	7S 39E	108
Darien, Gulf of	9N 77W	95
Darling (riv., Australasia)	35S 144E	124
Darling Range	33S 116W	124
Darwin, Northern Terr. (city, st. cap., Aust.)	12S 131E	123
Davis Strait	57N 59W	87
Deccan Plateau	20N 80E	115
DeGrey (riv., Australasia)	22S 120E	124
Delaware (st., US)	38N 75W	86
Delaware (riv., N.Am.)	38N 77W	87
Delhi, India (city)	30N 78E	108
Denmark (country)	55N 10E	100
Denmark Strait	67N 27W	101
D'Entrecasteaux Islands	10S 153E	124
Denver, Colorado (city, st. cap., US)	40N 105W	86
Derby, Australia (city)	17S 124E	123
Des Moines (riv., N.Am.)	43N 95W	87
Des Moines, Iowa (city, st. cap., US)	42N 92W	86
Desolacion Island	54S 73W	95
Detroit, MI (city)	42N 83W	86
Dhaka, Bangladesh (city, nat. cap.)	24N 90E	114
Dinaric Alps	44N 20E	101

Geographic Index

Name/Description	Latitude & Longitude	Page
Djibouti (country)	12N 43E	108
Djibouti, Djibouti (city, nat. cap.)	12N 43E	108
Dnepr (riv., Europe)	50N 34E	101
Dnipropetrovsk, Ukraine (city)	48N 35E	100
Dodoma, Tanzania (city)	6S 36E	108
Dominican Republic (country)	20N 70W	86
Don (riv., Europe)	53N 39E	101
Donetsk, Ukraine (city)	48N 38E	100
Dover, Delaware (city, st. cap., US)	39N 75W	86
Dover, Strait of	52N 0	101
Drakensberg	30S 30E	109
Dublin, Ireland (city, nat. cap.)	53N 6W	100
Duluth, MN (city)	47N 92W	86
Dunedin, New Zealand (city)	46S 171E	123
Durango (st., Mex.)	25N 108W	86
Durango, Durango (city, st. cap., Mex.)	24N 105W	86
Durban, South Africa (city)	30S 31E	108
Dushanbe, Tajikistan (city, nat. cap.)	39N 69E	114
Dvina (riv., Europe)	64N 42E	101
Dzhugdzhur Khrebet	58N 138E	115
East Cape (NZ)	37S 180E	124
East China Sea	30N 128E	115
Eastern Ghats	15N 80E	115
Ecuador (country)	3S 78W	94
Edmonton, Alberta (city, prov. cap., Can.)	54N 114W	86
Edward, Lake	0 30E	109
Egypt (country)	23N 30E	108
El Aaiun, Western Sahara (city)	27N 13W	108
El Djouf	25N 15W	109
El Paso, TX (city)	32N 106W	86
El Salvador (country)	15N 90W	86
Elbe (riv., Europe)	54N 10E	101
Elburz Mountains	28N 60E	115
Elbruz, Mt. 18,510	43N 42E	115
Elgon, Mt. 14,178	1N 34E	109
English Channel	50N 0	101
Entre Rios (st., Argentina)	32S 60W	94
Equatorial Guinea (country)	3N 10E	108
Erg Iguidi	26N 6W	109
Erie (lake, N.Am.)	42N 85W	87
Eritrea (country)	16N 38E	108
Erzegebirge Mountains	50N 14E	101
Espinhaco Mountains	15S 42W	95
Espiritu Santo (island)	15S 168E	123
Espiritu Santo (st., Brazil)	20S 42W	94
Essen, Germany (city)	52N 8E	100
Estonia (country)	60N 26E	100
Ethiopia (country)	8N 40E	108
Ethiopian Plateau	8N 40E	109

Geographic Index

Name/Description	Latitude & Longitude	Page
Euphrates (riv., Asia)	28N 50E	115
Everard, Lake	32S 135E	124
Everard Ranges	28S 135E	124
Everest, Mt. 29,028	28N 84E	115
Eyre, Lake	29S 136E	124
Faeroe Islands	62N 11W	101
Fairbanks, AK (city)	63N 146W	86
Falkland Islands (Islas Malvinas)	52S 60W	95
Farewell, Cape (NZ)	40S 170E	124
Fargo, ND (city)	47N 97W	86
Farquhar, Cape	24S 114E	124
Fiji (country)	17S 178E	123
Finisterre, Cape	44N 10W	101
Finland (country)	62N 28E	100
Finland, Gulf of	60N 20E	101
Firth of Forth	56N 3W	101
Fitzroy (riv., Australasia)	17S 125E	124
Flinders Range	31S 139E	124
Flores (island)	8S 121E	124
Florianopolis, Sta. Catarina (city, st. cap., Braz.)	27S 48W	94
Florida (st., US)	28N 83W	86
Florida, Strait of	28N 80W	87
Fly (riv., Australasia)	8S 143E	124
Formosa (st., Argentina)	23S 60W	94
Formosa, Formosa (city, st. cap., Argen.)	27S 58W	94
Fort Worth, TX (city)	33N 97W	86
Fortaleza, Ceara (city, st. cap., Braz.)	4S 39W	94
France (country)	46N 4E	100
Frankfort, Kentucky (city, st. cap., US)	38N 85W	86
Frankfurt, Germany (city)	50N 9E	100
Fraser (riv., N.Am.)	52N 122W	87
Fredericton, N.B. (city, prov. cap., Can.)	46N 67W	86
Fremantle, Australia (city)	33S 116E	123
Freetown, Sierra Leone (city, nat. cap.)	8N 13W	108
French Guiana (country)	4N 52W	94
Fria, Cape	18S 12E	109
Fuzhou, China (city)	26N 119E	114
Gabes, Gulf of	33N 12E	109
Gabes, Tunisia (city)	34N 10E	108
Gabon (country)	2S 12E	108
Gaborone, Botswana (city, nat. cap.)	25S 25E	108
Gairdiner, Lake	32S 136E	124
Galveston, TX (city)	29N 95W	86
Gambia (country)	13N 15W	108
Gambia (riv., Africa)	13N 15W	109
Ganges (riv., Asia)	27N 85E	115
Gascoyne (riv., Australasia)	25S 115E	124
Gaspé Peninsula	50N 70W	87
Gdansk, Poland (city)	54N 19E	100

Geographic Index

Name/Description	Latitude & Longitude	Page
Geelong, Aust. (city)	38S 144E	123
Gees Gwardafuy (island)	15N 50F	109
Genoa, Gulf of	44N 10E	101
Geographe Bay	35S 115E	124
Georgetown, Guyana (city, nat. cap.)	8N 58W	94
Georgia (country)	42N 44E	100
Georgia (st., US)	30N 82W	86
Germany (country)	50N 12E	100
Ghana (country)	8N 3W	108
Gibraltar, Strait of	37N 6W	101
Gibson Desert	24S 124E	124
Gilbert (riv., Australasia)	8S 142E	124
Giluwe, Mt. 14,330	5S 144E	124
Glasgow, Scotland (city)	56N 6W	100
Gobi Desert	48N 105E	115
Godavari (riv., Asia)	18N 82E	115
Godwin-Austen (K2), Mt. 28,250	30N 70E	115
Goiania, Goias (city, st. cap., Braz.)	17S 49W	94
Goias (st., Brazil)	15S 50W	94
Gongga Shan 24,790	26N 102E	115
Good Hope, Cape of	33S 18E	109
Goteborg, Sweden (city)	58N 12E	100
Gotland (island)	57N 20E	101
Grampian Mountains	57N 4W	101
Gran Chaco	23S 70N	95
Grand Erg Occidental	29N 0	109
Grand Teton 13,770	45N 112W	87
Great Artesian Basin	25S 145E	124
Great Australian Bight	33S 130E	124
Great Barrier Reef	15S 145E	124
Great Basin	39N 117W	87
Great Bear Lake (lake, N.Am.)	67N 120W	87
Great Dividing Range	20S 145E	124
Great Indian Desert	25N 72E	115
Great Namaland	25S 16E	109
Great Plains	40N 105W	87
Great Salt Lake (lake, N.Am.)	40N 113W	87
Great Sandy Desert	23S 125E	124
Great Slave Lake (lake, N.Am.)	62N 110W	87
Great Victoria Desert	30S 125E	124
Greater Khingan Range	50N 120E	115
Greece (country)	39N 21E	100
Greenland (Denmark) (country)	78N 40W	86
Gregory Range	18S 145E	124
Grey Range	26S 145E	124
Guadalajara, Jalisco (city, st. cap., Mex.)	21N 103W	86
Guadalcanal (island)	9S 160E	123
Guadeloupe (island)	29N 120W	87
Guanajuato (st., Mex.)	22N 100W	86

Geographic Index

Name/Description	Latitude & Longitude	Page
Guanajuato, Guanajuato (city, st. cap., Mex.)	21N 101W	86
Guangzhou, China (city)	23N 113E	114
Guapore, Rio (riv., S.Am.)	15S 63W	95
Guatemala (country)	14N 90W	86
Guatemala, Guatemala (city, nat. cap.)	15N 91W	86
Guayaquil, Ecuador (city)	2S 80W	94
Guayaquil, Gulf of	3S 83W	95
Guerrero (st., Mex.)	18N 102W	86
Guianas Highlands	5N 60W	95
Guinea (country)	10N 10W	108
Guinea, Gulf of	3N 0	109
Guinea-Bissau (country)	12N 15W	108
Guyana (country)	6N 57W	94
Gydan Range	62N 155E	115
Haiti (country)	18N 72W	86
Hakodate, Japan (city)	42N 140E	114
Halifax Bay	18S 146E	124
Halifax, Nova Scotia (city, prov. cap., Can.)	45N 64W	86
Halmahera (island)	1N 128E	115 inset
Hamburg, Germany (city)	54N 10E	100
Hammersley Range	23S 116W	124
Hann, Mt. 2,800	15S 127E	124
Hanoi, Vietnam (city, nat. cap.)	21N 106E	114
Hanover Island	52S 74W	95
Harare, Zimbabwe (city, nat. cap.)	18S 31E	108
Harbin, China (city)	46N 126E	114
Harer, Ethiopia (city)	10N 42E	108
Hargeysa, Somalia (city)	9N 44E	108
Harrisburg, Pennsylvania (city, st. cap., US)	40N 77W	86
Hartford, Connecticut (city, st. cap., US)	42N 73W	86
Hatteras, Cape	32N 73W	87
Havana, Cuba (city, nat. cap.)	23N 82W	86
Hawaii (st., US)	21N 156W	87 inset
Hebrides (island)	58N 8W	101
Helena, Montana (city, st. cap., US)	47N 112W	86
Helsinki, Finland (city, nat. cap.)	60N 25E	100
Herat, Afghanistan (city)	34N 62E	114
Hermosillo, Sonora (city, st. cap., Mex.)	29N 111W	86
Hidalgo (st., Mex.)	20N 98W	86
Himalayas	26N 80E	115
Hindu Kush	30N 70E	115
Ho Chi Minh City, Vietnam (city)	11N 107E	114
Hobart, Tasmania (city, st. cap., Aust.)	43S 147E	123
Hokkaido (island)	43N 142E	115
Honduras (country)	16N 87W	86
Honduras, Gulf of	15N 88W	87
Honiara, Solomon Islands (city, nat. cap.)	9S 160E	123
Honolulu, Hawaii (city, st. cap., US)	21N 158W	86 inset
Honshu (island)	38N 140E	115

Geographic Index

Name/Description	Latitude & Longitude	Page
Hormuz, Strait of	25N 58E	115
Horn, Cape	55S 70W	95
Houston, TX (city)	30N 95W	86
Howe, Cape	37S 150E	124
Huambo, Angola (city)	13S 16E	108
Huang (riv., Asia)	30N 105E	115
Huascaran, Mt. 22,133	8N 79W	95
Hudson (riv., N.Am.)	42N 76W	86
Hudson Bay	60N 90W	87
Hudson Strait	63N 70W	87
Hue, Vietnam (city)	15N 110F	114
Hughes, Aust. (city)	30S 130E	123
Hungary (country)	48N 20E	100
Huron (lake, N.Am.)	45N 85W	87
Hyderabad, India (city)	17N 79E	114
Ibadan, Nigeria (city)	7N 4E	108
Iceland (country)	64N 20W	100
Idaho (st., US)	43N 113W	86
Iguassu Falls	25S 55W	95
Illimani, Mt. 20,741	16S 67W	95
Illinois (riv., N.Am.)	40N 90W	87
Illinois (st., US)	44N 90W	86
India (country)	23N 80E	114
Indiana (st., US)	46N 88W	86
Indianapolis, Indiana (city, st. cap., US)	40N 86W	86
Indigirka (riv., Asia)	70N 145E	115
Indonesia (country)	2S 120E	114
Indus (riv., Asia)	25N 70E	115
Ionian Sea	38N 19E	101
Iowa (st., US)	43N 95W	86
Iquitos, Peru (city)	4S 74W	94
Iran (country)	30N 55E	114
Iraq (country)	30N 50E	114
Ireland (country)	54N 8W	100
Irish Sea	54N 5W	100
Irkutsk, Russia (city)	52N 104E	114
Irrawaddy (riv., Asia)	25N 95E	115
Irtysh (riv., Asia)	50N 70E	115
Ishim (riv., Asia)	48N 70E	115
Isla de los Estados (island)	55S 60W	95
Islamabad, Pakistan (city, nat. cap.)	34N 73E	114
Isles of Scilly	50N 8W	101
Israel (country)	31N 36E	100
Istanbul, Turkey (city)	41N 29E	100
Italy (country)	42N 12E	100
Jabal Marrah, 10, 131	10N 23E	109
Jackson, Mississippi (city, st. cap., US)	32N 84W	86
Jacksonville, FL (city)	30N 82W	86
Jakarta, Indonesia (city, nat. cap.)	6S 107E	114 inset

Geographic Index

Name/Description	Latitude & Longitude	Page
Jalisco (st., Mex.)	20N 105W	86
Jamaica (country)	18N 78W	86
James Bay	54N 81W	87
Japan (country)	35N 138E	114
Japan, Sea of	40N 135E	115
Japura, Rio (riv., S.Am.)	3S 65W	95
Java (island)	6N 110E	115 inset
Jaya Peak 16,503	4S 136W	124
Jayapura, New Guinea (Indon.) (city)	3S 141E	114 inset
Jebel Toubkal 13,665	31N 8W	109
Jefferson City, Missouri (city, st. cap., US)	39N 92W	86
Jerusalem, Israel (city, nat. cap.)	32N 35E	100
Joao Pessoa, Paraiba (city, st. cap., Braz.)	7S 35W	94
Johannesburg, South Africa (city)	26S 27E	108
Jordan (country)	32N 36E	100
Juan Fernandez (island)	33S 80W	95
Jubba (riv., Africa)	3N 43E	109
Jujuy (st., Argentina)	23S 67W	94
Jujuy, Jujuy (city, st. cap., Argen.)	23S 66W	94
Juneau, Alaska (city, st. cap., US)	58N 134W	86
Jura Mountains	46N 5E	101
Jurua, Rio (riv., S.Am.)	6S 70W	95
Kabul, Afghanistan (city, nat. cap.)	35N 69E	114
Kalahari Desert	25S 20E	109
Kalgourie-Boulder, Australia (city)	31S 121E	123
Kaliningrad, Russia (city)	55N 21E	100
Kamchatka Range	55N 159E	115
Kampala, Uganda (city, nat. cap.)	0 33E	108
Kanchenjunga, Mt. 28,208	30N 83E	115
Kano, Nigeria (city)	12N 9E	108
Kanpur, India (city)	27N 80E	114
Kansas (st., US)	40N 98W	86
Kansas City, MO (city)	39N 95W	86
Kara Sea	69N 65E	115
Karachi, Pakistan (city)	25N 66E	114
Karakorum Range	32N 78E	115
Karakum Desert	42N 52E	101
Kasai (riv., Africa)	5S 18E	109
Kashi, China (city)	39N 76E	114
Katherine, Aust. (city)	14S 132E	123
Kathmandu, Nepal (city, nat. cap.)	28N 85E	114
Katowice, Poland (city)	50N 19E	100
Kattegat, Strait of	57N 11E	101
Kazakhstan (country)	50N 70E	114
Kentucky (st., US)	37N 88W	86
Kenya (country)	0 35E	108
Kenya, Mt. 17, 058	0 37E	109
Khabarovsk, Russia (city)	48N 135E	114
Khambhat, Gulf of	20N 73E	115

Geographic Index

Name/Description	Latitude & Longitude	Page
Kharkiv, Ukraine (city)	50N 36E	100
Khartoum, Sudan (city, nat. cap.)	16N 33E	108
Kiev, Ukraine (city, nat. cap.)	50N 31E	100
Kigali, Rwanda (city, nat. cap.)	2S 30E	108
Kilimanjaro, Mt. 19,340	4N 35E	109
Kimberly, South Africa (city)	29S 25E	108
King Leopold Ranges	16S 125E	124
Kingston, Jamaica (city, nat. cap.)	18N 77W	86
Kinshasa, Congo Republic (city, nat. cap.)	4S 15E	108
Kirghiz Steppe	40N 65E	115
Kisangani, Congo Republic (city)	1N 25E	108
Kitayushu, Japan (city)	34N 130E	114
Klyuchevskaya, Mt. 15,584	56N 160E	115
Kobe, Japan (city)	34N 135E	114
Kodiak Island	58N 152W	87 inset
Kolyma (riv., Asia)	70N 160E	115
Kommunizma, Mt. 24,590	40N 70E	115
Komsomolsk, Russia (city)	51N 137E	114
Korea, North (country)	40N 128E	114
Korea, South (country)	3S 130W	114
Korea Strait	32N 130W	115
Kosciusko, Mt. 7,310	36S 148E	124
Krasnoyarsk, Russia (city)	56N 93E	114
Krishna (riv., Asia)	15N 76E	115
Kuala Lumpur, Malaysia (city, nat. cap.)	3N 107E	114
Kunlun Shan	36N 90E	115
Kunming, China (city)	25N 103E	114
Kuril Islands	46N 147E	115
Kutch, Gulf of	23N 70E	115
Kuwait (country)	29N 48E	100
Kuwait, Kuwait (city, nat. cap.)	29N 48E	100
Kyoto, Japan (city)	35N 136E	114
Kyrgyzstan (country)	40N 75E	114
Kyushu (island)	30N 130W	115
La Pampa (st., Argentina)	36S 70W	94
La Paz, Baja California Sur (city, st. cap., Mex.)	24N 110W	86
La Paz, Bolivia (city, nat. cap.)	17S 68W	94
La Plata, Argentina (city)	35S 58W	94
Laptev Sea	73N 120E	115
La Rioja (st., Argentina)	30S 70W	94
La Rioja, La Rioja (city, st. cap., Argen.)	29S 67W	94
Labrador Peninsula	52N 60W	94
Lachlan (riv., Australasia)	34S 145E	124
Ladoga, Lake	61N 31E	101
Lagos, Nigeria (city, nat. cap.)	7N 3E	108
Lahore, Pakistan (city)	34N 74E	114
Lake of the Woods	50N 92W	87
Lands End	50N 5W	101
Lansing, Michigan (city, st. cap., US)	43N 85W	86

Geographic Index

Name/Description	Latitude & Longitude	Page
Lanzhou, China (city)	36N 104E	114
Laos (country)	20N 105E	114
Las Vegas, NV (city)	36N 115W	86
Latvia (country)	56N 24E	100
Laurentian Highlands	48N 72W	87
Lebanon (country)	34N 35E	100
Leeds, UK (city)	54N 2W	100
Le Havre, France (city)	50N 0	100
Lena (riv., Asia)	70N 125E	115
Lesotho (country)	30S 27E	108
Leveque, Cape	16S 123E	124
Leyte (island)	12N 130E	115
Lhasa, Tibet (China) (city)	30N 91E	114
Liberia (country)	6N 10W	108
Libreville, Gabon (city, nat. cap.)	0 9E	108
Libya (country)	27N 17E	108
Libyan Desert	27N 25E	109
Lille, France (city)	51N 3E	100
Lilongwe, Malawi (city, nat. cap.)	14S 33E	108
Lima, Peru (city, nat. cap.)	12S 77W	94
Limpopo (riv., Africa)	22S 30E	109
Lincoln, Nebraska (city, st. cap., US)	41N 97W	86
Lisbon, Portugal (city, nat. cap.)	39N 9W	100
Lithuania (country)	56N 24E	100
Little Rock, Arkansas (city, st. cap., US)	35N 92W	86
Liverpool, UK (city)	53N 3W	100
Ljubljana, Slovenia (city, nat. cap.)	46N 14E	100
Llanos	33N 103W	95
Logan, Mt. 18,551	62N 139W	87
Logone (riv., Africa)	10N 14E	109
Lome, Togo (city, nat. cap.)	6N 1E	108
London, United Kingdom (city, nat. cap.)	51N 0	100
Londonderry, Cape	14S 125E	124
Lopez, Cape	1S 8E	109
Los Angeles, CA (city)	34N 118W	86
Los Chonos Archipelago	45S 74W	95
Louisiana (st., US)	30N 90W	86
Lower Hutt, New Zealand (city)	45S 175E	123
Luanda, Angola (city, nat. cap.)	9S 13E	108
Lubumbashi, Congo Republic (city)	12S 28E	108
Lusaka, Zambia (city, nat. cap.)	15S 28E	108
Luxembourg (country)	50N 6E	100
Luxembourg, Luxembourg (city, nat. cap.)	50N 6E	100
Luzon (island)	17N 121E	115
Luzon Strait	20N 121E	115
Lyon, France (city)	46N 5E	100
Lyon, Gulf of	42N 4E	101
Maccio, Alagoas (city, st. cap., Braz.)	10S 36W	94
Macdonnell Ranges	23S 135E	124

Geographic Index

Name/Description	Latitude & Longitude	Page
Macedonia (country)	41N 21E	100
Mackenzie (riv., N.Am.)	68N 130W	87
Macquarie (riv., Australasia)	33S 146E	124
Madagascar (country)	20S 46E	108
Madeira, Rio (riv., S.Am.)	5S 60W	95
Madison, Wisconsin (city, st. cap., US)	43N 89W	86
Madras, (Chennai) India (city)	13N 80E	114
Madrid, Spain (city, nat. cap.)	40N 4W	100
Magdalena, Rio (riv., S.Am.)	8N 74W	95
Magellan, Strait of	54S 68W	95
Maine (st., US)	46N 70W	86
Malabo, Equatorial Guinea (city, nat. cap.)	4N 9E	108
Malacca, Strait of	3N 100E	114
Malawi (country)	13S 35E	108
Malaysia (country)	3N 110E	114
Malekula (island)	16S 166E	124
Mali (country)	17N 5W	108
Malpelo Island	8N 84W	87
Malta (island)	36N 16E	101
Mamore, Rio (riv., S.Am.)	15S 65W	95
Managua, Nicaragua (city, nat. cap.)	12N 86W	86
Manaus, Amazonas (city, st. cap., Braz.)	3S 60W	94
Manchester, UK (city)	53N 2W	100
Mandalay, Myamar (city)	22N 96E	114
Manila, Philippines (city, nat. cap.)	115N 121E	114
Manitoba (prov., Can.)	52N 93W	86
Mannar, Gulf of	9N 79E	115
Maoke Mountains	5S 138E	124
Maputo, Mozambique (city, nat. cap.)	26S 33E	108
Maracaibo, Lake	10N 72W	94
Maracaibo, Venezuela (city)	11N 72W	94
Maracapa, Amapa (city, st. cap., Braz.)	0 51W	94
Maranhao (st., Brazil)	4S 45W	94
Maranon, Rio (riv., S.Am.)	5S 75W	95
Marseille, France (city)	43N 5E	100
Maryland (st., US)	37N 76W	86
Masai Steppe	5S 35E	109
Maseru, Lesotho (city, nat. cap.)	29S 27E	108
Mashad, Iran (city)	36N 59E	114
Massachusetts (st., US)	42N 70W	86
Massif Central	45N 3E	101
Mato Grosso	16S 52W	95
Mato Grosso (st., Brazil)	15S 55W	94
Mato Grosso do Sul (st., Brazil)	20S 55W	94
Mauritania (country)	20N 10W	108
Mbandaka, Congo Republic (city)	0 18E	108
McKinley, Mt. 20,320	62N 150W	87 inset
Medellin, Colombia (city)	6N 76W	94
Mediterranean Sea	36N 16E	101

Geographic Index

Name/Description	Latitude & Longitude	Page
Mekong (riv., Asia)	15N 108E	115
Melbourne, Victoria (city, st. cap., Aust.)	38S 145E	123
Melville, Cape	15S 145E	124
Memphis, TN (city)	35N 90W	86
Mendoza (st., Argentina)	35S 70W	94
Mendoza, Mendoza (city, st. cap., Argen.)	33S 69W	94
Merida, Yucatan (city, st. cap. Mex.)	21N 90W	86
Merauke, New Guinea (Indon.) (city)	9S 140E	123
Mexicali, Baja California (city, st. cap., Mex.)	32N 115W	86
Mexico (country)	30N 110W	86
Mexico (st., Mex.)	18N 100W	86
Mexico City, Mexico (city, nat. cap.)	19N 99W	86
Mexico, Gulf of	26N 90W	87
Miami, FL (city)	26N 80W	86
Michigan (st., US)	45N 82W	86
Michigan (lake, N.Am.)	45N 90W	87
Michoacan (st., Mex.)	17N 107W	86
Milan, Italy (city)	45N 9E	100
Milwaukee, WI (city)	43N 88W	86
Minas Gerais (st., Brazil)	17S 45W	94
Mindoro (island)	13N 120E	114
Minneapolis, MN (city)	45N 93W	86
Minnesota (st., US)	45N 90W	86
Minsk, Belarus (city, nat. cap.)	54N 28E	100
Misiones (st., Argentina)	25S 55W	94
Mississippi (riv., N.Am.)	28N 90W	87
Mississippi (st., US)	30N 90W	86
Missouri (riv., N.Am.)	41N 96W	87
Missouri (st., US)	35N 92W	86
Misti, Mt. 19,101	15S 73W	95
Mitchell (riv., Australasia)	16S 143E	124
Mobile, AL (city)	31N 88W	86
Mocambique, Mozambique (city)	15S 40E	108
Mogadishu, Somalia (city, nat. cap.)	2N 45E	108
Moldova (country)	49N 28E	100
Mombasa, Kenya (city)	4S 40E	108
Monaco, Monaco (city)	44N 8E	100
Mongolia (country)	45N 100E	114
Monrovia, Liberia (city, nat. cap.)	6N 11W	108
Montana (st., US)	50N 110W	86
Montorroy, Nuevo Leon (city, st. cap., Mex.)	26N 100W	86
Montevideo, Uruguay (city, nat. cap.)	35S 56W	94
Montgomery, Alabama (city, st. cap., US)	32N 86W	86
Montpelier, Vermont (city, st. cap., US)	44N 73W	86
Montreal, Canada (city)	45N 74W	86
Morelin, Michoacan (city, st. cap., Mex.)	20N 100W	86
Morocco (country)	34N 10W	108
Moroni, Comoros (city, nat. cap.)	12S 42E	108
Moscow, Russia (city, nat. cap.)	56N 38E	100

Geographic Index

Name/Description	Latitude & Longitude	Page
Mountain Nile (riv., Africa)	5N 30E	109
Mozambique (country)	19N 35E	108
Mozambique Channel	19N 42E	109
Munich, Germany (city)	48N 12E	100
Murchison (riv., Australasia)	26S 115E	124
Murmansk, Russia (city)	69N 33E	100
Murray (riv., Australasia)	36S 143E	124
Murrumbidgee (riv., Australasia)	35S 146E	124
Muscat, Oman (city, nat. cap.)	23N 58E	114
Musgrave Ranges	28S 135E	124
Myanmar (Burma) (country)	20N 95E	114
Nairobi, Kenya (city, nat. cap.)	1S 37E	108
Namibe, Angola (city)	16S 13E	108
Namibia (country)	20S 16E	108
Namoi (riv., Australasia)	31S 150E	124
Nan Ling Mountains	25N 110E	115
Nanda Devi, Mt. 25,645	30N 80E	115
Nanjing, China (city)	32N 119E	114
Nansei Shoto (island)	27N 125E	115
Naples, Italy (city)	41N 14E	100
Nashville, Tennessee (city, st. cap., US)	36N 87W	86
Nasser, Lake	22N 32E	109
Natal, Rio Grande do Norte (city, st. cap., Braz.)	6S 5W	94
Naturaliste, Cape	35S 115E	124
Nayarit (st., Mex.)	22N 106W	86
N'Djamena, Chad (city, nat. cap.)	12N 15E	108
Nebraska (st., US)	42N 100W	86
Negro, Rio (Argentina) (riv., S.Am.)	40S 70W	95
Negro, Rio (Brazil) (riv., S.Am.)	0 65W	95
Negros (island)	10N 125E	115
Nelson (riv., N.Am.)	56N 90W	87
Nepal (country)	29N 85E	114
Netherlands (country)	54N 6E	100
Neuquen (st., Argentina)	38S 68W	94
Neuquen, Neuquen (city, st. cap., Argen.)	39S 68W	94
Nevada (st., US)	37N 117W	86
New Britain (island)	5S 152E	124
New Brunswick (prov., Can.)	47N 67W	86
New Caledonia (island)	21S 165E	124
New Delhi, India (city, nat. cap.)	29N 77E	114
New Georgia (island)	8S 157E	124
New Guinea (island)	5S 142E	124
New Hampshire (st., US)	45N 70W	86
New Hanover (island)	3S 153E	124
New Hebrides (island)	15S 165E	124
New Ireland (island)	4S 154E	124
New Jersey (st., US)	40N 75W	86
New Mexico (st., US)	30N 108W	86
New Orleans, LA (city)	30N 90W	86

Geographic Index

Name/Description	Latitude & Longitude	Page
New Siberian Islands	74N 140E	115
New South Wales (st., Aust.)	35S 145E	123
New York (city)	41N 74W	86
New York (st., US)	45N 75W	86
New Zealand (country)	40S 170E	123
Newcastle, Aust. (city)	33S 152E	123
Newcastle, UK (city)	55N 2W	100
Newfoundland (prov., Can.)	53N 60W	86
Nicaragua (country)	10N 90W	86
Niamey, Niger (city, nat. cap.)	14N 2E	108
Nicobar Islands	5N 93E	115
Niger (country)	10N 8E	108
Niger (riv., Africa)	12N 0	109
Nigeria (country)	8N 5E	108
Nile (riv., Africa)	25N 31E	109
Nipigon (lake, N.Am.)	50N 87W	87
Nizhny-Novgorod, Russia (city)	56N 44E	100
Norfolk, VA (city)	37N 76W	86
North Cape (NZ)	36N 174W	124
North Carolina (st., US)	30N 78W	86
North Channel	56N 5W	101
North Dakota (st., US)	49N 100W	86
North Island (NZ)	37S 175W	124
North Saskatchewan (riv., N.Am.)	55N 110W	87
North Sea	56N 3E	101
North West Cape	22S 115W	124
Northern Territory (st., Aust.)	20S 134W	124
Northwest Territories (prov., Can.)	65N 125W	86
Norway (country)	62N 8E	100
Nouakchott, Mauritania (city, nat. cap.)	18N 16W	108
Noumea, New Caledonia (city)	22S 167E	123
Nova Scotia (prov., Can.)	46N 67W	86
Novaya Zemlya (island)	72N 55E	115
Novosibirsk, Russia (city)	55N 83E	114
Nubian Desert	20N 30E	109
Nuevo Leon (st., Mex.)	25N 100W	86
Nullarbor Plain	34S 125W	124
Nyasa, Lake	10S 35E	109
Oakland, CA (city)	38N 122W	86
Oaxaca (st., Mex.)	17N 97W	86
Oaxaca, Oaxaca (city, st. cap., Mex.)	17N 97W	86
Ob (riv., Asia)	60N 78E	115
Ohio (riv., N.Am.)	38N 85W	87
Ohio (st., US)	42N 85W	86
Okavongo (riv., Africa)	18S 18E	109
Okavango Swamp	21S 23E	109
Okeechobee (lake, N.Am.)	28N 82W	87
Okhotsk, Russia (city)	59N 140E	114
Okhotsk, Sea of	57N 150E	115

Geographic Index

Name/Description	Latitude & Longitude	Page
Oklahoma (st., US)	36N 95W	86
Oklahoma City, Oklahoma (city, st. cap., US)	35N 98W	86
Oland (island)	57N 17E	101
Olympia, Washington (city, st. cap., US)	47N 123W	95
Omaha, NE (city)	41N 96W	86
Oman (country)	20N 55E	114
Oman, Gulf of	23N 55E	115
Omdurman, Sudan (city)	16N 32E	108
Omsk, Russia (city)	55N 73E	114
Onega, Lake	62N 35E	101
Ontario (lake, N.Am.)	45N 77W	87
Ontario (prov., Can.)	50N 90W	86
Oodnadatta, Aust. (city)	28S 135E	123
Oran, Algeria (city)	36N 1W	108
Oregon (st., US)	46N 120W	86
Orinoco, Rio (riv., S.Am.)	8N 65W	95
Orizaba Peak 18,406	19N 97W	87
Orkney Islands	60N 0	101
Osaka, Japan (city)	35N 135E	114
Oslo, Norway (city, nat. cap.)	60N 11W	100
Ossa, Mt. 5,305 (Tasm.)	43S 145E	124
Ottawa, Canada (city, nat. cap.)	45N 76W	86
Otway, Cape	40S 142W	124
Ougadougou, Burkina Faso (city, nat. cap.)	12N 2W	108
Owen Stanley Range	9S 148E	124
Pachuca, Hidalgo (city, st. cap. Mex.)	20N 99W	86
Pacific Ocean	20N 115W	87
Pakistan (country)	25N 72E	114
Palawan (island)	10N 119E	115
Palmas, Cape	8N 8W	109
Palmas, Tocantins (city, st. cap., Braz.)	10S 49W	94
Pamirs	32N 70E	115
Pampas	36S 73W	95
Panama (country)	10N 80W	95
Panama, Gulf of	10N 80W	86
Panama, Panama (city, nat. cap.)	9N 80W	86
Papua, Gulf of	8S 144E	124
Papua New Guinea (country)	6S 144E	124
Para (st., Brazil)	4S 54W	94
Paraguay (country)	23S 60W	94
Paraguay, Rio (riv., S.Am.)	17S 60W	95
Paraiba (st., Brazil)	6S 35W	94
Paramaribo, Suriname (city, nat. cap.)	5N 55W	94
Parana (st., Brazil)	25S 55W	94
Parana, Entre Rios (city, st. cap., Argen.)	32S 60W	94
Parana, Rio (riv., S.Am.)	20S 50W	95
Paris, France (city, nat. cap.)	49N 2E	100
Pasadas, Misiones (city, st. cap., Argen.)	27S 56W	94
Patagonia	43S 70W	95

Geographic Index

Name/Description	Latitude & Longitude	Page
Paulo Afonso Falls	10S 40W	95
Peace (riv., N.Am.)	55N 120W	87
Pennsylvania (st., US)	43N 80W	86
Pernambuco (st., Brazil)	7S 36W	94
Persian Gulf	28N 50E	115
Perth, W. Australia (city, st. cap., Aust.)	32S 116E	123
Peru (country)	10S 75W	94
Peshawar, Pakistan (city)	34N 72E	114
Philadelphia, PA (city)	40N 75W	86
Philippine Sea	15N 125E	115
Philippines (country)	15N 120E	114
Phnom Penh, Cambodia (city, nat. cap.)	12N 105E	114
Phoenix, Arizona (city, st. cap., US)	33N 112W	86
Phou Bia 9,249	24N 102E	115
Piaui (st., Brazil)	7S 44W	94
Piaui Range	10S 45W	95
Pic Touside 10,712	20N 12E	109
Pierre, South Dakota (city, st. cap., US)	44N 100W	86
Pietermaritzburg, South Africa (city)	30S 30E	108
Pike's Peak 14,110	36N 110W	87
Pilcomayo, Rio (riv., S.Am.)	23S 60W	95
Pittsburgh, PA (city)	40N 80W	86
Plateau of Iran	26N 60E	115
Plateau of Tibet	26N 85E	115
Platte (riv., N.Am.)	41N 105W	87
Po (riv., Europe)	45N 12E	101
Point Barrow	70N 156W	87 inset
Poland (country)	54N 20E	100
Poopo, Lake	16S 67W	95
Popocatepetl 17,887	17N 100W	87
Port Elizabeth, South Africa (city)	34S 26E	108
Port Lincoln, Aust. (city)	35S 135E	123
Port Moresby, Papua N. G. (city, nat. cap.)	10S 147E	123
Port Vila, Vanatu (city, nat. cap.)	17S 169E	123
Port-au-Prince, Haiti (city, nat. cap.)	19N 72W	86
Portland, OR (city)	46N 123W	86
Porto Alegre, R. Gr. do Sul (city, st. cap., Braz.)	30S 51W	94
Porto Novo, Benin (city, nat. cap.)	7N 3E	108
Porto Velho, Rondonia (city, st. cap., Braz.)	9S 64W	94
Portugal (country)	38N 8W	100
Potomac (riv., N.Am.)	35N 75W	87
Potosi, Bolivia (city)	20S 66W	94
Prague, Czech Republic (city, nat. cap.)	50N 14E	100
Pretoria, South Africa (city, nat. cap.)	26S 28E	108
Pribilof Islands	56N 170W	87 inset
Prince Edward Island (prov., Can.)	50N 67W	86
Pripyat Marshes	54N 24E	101
Providence, Rhode Island (city, st. cap., US)	42N 71W	86
Puebla (st., Mex.)	18N 96W	86

Geographic Index

Name/Description	Latitude & Longitude	Page
Puebla, Puebla (city, st. cap., Mex.)	19N 98W	86
Puerto Monte, Chile (city)	42S 74W	94
Purus, Rio (riv., S.Am.)	5S 68W	95
Putumayo, Rio (riv., S.Am.)	3S 74W	95
Pyongyang, Korea, North (city, nat. cap.)	39N 126E	114
Pyrenees Mountains	43N 2E	101
Qingdao, China (city)	36N 120E	114
Quebec (prov., Can.)	52N 70W	86
Quebec, Quebec (city, prov. cap., Can.)	47N 71W	86
Queen Charlotte Islands	50N 130W	87
Queen Elizabeth Islands	75N 110W	87
Queensland (st., Aust.)	24S 145E	123
Querataro (st., Mex.)	22N 96W	86
Querataro, Querataro (city, st. cap., Mex.)	21N 100W	86
Quintana Roo (st., Mex.)	18N 88W	86
Quito, Ecuador (city, nat. cap.)	0 79W	94
Rabat, Morocco (city, nat. cap.)	34N 7W	108
Race, Cape	46N 52W	87
Rainier, Mt. 14,410	48N 120W	87
Raleigh, North Carolina (city, st. cap., US)	36N 79W	86
Rangoon, Myanmar (Burma) (city, nat. cap.)	17N 96E	114
Rapid City, SD (city)	44N 103W	86
Rawalpindi, India (city)	34N 73E	114
Rawson, Chubuy (city, st. cap., Argen.)	43S 65W	94
Recife, Pernambuco (city, st. cap., Braz.)	8S 35W	94
Red (of the North) (riv., N.Am.)	50N 98W	94
Red (riv., N.Am.)	42N 96W	94
Red Sea	20N 35E	109
Regina, Canada (city)	51N 104W	86
Reindeer (lake, N.Am.)	57N 100W	87
Repulse Bay	22S 147E	124
Resistencia, Chaco (city, st. cap., Argen.)	27S 59W	94
Revillagigedo Island	18N 110W	87
Reykjavik, Iceland (city, nat. cap.)	64N 22W	100
Rhine (riv., Europe)	50N 10E	101
Rhode Island (st., US)	42N 70W	86
Rhone (riv., Europe)	42N 8E	101
Richmond, Virginia (city, st. cap., US)	38N 77W	86
Riga, Gulf of	58N 24E	101
Riga, Latvia (city, nat. cap.)	57N 24E	100
Rio Branco, Acre (city, st. cap., Braz.)	10S 68W	94
Rio de Janeiro (st., Brazil)	22S 45W	94
Rio de Janeiro, R. de Jan. (city, st. cap., Braz.)	23S 43W	94
Rio de la Plata	35S 55W	95
Rio Gallegos, Santa Cruz (city, st. cap., Argen.)	52S 68W	94
Rio Grande (riv., N.Am.)	30N 100W	87
Rio Grande do Norte (st., Brazil)	5S 35W	94
Rio Grande do Sul (st., Brazil)	30S 55W	94
Rio Negro (st., Argentina)	40S 70W	94

Geographic Index

Name/Description	Latitude & Longitude	Page
Riyadh, Saudi Arabia (city, nat. cap.)	25N 47E	100
Roanoke (riv., N.Am.)	34N 75W	87
Roberts, Mt. 4,495	28S 154E	124
Rockhampton, Aust. (city)	23S 150E	123
Rocky Mountains	50N 108W	87
Roebuck Bay	18S 125E	124
Romania (country)	46N 24E	100
Rome, Italy (city, nat. cap.)	42N 13E	100
Rondonia (st., Brazil)	12S 65W	94
Roosevelt, Rio (riv., S.Am.)	10S 60W	95
Roper (riv., Australasia)	15S 135W	124
Roraima (st., Brazil)	2N 62W	94
Ros Dashen Terrara 15,158	12N 40E	109
Rosario, Santa Fe (city, st. cap., Argen.)	33S 61W	94
Rostov, Russia (city)	47N 40E	100
Rotterdam, Netherlands (city)	52N 4E	100
Ruapehu, Mt. 9,177	39S 176W	124
Rub al Khali	20N 50E	115
Rudolph, Lake	3N 34E	109
Russia (country)	58N 56E	100
Ruvuma (riv., Africa)	12S 38E	109
Ruwenzori Mountains	0 30E	109
Rwanda (country)	3S 30E	108
Rybinsk, Lake	58N 38E	101
S. Saskatchewan (riv., N.Am.)	50N 110W	87
Sable, Cape	45N 70W	87
Sacramento (riv., N.Am.)	40N 122W	87
Sacramento, California (city, st. cap., US)	39 121W	86
Sahara	18N 10E	109
Sakhalin Island	50N 143E	115
Salado, Rio (riv., S.Am.)	35S 70W	95
Salem, Oregon (city, st. cap., US)	45N 123W	86
Salt Lake City, Utah (city, st. cap., US)	41N 112W	86
Salta (st., Argentina)	25S 70W	94
Salta, Salta (city, st. cap., Argen.)	25S 65W	94
Saltillo, Coahuila (city, st. cap., Mex.)	26N 101W	86
Salvador, Bahia (city, st. cap., Braz.)	13S 38W	94
Salween (riv., Asia)	18N 98E	115
Samar (island)	12N 124E	115
Samara, Russia (city)	53N 50E	100
Samarkand, Uzbekistan (city)	40N 67E	100
San Antonio, TX (city)	29N 98W	86
San Cristobal (island)	12S 162E	124
San Diego, CA (city)	33N 117W	86
San Francisco, CA (city)	38N 122W	86
San Francisco, Rio (riv., S.Am.)	10S 40W	95
San Joaquin (riv., N.Am.)	37N 121W	87
San Jorge, Gulf of	45S 68W	95
San Jose, Costa Rica (city, nat. cap.)	10N 84W	86

Geographic Index

Name/Description	Latitude & Longitude	Page
San Juan (st., Argentina)	30S 70W	94
San Juan, San Juan (city, st. cap., Argen.)	18N 66W	86
San Lucas, Cape	23N 110W	87
San Luis Potosi (st., Mex.)	22N 101W	86
San Luis Potosi, S. Luis P. (city, st. cap., Mex.)	22N 101W	86
San Matias, Gulf of	43S 65W	95
San Salvador, El Salvador (city, nat. cap.)	14N 89W	86
Sanaa, Yemen (city)	16N 44E	108
Santa Catarina (st., Brazil)	28S 50W	94
Santa Cruz (st., Argentina)	50S 70W	94
Santa Cruz Islands	8S 168E	124
Santa Fe (st., Argentina)	30S 62W	94
Santa Fe de Bogota, Colombia (city, nat. cap.)	5N 74W	94
Santa Fe, New Mexico (city, st. cap., US)	35N 106W	86
Santa Rosa, La Pampa (city, st. cap., Argen.)	37S 64W	94
Santiago, Chile (city, nat. cap.)	33S 71W	94
Santiago del Estero (st., Argentina)	25S 65W	94
Santiago, Sant. del Estero (city, st. cap., Argen.)	28S 64W	94
Santo Domingo, Dominican Rep. (city, nat. cap.)	18N 70W	86
Santos, Brazil (city)	24S 46W	94
Sao Luis, Maranhao (city, st. cap., Braz.)	3S 43W	94
Sao Paulo (st., Brazil)	22S 50W	94
Sao Paulo, Sao Paulo (city, st. cap., Braz.)	24S 47W	94
Sarajevo, Bosnia and Herz. (city, nat. cap.)	43N 18E	100
Sardinia (island)	40N 10E	101
Sarmiento, Mt. 8,100	55S 72W	95
Saskatchewan (riv., N.Am.)	52N 108W	87
Saudi Arabia (country)	25N 50E	114
Savannah (riv., N.Am.)	33N 82W	87
Savannah, GA (city)	32N 81W	86
Sayan Range	45N 90E	115
Seattle, WA (city)	48N 122W	86
Seine (riv., Europe)	49N 3E	101
Senegal (country)	15N 15W	108
Senegal (riv., Africa)	15N 15W	109
Seoul, Korea, South (city, nat. cap.)	38N 127E	114
Sepik (riv., Australasia)	4S 142E	124
Sergipe (st., Brazil)	12S 36W	94
Sev Dvina (riv., Asia)	60N 50E	115
Severnaya Zemlya (island)	80N 88E	115
Shanghai, China (city)	31N 121E	114
Shasta, Mt. 14,162	42N 120W	87
Shenyang, China (city)	42N 123E	114
Shetland Islands	60N 5W	101
Shikoku (island)	34N 130E	115
Shiraz, Iran (city)	30N 52E	114
Sicily (island)	38N 14E	100
Sierra Leone (country)	6N 14W	108
Sierra Madre Occidental	27N 108W	87

Geographic Index

Name/Description	Latitude & Longitude	Page
Sierra Madre Oriental	27N 100W	87
Sierra Nevada	38N 120W	87
Sikhote Alin	45N 135E	115
Simpson Desert	25S 136E	124
Sinai Peninsula	28N 33E	109
Sinaloa (st., Mex.)	25N 110W	86
Singapore (city, nat. cap.)	1N 104E	114
Sitka Island	57N 125W	87
Skagerrak, Strait of	58N 8E	101
Skopje, Macedonia (city, nat. cap.)	42N 21E	100
Slovakia (country)	50N 20E	100
Slovenia (country)	47N 14E	100
Snake (riv., N.Am.)	45N 110W	87
Snowy Mountains	37S 148E	124
Sofia, Bulgaria (city, nat. cap.)	43N 23E	95
Solimoes, Rio (riv., S.Am.)	3S 65W	95
Solomon Islands (country)	7S 160E	124
Somalia (country)	5N 45E	108
Sonora (st., Mex.)	30N 110W	86
South Africa (country)	30S 25E	108
South Australia (st., Aust.)	30S 125E	123
South Cape, New Guinea	8S 150E	124
South Carolina (st., US)	33N 79W	86
South China Sea	15N 115E	124
South Dakota (st., US)	45N 100W	86
South Georgia (island)	55S 40W	95
South Island (NZ)	45S 170E	124
Southampton Island	68N 86W	87
Southern Alps (NZ)	45S 170E	124
Southwest Cape (NZ)	47S 167E	124
Spain (country)	38N 4W	100
Spokane, WA (city)	48N 117W	86
Springfield, Illinois (city, st. cap., US)	40N 90W	86
Sri Lanka (country)	8N 80E	114
Srinagar, India (city)	34N 75E	114
St. Elias, Mt. 18, 008	61N 139W	87
St. George's Channel	53N 5W	101
St. Helena (island)	16S 5W	108
St. John's, Nwfndlnd (city, prov. cap., Can.)	48N 53W	86
St. Louis, MO (city)	39N 90W	86
St. Lawrence (island)	65N 170W	87 inset
St. Lawrence (riv., N.Am.)	50N 65W	87
St. Lawrence, Gulf of	50N 65W	87
St. Marie, Cape	25S 45E	108
St. Paul, Minnesota (city, st. cap., US)	45N 93W	86
St. Petersburg, Russia (city)	60N 30E	100
St. Vincente, Cape of	37N 10W	101
Stanovoy Range	55N 125E	115
Stavanger, Norway (city)	59N 6E	100

Geographic Index

Name/Description	Latitude & Longitude	Page
Steep Point	25S 115E	124
Stockholm, Sweden (city, nat. cap.)	59N 18E	100
Stuart Range	32S 135E	124
Stuttgart, Germany (city)	49N 9E	100
Sucre, Bolivia (city)	19S 65W	94
Sudan (country)	10N 30E	108
Sulaiman Range	28N 70E	115
Sulu Islands	8N 120E	114
Sulu Sea	10N 120E	114
Sumatra (island)	0 100E	115 inset
Sumba (island)	10S 120E	124
Sumbawa (island)	8S 116E	124
Sunda Islands	12S 118E	124
Superior (lake, N.Am.)	50N 90W	87
Surabaya, Java (Indonesia) (city)	7S 113E	114 inset
Suriname (country)	5N 55W	94
Svalbard Islands	75N 20E	115
Swan (riv., Australasia)	34S 115E	124
Sweden (country)	62N 16E	100
Sydney, N.S.Wales (city, st. cap., Aust.)	34S 151E	123
Syr Darya (riv., Asia)	36N 65E	115
Syria (country)	37N 36E	100
Tabasco (st., Mex.)	16N 90W	86
Tabriz, Iran (city)	38N 46E	114
Tahat, Mt. 9,541	23N 8E	109
Taipei, Taiwan (city, nat. cap.)	25N 121E	114
Taiwan (country)	25N 122E	114
Taiwan Strait	25N 120E	115
Tajikistan (country)	35N 75E	114
Takla Makan	37N 90E	115
Tallahassee, Florida (city, st. cap., US)	30N 84W	86
Tallinn, Estonia (city, nat. cap.)	59N 25E	100
Tamaulipas (st., Mex.)	25N 95W	86
Tampico, Mexico (city)	22N 98W	86
Tanganyika, Lake	5S 30E	109
Tanzania (country)	8S 35E	108
Tapajos, Rio (riv., S.Am.)	5S 55W	95
Tarim Basin	37N 85E	115
Tashkent, Uzbekistan (city, nat. cap.)	41N 69E	114
Tasman Sea	38S 160E	124
Tasmania (st., Aust.).	42S 145E	123
Tatar Strait	50N 142E	115
Tbilisi, Georgia (city, nat. cap.)	42N 45E	100
Teguicigalpa, Honduras (city, nat. cap.)	14N 87W	86
Tehran, Iran (city, nat. cap.)	36N 51E	114
Tel Aviv, Israel (city)	32N 35E	100
Tennant Creek, Aust. (city)	19S 134E	123
Tennessee (st., US)	37N 88W	86
Tennessee (riv., N.Am.)	32N 88W	87

Geographic Index

Name/Description	Latitude & Longitude	Page
Tepic, Nayarit (city, st. cap., Mex.)	22N 105W	86
Teresina, Piaui (city, st. cap., Braz.)	5S 43W	94
Texas (st., US)	30N 95W	86
Thailand (country)	15N 105E	114
Thailand, Gulf of	10N 105E	115
Thames (riv., Europe)	52N 4W	101
The Hague, Netherlands (city, nat. cap.)	52N 4E	100
The Round Mountain 5,300	29S 152E	124
Thimphu, Bhutan (city, nat. cap.)	28N 90E	114
Tianjin, China (city)	39N 117E	114
Tibest Massif	20N 20E	109
Tien Shan	40N 80E	115
Tierra del Fuego	54S 68W	95
Tierra del Fuego (st., Argentina)	54S 68W	94
Tigris (riv., Asia)	37N 40E	101
Timor (island)	7S 126E	115
Timor Sea	11S 125E	123
Tirane, Albania (city, nat. cap.)	41N 20E	100
Titicaca, Lake	15S 70W	95
Tlaxcala (st., Mex.)	20N 96W	86
Tlaxcala, Tlaxcala (city, st. cap., Mex.)	19N 98W	86
Toamasino, Madagascar (city)	18S 49E	108
Tocantins (st., Brazil)	12S 50W	94
Tocantins, Rio (riv., S.Am.)	5S 50W	95
Togo (country)	8N 1E	108
Tokyo, Japan (city, nat. cap.)	36N 140E	114
Toliara, Madagascar (city)	23S 44E	108
Tolima, Mt. 17,110	5N 75W	95
Toluca, Mexico (city, st. cap., Mex.)	19N 100W	86
Tombouctou, Mali (city)	24N 3W	108
Tomsk, Russia (city)	56N 85E	114
Tonkin, Gulf of	20N 108E	115
Topeka, Kansas (city, st. cap., US)	39N 96W	86
Toronto, Ontario (city, prov. cap., Can.)	44N 79W	86
Toros Mountains	37N 45E	115
Torrens, Lake	33S 136W	124
Torres Strait	10S 142E	124
Townsville, Aust. (city)	19S 146E	123
Transylvanian Alps	46N 20E	101
Trenton, New Jersey (city, st. cap., US)	40N 75W	86
Tricara Peak 15,584	4S 137E	124
Trinidad and Tobago (island)	9N 60W	95
Tripoli, Libya (city, nat. cap.)	33N 13E	108
Trujillo, Peru (city)	8S 79W	94
Tucson, AZ (city)	32N 111W	86
Tucuman (st., Argentina)	25S 65W	94
Tucuman, Tucuman (city, st. cap., Argen.)	27S 65W	94
Tunis, Tunisia (city, nat. cap.)	37N 10E	108
Tunisia (country)	34N 9E	108

Geographic Index

Name/Description	Latitude & Longitude	Page
Turin, Italy (city)	45N 8E	100
Turkey (country)	39N 32E	100
Turkmenistan (country)	39N 56E	100
Turku, Finland (city)	60N 22E	100
Tuxtla Gutierrez, Chiapas (city, st. cap., Mex.)	17N 93W	86
Tyrrhenian Sea	40N 12E	101
Ubangi (riv., Africa)	0 20E	109
Ucayali, Rio (riv., S.Am.)	7S 75W	95
Uele (riv., Africa)	3N 25E	109
Uganda (country)	3N 30E	108
Ujungpandang, Celebes (Indon.) (city)	5S 119E	114 inset
Ukraine (country)	53N 32E	100
Ulan Bator, Mongolia (city, nat. cap.)	47N 107E	114
Uliastay, Mongolia (city)	48N 97E	114
Ungava Peninsula	60N 72W	87
United Arab Emirates (country)	25N 55E	114
United Kingdom (country)	54N 4W	100
United States (country)	40N 100W	86
Uppsala, Sweden (city)	60N 18E	100
Ural (riv., Asia)	45N 55E	115
Ural Mountains	50N 60E	115
Uruguay (country)	37S 67W	94
Uruguay, Rio (riv., S.Am.)	30S 57W	95
Urumqui, China (city)	44N 87E	114
Utah (st., US)	38N 110W	86
Uzbekistan (country)	42N 58E	100
Vaal (riv., Africa)	27S 27E	109
Valdivia, Chile (city)	40S 73W	94
Valencia, Spain (city)	39N 0	100
Valencia, Venezuela (city)	10N 68W	94
Valparaiso, Chile (city)	33S 72W	94
van Diemen, Cape	11S 130E	124
van Rees Mountains	4S 140E	124
Vanatu (country)	15S 167E	124
Vancouver, Canada (city)	49N 123W	86
Vancouver Island	50N 130W	87
Vanern, Lake	60N 12E	101
Vattern, Lake	56N 12E	101
Venezuela (country)	5N 65W	94
Venezuela, Gulf of	12N 72W	95
Venice, Italy (city)	45N 12E	100
Vera Cruz (st., Mex.)	20N 97W	86
Vera Cruz, Mexico (city)	19N 96W	86
Verkhoyanskiy Range	65N 130E	115
Vermont (st., US)	45N 73W	86
Vert, Cape	15N 17W	109
Vestfjord	68N 14E	101
Viangchan, Laos (city, nat. cap.)	18N 103E	114
Victoria (riv., Australasia)	15S 130E	124

Geographic Index

Name/Description	Latitude & Longitude	Page
Victoria (st., Aust.)	37S 145W	123
Victoria, B.C. (city, prov. cap., Can.)	48N 123W	86
Victoria, Lake	3S 35E	109
Victoria, Mt. 13,238	9S 137E	124
Victoria Riv. Downs, Aust. (city)	17S 131E	123
Viedma, Rio Negro (city, st. cap., Argen.)	41S 63W	94
Vienna, Austria (city)	48N 16E	100
Vietnam (country)	10N 110E	114
Villahermosa, Tabasco (city, st. cap., Mex.)	18N 93W	86
Vilnius, Lithuania (city, nat. cap.)	55N 25E	100
Virginia (st., US)	35N 78W	86
Viscount Melville Sound	72N 110W	87
Vitoria, Espiritu Santo (city, st. cap., Braz.)	20S 40W	94
Vladivostock, Russia (city)	43N 132E	114
Volga (riv., Europe)	46N 46E	101
Volgograd, Russia (city)	54N 44E	100
Volta (riv., Africa)	10N 15E	109
Volta, Lake	8N 2W	109
Vosges Mountains	48N 7E	101
Wabash (riv., N.Am.)	43N 90W	87
Walvis Bay, Namibia (city)	23S 14E	108
Warsaw, Poland (city, nat. cap.)	52N 21E	100
Washington (st., US)	48N 122W	86
Washington, D.C., United St.s (city, nat. cap.)	39N 77W	86
Wellington Island	48S 74W	95
Wellington, New Zealand (city, nat. cap.)	41S 175E	123
Weser (riv., Europe)	54N 8E	101
West Cape Howe	36S 115E	124
West Indies	18N 75W	87
West Siberian Lowland	60N 80E	115
West Virginia (st., US)	38N 80W	86
Western Australia (st., Aust.)	25S 122W	123
Western Ghats	15N 72E	115
Western Sahara (country)	25N 13W	108
White Nile (riv., Africa)	13N 30E	109
White Sea	64N 36E	101
Whitney, Mt. 14,494	33N 118W	87
Wichita, KS (city)	38N 97W	86
Wilhelm, Mt. 14,793	4S 145E	124
Windhoek, Namibia (city, nat. cap.)	22S 17E	108
Winnipeg (lake, N.Am.)	50N 100W	87
Winnipeg, Manitoba (city, prov. cap., Can.)	53N 98W	86
Wisconsin (st., US)	50N 90W	86
Wollongong, Aust. (city)	34S 151E	123
Woodroffe, Mt. 4,724	26S 133W	124
Woomera, Aust. (city)	32S 137E	123
Wrangell (island)	72N 180E	115
Wuhan, China (city)	30N 114E	114
Wyndham, Australia (city)	16S 129E	123